Travels Without Charlie – the Natives Ate Him

Exploits and Misadventures of an Exploration Geologist

Bob Cuffney

Splash Books

Travels without Charlie – the Natives Ate Him
Copyright 2022 @ Bob Cuffney

LeRue Press, LLC
Reno, Nevada
Splash Books

All rights reserved. No part of this book may be reproduced in any form or by any electronic or mechanical means, including information storage and retrieval systems, without permission in writing from the author.

This book is an attempt at a memoir. It reflects the author's present and sometimes warped recollections of experiences over many decades. Some names and characteristics have been changed; some dialogue has been recreated. In some instances, the author has been liberal with adjusting both time and space of the events.

First Edition

Library of Congress Control Number 2021919706

ISBN: 978-1-938814-37-2

Cover design by Peg O'Malley

Printed in the United States

For Eric and Craig, the Fruit Loops of my loins.

No, sons, these stories were not, "made up on the long trip home," as you once suggested. They are true stories, as best as my fading memory can recollect. I hope you enjoy them and have many an adventure of your own during your time on this crazy rotating rock.

Contents

		Page
Acknowledgments		vii
Index Maps to Chapters		viii
Introduction		1
Section I	**Preparing for Misadventure**	5
Chapter 1	Rocky Start (1956-1967)	7
Chapter 2	Rambling Wrecks (1967-1975)	12
Section II	**The Great North**	31
Chapter 3	Doing Strange Things in the Midnight Sun (1974-1981)	33
Chapter 4	In Search of Smokes (1980)	55
Chapter 5	Fear of Flying (1974-1981)	68
Section III	**Australasia**	91
Chapter 6	Into the Soup Pot (1981)	93
Chapter 7	Tarzan of Sulawesi (1994-1996)	99
Chapter 8	The Drowning of Regex (1995)	124
Chapter 9	Wild Man of Borneo (2007)	135
Chapter 10	Bugger Me! (2007-2011)	139
Chapter 11	The First Cut is the Deepest (2011)	147
Chapter 12	Safety First (2012)	150
Section IV	**Asia**	159
Chapter 13	North of the Great Wall (2007-2011)	161
Chapter 14	One Day of Mongolia (2007)	175
Chapter 15	Where Reindeer Fly (2011)	188
Chapter 16	South of the Great Wall (2008)	204
Chapter 17	Kung Fu Food (2008)	224
Section V	**Closer to Home**	229
Chapter 18	Rocky (1978)	231
Chapter 19	White Knight (1996-2007)	238
Chapter 20	Highway to Hell (2016)	244
Chapter 21	Living Hell (2016-2017)	257
Chapter 22	Ring of Fire (2017)	265
Chapter 23	Odd Quest (1990)	268
Chapter 24	Kay Mine Disaster (1984)	273

Chapter 25	Radioactive (1975-1978)	277
Chapter 26	PokIng the Bear (1975-2020)	280
Chapter 27	Four-Wheel Misadventures (1970-2020)	289
Chapter 28	Illegal Alien (2011-2017)	299
Chapter 29	Office Mates (1974-2020)	303
Chapter 30	Lions, and Tigers, and Bears...and Anoas (1974-2020)	310
Chapter 31	Scatological Musings	327
Chapter 32	Apps, please (2018)	332
Chapter 33	Wandering Star	334

ACKNOWLEDGMENTS

Many people, both extant and deceased, contributed to the creation of this masterpiece of literature – willingly, or not. I am eternally grateful to all of them.

For starters, I must acknowledge the ancient Phoenicians for inventing the first rudimentary alphabet, and my long-lost relatives, the Anglo Saxons, for finishing the job. My writing efforts would have been in vain without the essential 26 building blocks they created. I thank Noah Webster, my favorite lexicographer, for compiling those letters into some 470,000 words in his *American Dictionary of the English Language,* from which I borrowed liberally. Sadly, I found his work inadequate. I was compelled to make up a few words of my own, which drove Spell Check and the editor crazy.

Writing accurate nonfiction requires a lot of fact checking. That can be difficult when you are trying to remember events and places you have not thought of for decades – *Where was I, really?* In the past, such investigative work would have required countless hours searching through books, encyclopedias, and maps at the local library. I thank Senator Al Gore for inventing the internet, allowing me to do my research from the comfort of my high-country cabin. Many thanks to Mr. Google for putting lots of misinformation and occasional snippets of truth on his websites. I have included many tidbits of information from his search engine, but disclaim responsibility for any inaccurate information derived therefrom.

I thank my high school and college buddies, colleagues, associates, native crews, and the many wild beasts that are featured in these stories, without whose participation the misadventures would not have happened. Special thanks to my colleague and good friend, John Carden, for participating in and surviving (barely) many of the misadventures.

These stories were written over a period of more than 15 years, partly because I am a slow writer, and partly because misadventures were still happening. Like many authors, I suffered from periods of inactivity when I lacked interest in finishing the work – times when my literary mojo was running low, or I decided my writing just plain sucked. It was only through the efforts of a few kind individuals, who reviewed early drafts of stories and provided much needed encouragement (OK, they lied), that these stories have come together in print. I am deeply grateful to Carrie Hatcher, Pat Harris, and Donna Robert for their helpful comments and encouragement.

Ernest Hemingway is credited (rightly or wrongly) with this sage advice, "Write drunk, edit sober." I tried following his advice and found that I was good at the first part, but not the second. Rather than endure long periods of sobriety, I opted to farm out the editing. I am forever indebted to Cynda Green, investigative reporter par excellence, turned editor, for editing and critical review – including tempering my bawdy language and my politically and socially incorrect rants (you should have seen the early drafts) – and for her never-wavering support and encouragement.

Many thanks to Peg O'Malley, fellow geologist, draftsperson, and artist for coming out of retirement to work her magic on the cover design.

I owe a debt of gratitude to Ian Macky for the wonderful open-source maps on his website PAT. They saved me countless hours trying to draft my own ugly background maps.

Finally, a tip of the ole hat to Chairman Xi Jinping, the Chinese Communist Party, and the Wuhan Institute of Virology for sending us the lovely Wuhan Virus, aka COVID-19. Without the ensuing lockdowns and bar closures, I surely would have never completed this project. I mean, what the hell, there was NOTHING else to do!

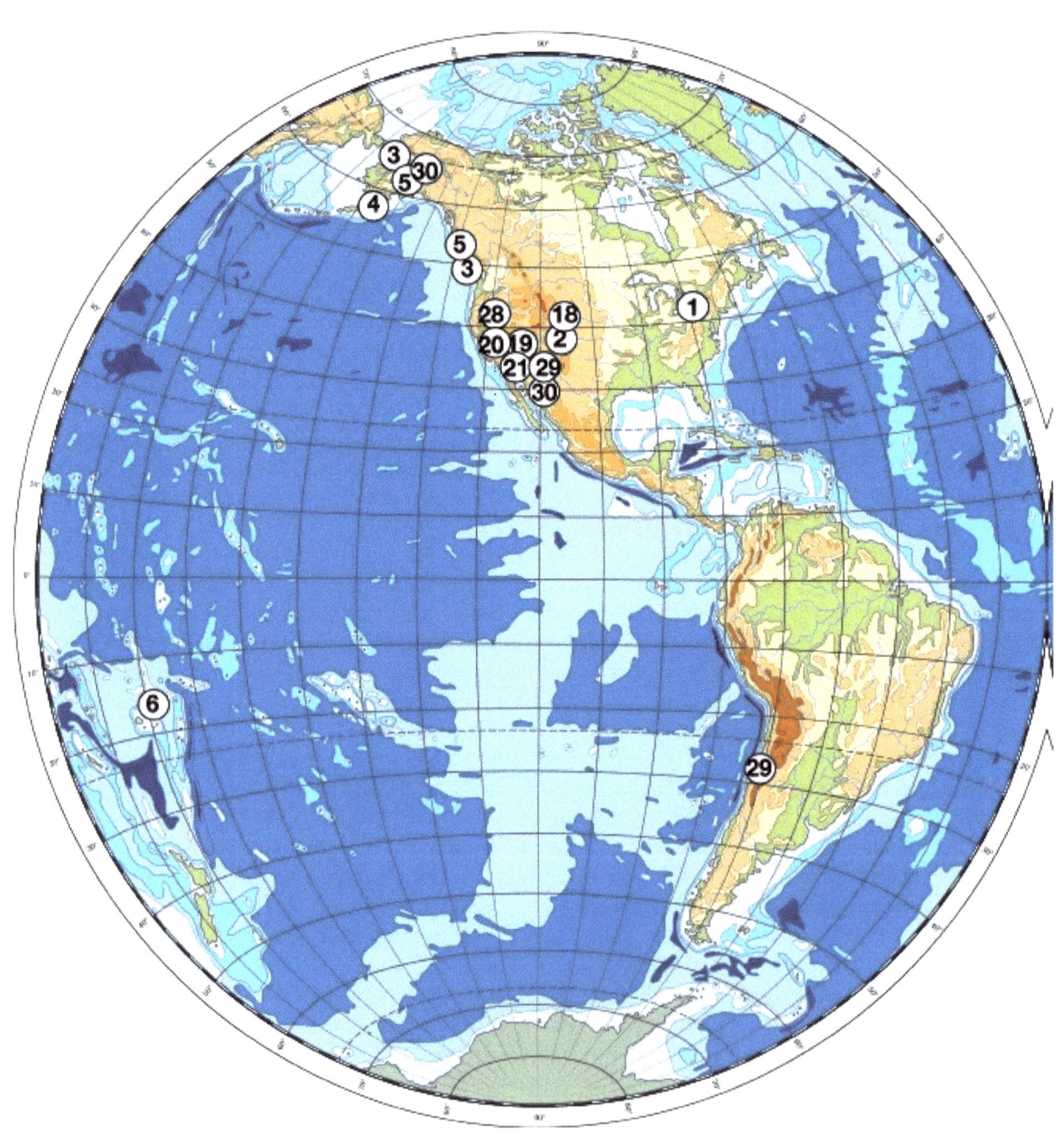

Index to Chapters – North and South America

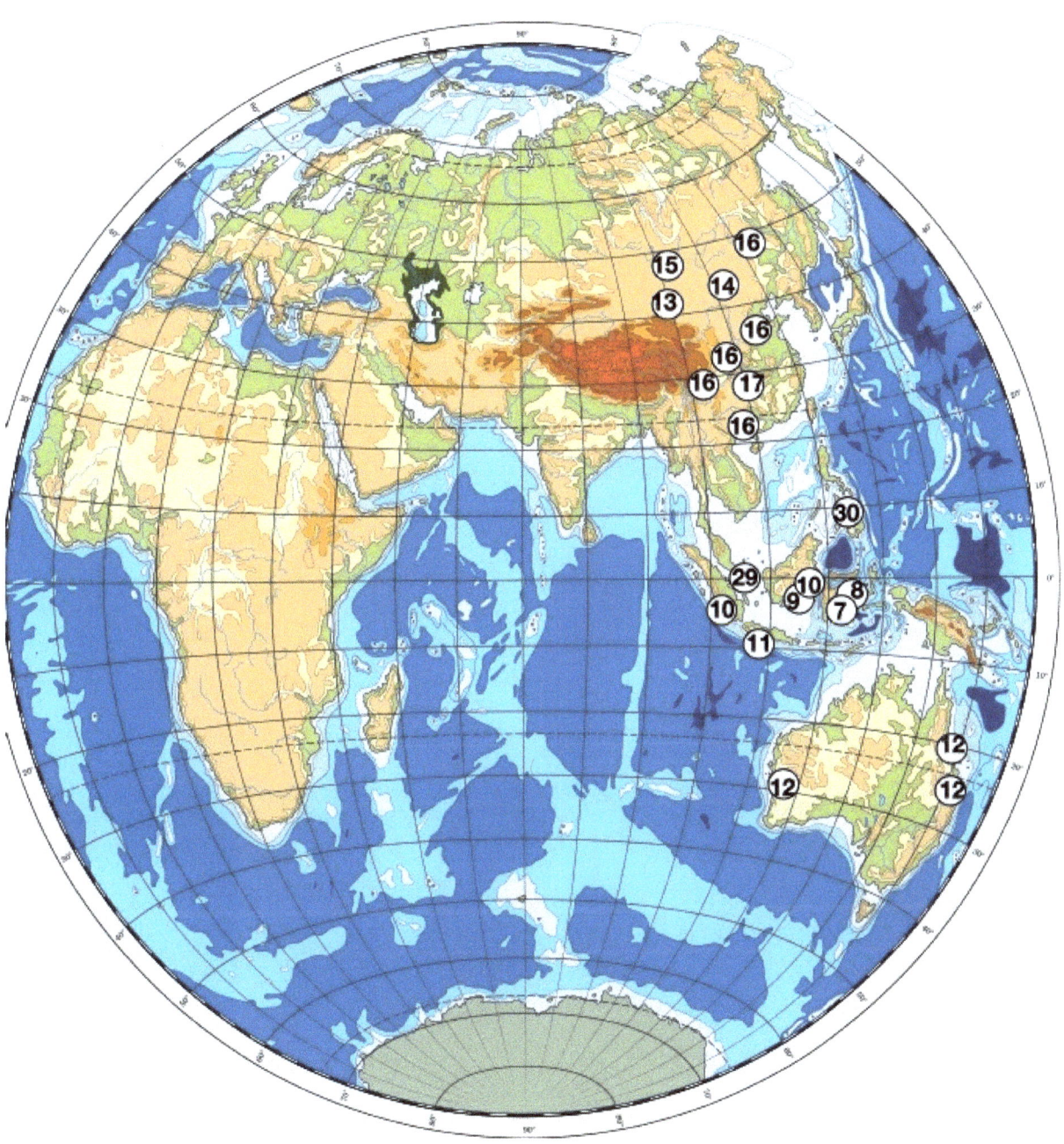

Index to Chapters – Asia and Australasia

INTRODUCTION

"Geology is not a real science!"
Sheldon Cooper — *The Big Bang Theory*

Some academics, nerdy physicists, for instance, may question the legitimacy of the noblest of all sciences– geology – claiming it is not a real science. Hah! Poppycock! Geology may be different, but it is just as much a real science as are political science and Scientology. And, unlike pseudo-science such as flip-flopping Fauci's flaky virology, geology is based on rock-solid principles. If you are going to "follow the science" – follow geology.

Just because it takes a few million years to test a geological hypothesis and a few million more to repeat the experiment does not preclude application of the scientific method. We need a little more time to get the job done than do other scientists. Consider it long-term job security – the only security we get.

So-called "normal" people may consider geologists to be, and it pains me to say, a little weird because we talk to rocks, and they talk back to us, telling stories of how the earth formed and the many traumatic events it has gone through since its birth. People also think we are odd because we lick rocks. "Ooh, gross!" is often the reaction I get. It may not help my case by saying we also smell, scratch, fondle, and drop acid on our stony subjects. It takes all five senses to entice rocks to reveal their deepest, darkest secrets.

There are many branches of the geological sciences and careers to accompany each "-ologist" specialty, including mineralogists, petrologists, paleontologists, geomorphologists, volcanologists, and engineering geologists, to name a few. One very special category of geologists stands out – Exploration Geologists – the true Men of the geological profession, those hardy souls who risk life, limb, and liver roaming the deserts, mountains, jungles, and bars of the world in search of the essential building blocks of our society: metals such as copper, nickel, zinc, gold, and silver; energy resources: coal, uranium, lithium, oil, and gas. Finally, there is the largest group of geologists – Unemployed Geologists. I have spent the better part of five decades bouncing between the latter two categories.

If anybody has picked up this book hoping to read an intriguing autobiography, put it down now. Trust me, this is not an autobiography. Such an opus would bore me, let alone the poor reader, to death. I am not fond of autobiographies. They tend to be somewhat biased works, often wildly self-aggrandizing platforms for the author to write glowingly about his or her many wonderful accomplishments and successes, be they real or imaginary. I look at them as expanded Facebook pages – focusing on the good things and none of the bad. They remind me of those awful Christmas letters you receive from your friends and relatives who pretend to have perfect families, like cousin Gary from New Jersey, the one you hope doesn't show up at family reunions and who fills the annual missive with sugary gems like: "Johnnie started 9th grade this year at Hoboken High, but the school curriculum didn't present enough of a challenge. He transferred to a smaller school on the south side, where he could get more one-on-one attention and pursue his varied interests." Translation: "Johnny got busted smoking crack in the boys' bathroom. He was expelled and is now in the Juvenile Detention Center, where he is learning useful skills like lock-picking, carjacking, and the finer points of manufacturing meth."

The closest thing to an autobiography I bothered to read in recent years was when the big, bad, bombastic, belligerent, boisterous, braggadocious, billionaire bully, Donald Trump, was campaigning for the Republican nomination for president. I took the time to read his book *Crippled America: How to Make America Great Again*, intrigued by the idea of a non-politician, an outsider owing nothing to special interests, a man with over-size cojones, running the country.

I read The Donald's book with great anticipation, eager to learn about his grandiose plans to make America great again. Much to my disappointment, it was short on details but certainly not short on self-aggrandizing, referencing his many accomplishments on nearly every page. But Donald had a moment of literary genius. In the back of the book, on pages 191-193, was a concise list of the magnificent buildings his calloused hands toiled to build. You could skip straight to the end to learn all you need to know.

Fantastic, I thought, *I should do the same thing*. Although I have no intention of using this book to tout the many accomplishments, discoveries, accolades, and prestigious awards of my distinguished career, I feel compelled to at least hit the highlights, so you don't come away thinking I am unqualified to write a book about exploration geologists. Donald's idea of a list at the back of the book is superb. The problem is – if I put a list of my wonderful deeds in the back of the book, many (most?) readers might get bored, toss the book, and never get to the end to peruse my list of outstanding achievements. Plus, I had other plans for the end of the book. I wanted to conclude with a bang, leaving the reader inspired, deep in thought, pondering philosophical conundrums, like "What is the meaning of life?", "Who wrote *The Book of Love?*" or *"Why are farts funny?"* Therefore, I am going to present my list of accomplishments upfront, lest they be ignored. Bear with me; here they are:

LIST OF MR. CUFFNEYS' GREAT PROFESSIONAL ACHIEVEMENTS, DISCOVERIES, PUBLICATIONS, CONTRIBUTIONS TO THE SCIENCE OF GEOLOGY, ACCOLADES, AND PRESTIGIOUS AWARDS

Page left **Un**intentionally Blank

Introduction

Yes, there they are: zip, nada, squat – no great honors, awards, or rewards. I take solace in one of Mark Twain's famous and wholly applicable quotes, "It is better to deserve honors and not have them than to have them and not deserve them."

With that out of the way, I can get down to what this literary masterpiece is all about – entertainment. This is a somewhat disjointed collection of short (and not so short) sometimes rambling stories based on the crazy, funny, stupid, sometimes harrowing, and all too often embarrassing happenings of my geological career and the rocky road on the way to it.

These stories span some 60 years, from collecting rocks and fossils as a boy growing up in the 1950s and 1960s, through the crazy juvenile antics and tomfoolery of my high school and college years, on to the crazy juvenile antics and tomfoolery of a career as a geologist exploring for mineral and energy resources on four continents and a bunch of remote islands.

Settings for the stories range from the fields and streams of upstate New York to the high country of the Rocky Mountains and Alaska; the sizzling Sonoran, Mohave, Gobi, and Atacama deserts; the steaming Jungles of Indonesia, the Philippines, Fiji, and China; the mosquito-infested tundra of Alaska, the windswept Steppes of Mongolia, the Outback of Australia – and, of course, the bars and taverns of all the aforementioned places.

Astute readers may notice the frequency with which alcohol, in its various and sundry forms, plays a role in these stories. That may be true (statistically speaking, it is – the Pearson Correlation Coefficient is 0.82). Before assuming exploration geologists, let alone this one, are drunken sots by nature, cause and effect must be considered. Admittedly, some misadventures may have been fueled by an ill-fated mixture of liquor and poor judgment. Other adventures merely conclude with well-earned adult beverages to celebrate the occasion or libations for medicinal purposes to ease the pain or to soften the psychological trauma of the ordeal. Trust me, there were sober moments between the alcohol-fueled ones. Few of those are worth remembering, let alone writing about.

The stories are based on actual events. I have strived to describe them as accurately as possible but admit to some errors and omissions, given the fading state of my addled brain. Perhaps I should have started this project earlier, before I slipped into hopeless mental decrepitude. I mistakenly thought senility was something that came on suddenly and not until one's late 80s. For me, it has been a surprisingly smooth transition, starting in my 30s. Besides likely having mixed up a few details, I must confess to adding a wee bit of embellishment here and there to liven things up.

For the most part, the characters' names in the stories are those of actual individuals. If they were foolish enough to participate in the misadventures, they deserve credit for their contributions. A few exceptions apply to certain individuals who might have the motive and resources (like cousin Vinnie from Philly) to seek revenge. I made up the names of a few foreign fellows because I never knew their names, forgot them, or couldn't begin to pronounce or spell them.

This is also not a treatise on geology. Some geological rambling appears here and there where appropriate. I have tried to keep the geological jargon to a minimum, but I admit to having slipped in a few words ending in "-ite" occasionally.

These stories are written in the vernacular of the field geologist and contain adult language and humor. Some readers, especially young cancel-culture snowflakes, who may self-identify as someone or something that they are not, may find my exhilarating prose rough, perhaps rude, crude, irreverent, at times socially and politically incorrect. I must apologize in advance because the last thing I want is to offend anyone who is thin-skinned or overly sensitive.

I said offending overly sensitive people was the *last* thing I wanted to do – not that I wouldn't do it. It comes naturally, albeit last.

One of my great heroes, Roy Chapman Andrews, who led the incredibly challenging and superbly successful Central Asiatic Expeditions to Inner and Outer Mongolia in 1928-1930, wrote in *This Business of Exploration*, "Adventures, of course, are always associated with exploration. Yet they are the one thing which a real explorer tries to guard against." He then cites what he calls Steffansson's dictum, "Adventures are a mark of incompetence." I particularly liked that quote, although I had no idea who Steffansson was, and Mr. Andrews did not provide any insight. Perhaps he was a famous Scandinavian explorer who discovered the North Pole, the Inside Passage, or perhaps the long-lost Outside Passage. I googled him. Vilhjam Steffansson was indeed an Arctic explorer, a Canadian of Icelandic descent, which certainly should have qualified him for adventures in cold, bleak places. He led several expeditions to the Canadian and Alaskan Arctic in the early 1900s, most of which turned into unmitigated disasters. In 1913, his flagship, the *Kaluk*, became frozen in the Arctic pack ice. Vilhjam cleverly abandoned the ship with the excuse of going hunting to score some walrus meat to feed the crew. The problem was he boogied south to terra firma and didn't return. The ship was carried off with the shifting ice and eventually was crushed to splinters. Thirteen of the 24 starving men aboard perished despite desperate attempts to survive the winter on various barren islands they trekked across the ice to reach.

Undaunted by that experience, in 1921, ole Vil organized an expedition to Wrangell Island, a lovely bare, windswept frozen rock in the Chukchi Sea betwixt Siberia and Alaska. His intent was to establish a tourism business on the island. What tourists would do there remains a mystery to this day. The expedition comprised four men, an Inuk Eskimo lady, and a cat. Vilhjam wisely did not accompany the team. Only the Eskimo lady survived, likely because she was the only one fond of cat stew.

If Andrews is right about Steffansson's dictum, I must plead guilty to incompetence. I have had many adventures and misadventures, too many of the latter self-inflicted, during my 50-some years as an exploration geologist. After telling stories of my adventures to friends and family for years, it dawned on me I should put some of them in writing to be shared with a broader audience. As John Steinbeck wrote in his famous travelogue, *Travels with Charley, In Search of America*, describing his year-long trek across America with his dog, Charley, "One goes, not so much to see, but to tell afterward."

Stories of the crazy adventures and misadventures are what I wish to tell, not the boring times when things were actually going right. My incompetence is your key to armchair adventure.

Enjoy.

SECTION I

Preparing for Misadventure

Sneaking the keg into the football stands – Homecoming 1971.

Chapter 1

Rocky Start

"The desire to see new places, to discover – the curiosity of life always has been a restless driving force in me."

This Business of Exploring   Roy Chapman Andrews

"Ouch!" I exclaimed as a wooden yardstick came crashing down on my tender young knuckles. Looking up, I saw the source of the abuse – the black-and-white-framed scowl of an angry nun towering over me. *What the heck, why did I deserve that?* I was minding my own business – gazing out the window, watching the chirping birds play amongst the budding maple leaves, and daydreaming about exploring the woods and streams of upstate western New York.

I was one of those fidgety kids, the ones that can't sit still and don't do well in captivity. Daily class activities, especially lectures on some holy sacrament or another, tended to bore me. While the class was reading about the latest adventures of Dick and Jane ("See Spot run. Run Spot, run!"), I often had a book on dinosaurs, snakes, or lizards in my lap that I was secretly trying to read. Those nuns were strict and mean. They had a habit (bad pun, I know) of sneaking up behind me and catching me in the act. Rather than complimenting me on my advanced reading skills, the sisters took delight in physically abusing me. Sister Teresa Marie was the worst. She would grab my hair and jerk me right out of my chair. My solution was to get a buzz cut, leaving her no follicle purchase. *Take that, Sister!* Unfazed, she went straight for my ears. I attribute my pointy Dr. Spock ears and my penguin-phobia, which lingers to this day, to the nasty nuns of Immaculate Conception School.

I was a bit of a nerd in my younger years, although some people would argue that nothing has changed. I was not the modern-day video-game-playing/comic book-reading type, but I do confess to having been a follower of Superman and his arch-foe, Lex Luthor. I was more of a shy, introverted nature-boy type, who was more comfortable alone in the woods with his dog than socializing with his fellow humanoids.

While the other kids were playing hopscotch on the playground at recess, and the cool boys were trying to steal kisses from the girls or were jostling for position to look up their skirts on the jungle gym, I was patrolling the perimeter of the playground, digging up earthworms to take fishing when I got home. My mother was not happy when a few of my fishing expeditions got canceled, and the worms in my pockets went through the washer and dryer.

When I wasn't anxiously waiting at the mailbox for my Captain Midnight magic decoder ring and Secret Squadron membership to arrive in the mail, (it took months!), I was outdoors exploring the woods, creeks, and ponds of my rural surroundings: fishing or catching frogs, snakes, snapping turtles, and anything else that moved and fit into my 2-gallon bucket. Summer days found me wandering the woods with my best friend, Lucky, by my side, my ballcap pulled low over my eyes and a sweet stick of grass between my teeth; exploring; taking the time to smell the roses, the pines, and – dang it – the dog doo-doo I stepped in behind the Smith's barn.

Like many boys, dinosaurs fascinated me, as did expeditions of any kind (but best if dinosaurs were encountered). Roy Chapman Andrews soon became my hero with his book, *All About Dinosaurs*, and his tales of discovering dinosaurs and their eggs during his incredible expeditions to the Gobi Desert.

I was a big fan of Tarzan. I spent a lot of time swinging from the limbs of the maple trees in our front yard, clad in a leopard-patterned loincloth, attempting a Tarzan yodeling call – "aahuaaa uaaa uaaa" (or something like that), but never even coming close to succeeding.

Rocks were not initially of great importance. They were for dodging when the big kids threw them at me. Dirt clods were much more fun for friendly wars with your buddies; the return fire didn't hurt so much. But the big bullies were playing with hard ammo. I picked up their rocks and slung them back. They just laughed since I was not a good pitcher. One fateful day while being pelted by the bullies, I picked up a rock to throw it back. It had bright, sparkly specks in it. I picked up another one of their missiles. It had what looked like tiny shells in it. I pocketed the rocks, much to the bewilderment of my tormentors. The next time I came under fire, I put all the rocks in my Roy Rogers lunch pail and yelled at the big boys, "Come on, is that all you got? I need more rocks for my collection." That took all the fun out of it for them. I ended the bullying, started a rock collection, and unwittingly launched a future career in the process.

We had few interesting native rocks within my radius of wandering as a kid. Most of the good stuff came from our neighbor to the north, Canada. Massive glaciers, up to two miles thick, scooped up sparkly crystalline rocks like granite, schist, and gneiss from the Canadian Shield, carried them south across what would become the Great Lakes, and dumped them in my backyard when the ice melted some 15,000 years ago. All that was accomplished without a single permit, let alone an Environmental Impact Statement, and long before Al Gore figured out how to get filthy rich off of global warming alarmism. Under the glacial till with its illegal alien rocks was a thick section of black shale, deposited in the deep Devonian Sea about 350 million years ago. The shale was boring, but it made good skipping stones.

The best Christmas present Santa ever slid down our chimney and deposited under the tree was an Estwing rock pick, the old style with a handle made from shellacked rings of leather, stained to look like wood grain. Since it was the dead of winter in the western New York snowbelt, my shiny prize held a position of honor on my bookcase, waiting impatiently for several feet of snow to melt and expose my quarry. As soon as the snow was gone, my shiny rock hammer, my dog Lucky, and I spent every free minute outdoors searching for rocks. Nary a cobble or boulder escaped the wrath of my hammer. Mightier than Thor's Hammer it was.

My geological world expanded considerably as a young teenager, when I learned about the fossil-rich Devonian rocks of 18-mile Creek, thanks to A. W. Grabau's epic treatise, *The Paleontology of Eighteen Mile Creek and the Lake Shore Sections of Erie County, New York*, published by the Buffalo Society of Natural Sciences in 1899. On summer days, my dad would drop off my faithful companion and me at the bridge over 18-mile Creek on his way to work and pick us up on his return. My best buddy and I spent the day searching for brachiopods, crinoids, trilobites, and other extinct invertebrate critters; and sharing my PB&J lunch. We had the creek and the Wanakah Cliffs along Lake Erie's shore to ourselves, with the occasional visit from an angler or two hoping to catch a rare giant sturgeon in the lower reaches of the creek. It was a magical time in a young nerd's life.

Lucky guarding the trilobite beds of the Ludlowville Formation, Eighteen Mile Creek, 1964.

Everyone has a particular teacher who stands out as a mentor, that one dedicated educator who helps shape your path through life. For me, it was my 8th-grade science teacher, Bernard Petrusky. Mr. Petrusky's classroom was a combination of a museum and a zoo. Cages with rabbits, mice, snakes, frogs, and turtles lined the walls, sharing space with odd plants, bones, skulls, and rocks. There was even a resident alligator living in the sink. This was my kind of classroom and my homeroom as well. I was in nerd heaven. Because 8th grade science class was general science, Mr. Petrusky had no set curriculum. We never knew what was going to happen in class. One day we might study light refraction using prisms. The next day we could dissect chicken eyeballs (which usually happened the Monday after his neighbor had a big barbeque).

Bernie was driving to school one morning when he came across fresh road kill – a young calf on the side of the road. The average person would have driven on, perhaps glancing at the poor deceased bovine and maybe even feeling sorrow for its early demise, sad it had been deprived of a few years of grazing before its ultimate destruction at the hands of a burly man with a hammer; then cleaved into steaks and ground into hamburger. We are not talking about an average person here. Not to miss an opportunity for educating his students (and grossing out the squeamish ones), Bernie loaded the calf's carcass into the trunk of his car, with its legs hanging limply out the side, and brought it to school. When I walked into homeroom, the calf was laid out belly-up on the second-row table, ready for the day's classes to dive into the glistening wonders of its inner workings. It took two days for the various classes to complete the autopsy. By then, the other teachers on the second floor, even those at the far end, were complaining about the overwhelming stench that emanated from room 204.

I think it was Mr. Petrusky who told my junior high school guidance counselor I was interested in science. It certainly wasn't me because I never went out of my way to talk to such extraneous people. I was required to learn a foreign language. Any good counselor with a crystal ball would have recommended I study Spanish. But back then, we were too far north to feel the impact of the wave of immigrants from south of the border. You could order a 15-cent McDonald's burger without even a rudimentary knowledge of *Espanol*. Learning Polish would have been a good choice, given the large Polish population of the Buffalo area. Had I done that, I might have been able to sweet-talk Mrs. Bojarczuk into letting me take Eva, the smartest student in my class, to the next Polka dance. Eva wasn't a bad-looking gal. She had that exotic foreign look to her – with long braided blonde hair – both on her head and under her arms. Sadly, it was not to be.

"So, you are interested in science?" the counselor posed.

"Yes," I replied sheepishly, not knowing what she was up to.

"Then you should study Latin."

"Latin, really? Why should I study Latin?" I inquired, thinking she was trying to trick me into qualifying to be an Altar Boy or at least to be able to understand what the heck the priests were mumbling at High Mass.

"Latin is the root of all scientific nomenclature. It would be helpful for you to learn it if you want to pursue a career in science."

Deprived of the chance of a date with Eva, I wasted four years studying a dead language, one that I soon learned truly deserved to die. To make matters worse, to this day, I cannot converse with fast-food workers. Oh, the lost opportunities of youth!

Once I finished four freaking years of Latin, my counselor recommended I learn German. "Why German?" I asked. "Didn't we beat the crap out of the Krauts in World War Two?" The counselor's argument was that German is also a language used extensively in science. That is true, but only German scientists use it. Even Wernher von Braun gave up on German and learned English after he surrendered to the US Army and joined NASA to make up for killing tons of people with his V-2 rockets. I dutifully studied *Deutsch* for two years and have not used it since. As it turns out, Germans today speak better English than half the American population, although they don't speak Spanish nearly as well as that half.

Despite being a good student, I admit to having been somewhat of a miscreant who didn't quite fit in socially and got into a fair amount of trouble in junior high and high school. Perhaps I was bored, or a wee bit rebellious (my Scotch-Irish heritage may have contributed to the latter), or both. It didn't help that

I have always had a problem with authority figures. Long before speed dial was invented, the principal had my parent's phone number memorized. I don't recall the specific reasons for most of my detentions. However, one stands out clearly in my mind. It was 10th grade – fourth-period biology class. We were dissecting leopard frogs. Jerry Cully, who sat beside me in most classes (likely having something to do with our surnames), was my partner. We sliced and diced our beautiful *Lithobates pipiens,* which reminded me of several I saw in the pond behind the Koningisor's house that summer *"Excuse me, Mr. Frog, have we met before?"* We finished the autopsy long before the rest of the class, pinning the frog's guts to a board with all the proper labels. This time I was bored. I surveyed the room and Mrs. Eller, a short, stocky blonde (I am sure she shared DNA with Frodo and Sam), was nowhere to be found. She probably snuck out for a smoke break, the other possible reason for her stunted growth. I stuck one of those long metal probes with a wooden handle into poor Kermit's webbed foot and held it aloft in triumph to celebrate our record dissection time.

Before long, with Jerry's encouragement, Kermit's formaldehyde-soaked body, sans innards, was spinning around the end of the stick. Next thing you know, my biology lesson turned into a physics demonstration. As the frog's body spun around the axis of the sticker, it slowly slid up until it came off the end. Suddenly, the kinetic energy of the frog's spinning centrifugal force converted to centripetal force. Kermit was flying. It would not have been so bad if he flew behind me and splatted against the wall, or if he took a short hop a couple of rows to the northeast and landed in Sally Cotton's hair (I had been trying to work up the courage to talk to her – any excuse would do). No such luck. My frog flew straight forward, zoomed over five rows of budding biologists busily searching for their frogs' hearts, livers, and kidneys, until gravity took control and he landed with a big splash – *kerplop* – in the fish tank that sat on Mrs. Eller's workbench, just as that dwarf of a biologist stood up from looking for supplies in the cabinet under the bench. I sat there in stunned silence, my frog poker raised in salute to Kermit, the first amphibious aircraft to grace the classrooms of East Aurora High. Busted, I was marched off to room 118, the dreaded lair of my archenemy, Mr. Lawson, the high school principal. Yet another afternoon of learning opportunity was lost to detention.

I got in serious trouble in Math 12b, Analytic Geometry/Calculus class, in my senior year, but luckily avoided another visit with Mr. Lawson. Charlie Pfleeger, a withered old bald guy who should have retired years before, was the math teacher. He never smiled, perhaps because a life dealing with unsolvable equations containing too many variables left him feeling frustrated, or maybe he had dreams of working for NASA – calculating the flight of the first rocket to the moon – only to realize he didn't have a chance of working there since he never studied German and couldn't communicate with Herr Professor Braun, who was still working on his "English as a Second Language" course. Charlie's career was just about finished, having peaked with trying to teach numbskulls the difference between a definite integral and an antiderivative. He always came late to class. No one was sure why. My theory was he dreaded dealing with us rowdy seniors.

Mr. Pfleeger used an overhead projector instead of the blackboard. After greeting the class with a sneer, he sat down at the projector and started drawing x and y axes and writing formulae. We were supposed to construct squiggly lines based on the formulae: parabolas, hyperbolas, asymptotic curves, voluptuous curves! While waiting for our beloved purveyor of algebraic equations and graphs to arrive, we amused ourselves by writing a joke or drawing a cartoon on the projector screen. Our artwork popped up when he turned on the projector. Charlie would snarl, grab his bottle of eraser fluid, and quickly wipe the overhead clean. It was a fun way to start a boring class.

One cloudy October day, Mr. Pfleeger came to class with an even sourer look than usual on his tired face. He was in a particularly foul mood. Perhaps his hemorrhoids were itching, and he ran out of Preparation H. We didn't ask. He flipped on the projector. A nonsensical equation with a graph that looked like small intestines during a diarrhea attack appeared on the screen. Charlie was not amused. He had enough of this impertinence. His beady eyes scanned the room, focusing on the back row where I was sitting between Chuck and Ed. Charlie looked directly at me and shouted, "You, Cuffney, get out of my class, and don't come back!" I was wrongly accused; the cartoon of the day was not my work (yesterday's was, but not today's). This beauty was the creation of Chuck and Ed. I could have argued my innocence, noted that

I paid attention to his lectures and would never write such an improbable equation, that dx/dy should have been written as dy/dx; then turned in the guilty parties. But you don't rat on the star fullback and tight end of the Blue Devils football team. Plus, it is not advisable to argue with a crazed mathematician armed with a wet-erase felt-tip pen. I took one for the team. I gathered up my books, walked out the door, and waved farewell to the somewhat shocked class.

Through my sacrifice, Chuck and Ed would retain their eligibility; their reputations would be unsullied; they would go on to score more touchdowns; the pretty cheerleaders would fawn over them. In exchange, I got seventh period to aimlessly wander the halls of East Aurora High, belonging nowhere – like the man without a country. Mr. Pfleeger, no doubt in a fit of rage-inflamed dementia, neglected to send me to visit Mr. Lawson; he also failed to notify the admin office that I was no longer attending his class. I received a token bonus for my sacrifice – a degree of notoriety and newly found respect from my fellow students. Some of the cute girls even started saying hello to me.

A couple of cross-country camping trips with my family opened this eastern lad's eyes to the wonders of the west – real mountains – and rocks everywhere. I wanted to bring them all home and tried hard. My dad never forgave me when the car-top carrier broke under the weight of my bootlegged rock collection. If I couldn't bring the rocks home, I would have to go to them. The decision was made; I was going to head west, young man, and become a geologist. The decision of which college to attend was almost a difficult one. I considered South Dakota School of Mines since I was interested in paleontology and had visited the fantastic collection of Badlands vertebrate fossils in the school's museum. On one of our cross-country trips, my dad, my younger brother, and I spent a hot July afternoon searching (successfully, I might add) the Badlands for oreodont and titanothere bones just outside of the park, while my mother and my sisters roasted in the 100-degree heat of the campground, a sacrifice for which they should be sainted. The Black Hills offered opportunities for outdoor escapes and decent skiing. Rolla School of Mines in Missouri was a consideration but was a long way from mountains and skiing. Colorado School of Mines (CSM) had access to great skiing AND a brewery in town. I made the decision to enroll at CSM if they would accept me. I applied for early admission, hoping the admissions people would not notice I was no longer attending Mr. Pfleeger's math class, a prerequisite for admission. My deception worked. I was accepted and was off to Golden, Colorado to attend the Foremost School of Mineral Engineering (not that there is much competition for second place), and the torture of engineering classes (alleviated by copious quantities of cold Coors) on the way to becoming a Geological Engineer. I had no clue what it entailed to become one or what one really did for a living, but was soon about to learn.

Chapter 2

THE RAMBLING WRECKS

I wish I had a barrel of rum and sugar three hundred pounds,
The college bell to mix it in and clapper to stir it 'round.
Like every honest fellow, I take my whiskey clear
I'm a rambling wreck from Golden Tech, a helluva engineer.

"The Mining Engineer" – Anonymous (1937)

Leaving family, friends, home, and most importantly, man's best friend to find a new beginning at a far-away college can be a traumatic experience for a young lad. In late August 1967, I apprehensively boarded a United Airlines DC-8 at Buffalo International Airport, headed to Chicago, connecting to Denver. I was off to Golden, Colorado, to attend the Colorado School of Mines. Golden lies about 15 miles due west of downtown Denver, tucked away in a verdant valley along Clear Creek, between the eastern edge of the Front Range and the two basalt-capped buttes of North and South Table Mountain. Entering town for the first time, I was greeted by the big arch over Washington Avenue announcing the town's western hospitality, "Howdy Folks, Welcome to Golden, Where the West Remains" and the equally welcoming aroma of hops and barley emanating from the Coors Brewery on the northeast side of town. My apprehension faded. Something told me I was going to like this place.

Mines, as we affectionately called our home/prison for the next four (or more) years, was a real eye-opener for this eastern lad. The brainiac engineering students thought it was cool to carry slide rulers on their belts, like a swinging badge of honor. For once, I wasn't the nerdiest student. I was surrounded by nerds of all kinds: big ones, little ones, skinny ones, fat ones, domestic ones, and foreign ones. In that crowd I almost could have passed as a cool dude. I might have even become a stud, the college heartthrob, impressing the cute gals with my suave and debonair personality. There was only one problem, other than my lack of a suave and debonair personality. My school selection, which I based on the exceedingly important elements of skiing and beer, left out one critically important factor – girls. My freshman class comprised something like 365 students, of which only 12 were females. Five of the coeds were, shall we say, less than attractive; three were man-hating women's libbers; we were uncertain about the sexual orientation of a couple of the others. That left Nina and Anita for 353 guys to fight over.

My younger son, Craig, was much smarter in his college selection criteria. He was interested in either architecture or engineering. When it was time for him to apply to universities, we took a trip to visit Cal Poly in San Luis Obispo and UC San Diego. Since we passed through Santa Barbara, we stopped at UCSB to check out the campus. It was a warm, sunny fall afternoon. Kids were playing soccer on well-manicured grass fields. Girls in short shorts and halter tops were jogging. Guys were surfing the waves of the bay. Gorgeous, impossibly tall girls in skimpy bikinis were playing beach volleyball. Bronzed goddesses with sumptuous, perky, round breasts bursting out of itsy-bitsy bikinis were sunbathing on the beach below the dorms. Craig and I looked at the college's catalog – no engineering or architecture classes, let alone degrees. Undaunted, Craig allowed he would really like to go to UCSB, despite that insignificant flaw. "I could get a Liberal Arts degree here, then go to grad school for architecture," he argued. I knew where this newfound interest in UCSB was coming from. Heck, I was thinking about enrolling in a few classes myself. I was not taking the bait and stuck to my guns. Then the testosterone-charged teenage logic kicked in.

"Dad, I totally *have* to go to school here," my darling son insisted as he perused a college brochure, he picked up at the information kiosk.

"Why? We've already discussed this. The college doesn't offer the courses you want. I am not going to pay for you to get a degree in underwater basket weaving."

"Dad, it's the 'Leftover Factor'."

"The what?"

"The Leftover Factor. It's huge."

"What is that?"

"It's over 2,000! That's the most of any college I have looked at."

"I mean, what the heck is the Leftover Factor?"

"There are 2,000 more girls than guys here. Do you know how easy it would be to get a date?"

Well, it appears that kids today choose the education that will shape their lives and careers and leave their parents broke based on leftover co-ends. If only I used the same criteria, I might have had a date or two in college. Instead, I was stuck drinking beer, skiing, exploring old mines, lighting farts, and "murdering" my frat brothers. I did not realize it at the time, but those were all essential skills to master on the way to becoming an exploration geologist.

Colorado School of Mines was about as "old school" as you could get. The first week of school was dubbed Freshman Agitation Week, officially advertised as, "A time-honored tradition, designed to instill in the freshmen some of the history and spirit of the school." Agitation, my ass, it was brutal hazing. As for spirit, one must assume sadomasochists were in charge of the school's spirit. For an entire week, we had to wear our clothes backward. I guess that was supposed to help roommates and classmates bond – "Hey Ralph, can you zip me up here?" Any upperclassman could stop you between classes and shout, "Button up and sound off!" The proper reaction was to hold your books over your head and run around in a circle like a headless chicken babbling some ridiculous saying about the next football game. I did that for a couple of days, then took another approach – run like hell to the next class. That strategy got me out of the button-up thing, but I paid the price later. The freshmen were required to report to the intramural field after class on Friday afternoon. Two parallel lines of upperclassmen holding wooden and leather paddles were assembled there. We had to run between the lines of the gauntlet while the upperclassmen took swats at our behinds. The amount and strength of swatting were largely a function of one's compliance with the rules and regulations of Agitation. Guys were looking for the rebels who refused to button up and sound off. Of course, my name was on that list. If there ever was a time when speed was paramount, this was it. Ever try to sprint with your pants on backward? *Run, Forest, run!*

There were some pleasant aspects of Agitation Week, like the midnight songfest. The frosh assembled on the football field to hear Kenny Walker, one of the few All American football players that the Orediggers team ever produced, serenade them at the stroke of midnight with a rousing rendition of "Charlotte the Harlot," which lacking any alumni who became famous songwriters and could produce something better, was the unofficial school song. I remember the chorus well:

She's dirty, she's vulgar, she spits in the street
Why whenever you see her, she's always in heat
She'll lay for a dollar, take less or take more
The pride of the prairie, the cowpuncher's whore.

The song went on for several verses, coming to a rather sad end when the lovely lady had a run-in with a rattlesnake and a cowboy with a pistol. I could go on, but you get the picture.

Agitation Week ended on Saturday with a fun outing in the mountains. Prior to that event, each freshman had to collect a rock weighing a minimum of ten pounds. For my first geology project, I chose a rounded cobble of sparkly Silver Plume granite that washed down Clear Creek from the high mountains in some ancient flood. The class was transported en masse on flatbed trucks up the Grapevine Road, whose switchbacks wound their way up the face of Mount Zion, overlooking the campus and the town of Golden. The destination was the M, a massive pile of whitewashed rocks laid out on the steep side of the mountain in the shape of, you guessed it – an M. The M, built by the senior class of 1908, is the iconic symbol of Colorado School of Mines and the greatest source of pride among the students. Part of that may be because it is illuminated at night and is visible from just about anywhere in the Denver area, a veritable lighthouse for lost Mines students, beckoning them home from their drunken off-campus excursions. Our mission was for each of us to add a rock to the M, thus earning pride in ownership. Then we gave the giant emblem a fresh coat of whitewash. The formula was two parts whitewash for the M and one part for your classmates – not a good time to wear those new jeans frontwards for the first time. Four years later, we repeated the process on Senior Day.

Things did not get much better after Agitation Week. CSM was a land grant school, which meant two years of Army Reserve Officers' Training (ROTC) was mandatory for all male students, essentially the entire class – minus the handful of ladies and the foreign students. I should have claimed I was a poor refugee from Kenya. That trick has provided really nifty privileges for some people, and would surely have spared me the pain of Military Science classes. As pseudo-Army recruits, we were required to have short military-style haircuts – really? – in the middle of the 1960s, the age of hippies and free love, not to mention the age of long hair? Despite that, I actually enjoyed the classes. We learned about maps and orienteering, jungle survival, exotic diseases (with a focus on the venereal brand), breaking down and reassembling rifles, and making and using explosives – all very useful skills for freshmen nerds or anyone else bent on mass murder, especially those crazed kids jumping up and down at recruitment centers singing a verse from "Alice's Restaurant."

My worst problem with ROTC came on Monday afternoons. We polished our brass insignia, which clearly advertised our status as lowly Privates, with Brasso; spit polished our terribly uncomfortable Army-issue dress shoes until they shined like black mirrors; donned our ugly olive-green uniforms; then headed to the armory. There we picked up WWII-era M-1 rifles, shouldered them, and marched down to the football field next to Clear Creek where we were supposed to perform various marches and drills. The drill instructors soon learned that marching was not my thing, especially when the wind blew up the valley from the east, the site of the Coors brewery. The smell of hops and barley wafting on the breeze was too much for me. My mind wandered to thoughts of taking the "short tour" of the brewery – straight to the tasting bar, for a couple glasses of ice-cold sweet elixir of the gods. On more than one Monday afternoon, the instructor barked, "Column leffffft!" but I mistakenly turned right, heading solo toward the source of the enticing aroma and marching smack into the rest of the column.

Mines had a few other odd traditions. In addition to underclassmen being forced to have short haircuts for two years, they could not sport beards or mustaches. Those manly adornments were reserved for seniors. Imagine my disappointment when after three long years of waiting, I finally became a senior only to realize I couldn't grow a beard worth a crap. My best effort looked like my crotch crawled up and established a new home under my chin. Seniors were the only ones allowed to wear Stetson hats. I took advantage of that tradition, in part to make up for my lack of whiskered prowess.

I should have used my son's criteria for selection of an institution of higher learning. The dearth of females at Mines was a killer. The rich kids who had wheels could travel to Denver to pick up chicks at the all-women colleges: Colorado Women's College and Loretta Heights College, where the gals were as desperate for dates as we were. Or they could drive a mere 70 miles north to Colorado State College in Greeley, which had a favorable Leftover Factor. Colorado University at Boulder had a surplus of hot chicks. But they were all dope-smoking hippies, who were busy protesting the ongoing war in Vietnam and wanted nothing to do with short-haired, gun-totin', beer-drinking redneck Miners. Traveling south to Colorado Springs for dates was not an option. The Zoomies at the Air Force Academy had a lock on that market. Guys like me, who lacked transportation, were stuck with the rather limited local talent from Golden High. A partial solution to this dilemma was to join a fraternity and at least be able to have an occasional double date with a frat brother.

Fraternities at Mines were not like the elite, snobby social clubs of large universities. They were groups of ordinary guys, who joined together to tackle the challenges of earning an engineering degree, while having some fun and comradeship along the way. I joined Sigma Phi Epsilon fraternity, the Sig Eps, the only fraternity with low-enough standards to accept me. Once again, I found myself back in the world of hazing. Oh, the price one pays for the chance to get a little nooky!

Fraternity Pledges were treated as second-class citizens. We waited upon the Brothers and did all the house chores. That was fine; we had to earn our way into the society. It was the crazy hazing that was a challenge. About once a month, out of the blue, the Brothers decided that one of the Pledges (more often than not, me) had screwed up. The entire pledge class needed to be disciplined. They hunted us down in our dorm rooms in the middle of the night and dragged us back to the frat house for a "Fire Drill." A crackling fire was burning in the chapter room fireplace. Stripped down to our skivvies, we were required to form a circle and crawl on our hands and knees past a big bowl of slimy, gross, foul-tasting liquid – "fire retardant" – and suck up a mouthful. We crawled over to the fireplace and spit the evil fluid on the fire to put it out. It didn't work very well, since the two main ingredients of the fire retardant were chili pepper and kerosene. Of course, the Brothers employed paddles to spur on the firefighters.

Instead of spending our semester break at home visiting family and friends, or taking a ski vacation, maybe even partying on Florida's beaches like other college kids, we returned to Golden a week early to partake in Hell Week, aptly named for seven sleep-deprived days of doing house chores, repairs, and other slave duties. The event kicked off at midnight with the Pledges lined up for a good paddling – just to get things off to the proper start. We were stripped of our names in classic POW style, and given "Scummy" numbers; I was Scummy #4. The actives took turns supervising us 24 hours a day. Anyone who was caught sleeping was hauled off for butt paddling. It didn't take long before we worked out a system where we took turns getting 20 minutes or so of sleep, usually crammed into a closet or behind a piece of furniture. When the Brothers started looking for a certain missing scummy, one of us would sneak off and rouse the culprit before he got caught.

The week culminated in Hell Night, also aptly named. It began with the Mother of All Fire Drills, followed by a series of silly, gross, and/or humiliating events. We were, of course, naked throughout the ordeal and blindfolded for much of it. One of the more perverted but funnier events was the Olive Race. Split up into teams of four, we lined up on the dining room floor, where piles of green olives were at one end of the room; dixie cups sat at the other end. The task was to deliver the olives to the Dixie cups – without using your hands. We assumed a crab walking position and attempted to grab an olive in our cheeks; I am not talking about the ones with teeth between them. Once an olive was secured, the challenge was to crab walk it across the room and deposit it in the cup. Many olives succumbed to the rigors of the event, ending up smashed on the dining room floor. A few never made it across the room, yet were never seen again. Where or where did those go? A few hearty olives made the trip successfully and were deposited in the dixie cups. To this day I am amazed at the dexterity of the derrieres of some of my Brothers.

The next event was truly gross. The head chef spent the evening cooking up a combination of sausage, corn, peanut butter, and Limburger cheese. This he rolled into little turd-shaped pieces, which were strategically placed in the downstairs commode (which had been scrubbed and sterilized). Each of us was led blindfolded to the head. At the edge of the crapper, our blindfold was removed. We stared into what certainly looked and smelled like a nasty case of number two. We were told one of our pledge mates snuck into the bathroom, took a crap and did not flush the toilet. Our penance was to take a turd and eat it. Needless to say, it was a good thing the chef made a lot of artificial turds. Several of us promptly filled the bowl with the afternoon's lunch, requiring cleaning and restocking of the commode. Fortunately, my mother was a big fan of Limburger cheese. I was used to having food that smelled like excrement on the dinner table, so the stench of the little turds didn't bother me.

Perhaps the most perverted station was the Brick Drop. Blindfolded and buck naked, we sat on the upper bunk of a bed with our hands outstretched, palms upward. A string was tied around our family jewels, and a brick with a string tied around it was placed in each hand. We were told that one brick was tied to our pride and joy with a string half the length of the height of the bed; the string on the other brick was not attached. Some nonsensical clues were given to assist in choosing which brick was which; then we had to drop one brick. In fact, neither brick was attached, but that was a well-guarded secret. Gary Garlough, who was a Pledge the year before me, was an exceedingly horny young fellow, who was exceptionally proud of his gonads. Gary sat on the top bunk for a while, pondering what to do. His ingenious solution was to jump off the bed, simultaneously throwing both bricks in the air, hoping to beat the bricks to the floor. He succeeded, but one brick landed on Tom Roberts' knee, giving him a nasty bruise. Gary's gonnies were saved, but his behind suffered a few extra swats for that move.

Did I mention sadomasochists at Mines?

The hazing during Freshmen Agitation Week, the fraternity's fire drills, and the crazy antics of Hell Week may seem like sick and perverted torture by today's standards. But they all had a purpose – creating a bond among the participants – a lasting one at that. The hazing prevalent in fraternities in the '50s and '60s was not like the heavy drinking contests of later years, which unfortunately resulted in deaths and nearly put an end to the fraternity system. Based on military boot camp (many college students in the late 40s and 50s were WWII and Korean War vets), the hazing was designed as a group exercise for the participants to face unusual and difficult challenges, and pull through as a unit. It worked. My pledge Brother, Larry Fischer, and I attended the national convention of Sigma Phi Epsilon in Memphis during the summer of 1969. The big topic of the meeting was reducing or eliminating hazing. We came back full of enthusiasm and convinced the Brothers to change the system. Guess what?

Halfway through the semester, the Pledges revolted. They saw the incredible bond shared by previous pledge classes and realized they did not have that. They demanded we restore hazing!

To this day, most of my true friends are my Sig Ep Brothers. Although we are spread out across the country and the world, we keep in touch regularly. We have held six reunions over the years, with as many as 60 Brothers, wives, and offspring attending – including our 50th held in Golden in July, 2021 (some of the *other* guys were looking a little wrinkled at that one). We went through a lot together, when we were lowly scummies, and afterwards as Brothers working and studying together. We will not forget our unique bond, and how we gained it.

The Sigma Phi Epsilon mob - 1971; the author is in there somewhere (like under the heart with a bowtie)

Reno reunion & pool party;   *Representatives of 1967 Pledge Class at the 2006 reunion: l to r – Joe Rousseau, Rick Ramert, the author, Charlie McNeil, Larry Fischer*

CSM had high admission standards. We freshmen were about to learn why. It was even more difficult to stay in school than to be accepted there. Taking a full load of engineering, math, physics, and chemistry classes was difficult. Attrition took a toll on the freshman class. Organization, focus, dedication, and good study habits were required to keep one's grades above the 2.0 grade-point average needed to stay in school. I lacked all four essential qualities and almost became a casualty. In the middle of the first semester, I was failing every class except one, Geology 101, in which I had an A. I was having fun, enjoying the freedom of being 1,500 miles from home and living in the shadow of a brewery. Fortunately, Norm Lewis, my Big Brother in the fraternity, hauled me kicking and screaming to his room and sequestered me there, forcing me to study five nights a week. It worked. In the second half of the semester, I pulled my grades up to Bs, with two exceptions: an A in Geology and a compensating F in BE 105A, Basic Engineering. It didn't help that the engineering professor was a particularly hard nut to crack. If you were brave enough to ask Professor Preston a question, he would usually answer, "You're the engineer, you figure it out." Wonderful advice for a neophyte freshman! That attitude characterized Mines. If nothing else, those students who survived to graduate left with a strong sense of self-reliance, an essential trait for an exploration geologist who would spend the better part of his career working alone.

Studying hard five days a week can make Jack a dull boy. Weekends were for blowing off steam. The lack of loose women to cavort with required inventing different distractions. Water fights were one, as were cherry bomb and pop bottle rocket wars. The former is almost normal, involving balloons, buckets, and lots and lots of water. The latter two require some explanation. Cherry bomb wars consisted of dueling parties of two warriors armed with slingshots and cherry bombs, better yet M-80s when we could get them. The Sig Ep warriors attacked the Kappa Sigma house across the street. One attacker loaded the round and pulled back on the slingshot while his partner lit the fuse and yelled, "Fire in the hole!" Volleys of miniature dynamite charges flew back and forth, followed by loud booms and whooping, hollering, and cussing. I will never forget when Randy Parsley was firing at us from the roof of the Kappa Sig house. One of our mortars landed on the sloped roof and rolled toward him, chasing him up into the overhang of the second floor. It went off there – 200 decibels in a confined space! Poor Randy was half-deaf for the better part of a week.

Pop bottle rocket wars were much more organized, perhaps a function of our two years of tactical training in ROTC class. Teams were formed and strategies planned. Armed with plenty of pop bottle rockets and employing coke bottles and cut-off ski poles as launchers, we made nighttime attacks on the enemy lines and were counter-attacked. The inky darkness of the night was suddenly filled with the tracks of dozens of rockets whizzing through the air and bright bursts of varicolored light as they exploded. The rules required wearing ski goggles or safety glasses to prevent eye injuries, without which there would have been many blind combatants. We added the additional requirement of earmuffs after a pop bottle rocket stuck in Tom Palmer's left ear and exploded. Tom was never quite right after that. It was no big deal – he wasn't right to begin with.

Water fight: Mark (DB) Frauenhoff and Dennis (OD) LaPrairie firing away
... and the rockets' red glare, bombs bursting in air.

Of course, we deny any connection to the rocket that flew into an open window of the SAE house and set the drapes on fire. That surely was not one of our missiles.

The subject of cherry bombs reminds me to digress a bit and introduce Bruce Peers, the biggest nerd in the Sig Ep house. Being nerds of some degree ourselves, we generally accepted our kin. But Bruce, a skinny little guy with thick glasses, was loudmouthed, obnoxious – just downright irritating. He was the human embodiment of fingernails on a chalkboard, that itch you can't reach, the zit on your forehead that refuses to go away before the senior prom despite three attempts to pop the damn thing. To make things worse, Bruce had a girlfriend, Monica ("Monica Hotbox" as she was known to everyone but Bruce). Every evening Bruce called Ms. Hotbox from the pay phone in the hall to profess his undying love for her and to whine about some petty personal problem, like the wedgie he got the previous evening for tying up the phone for an hour while other guys waited impatiently to order a pizza, place a bet on the next Bronco's game with their bookie, or call the suicide hotline in desperation after flunking the latest Physics exam.

The phone booth was tiny, just big enough for one person. It had a bench to sit on and a small shelf, above which was the pay phone. Although it had a solid wooden door, conversations were far from private owing to the six-inch-high open slot above the door. I am not sure why that was there – for ventilation, perhaps. Regardless of its intended function, we came up with a novel use for it. I don't recall who hatched the plan (not me, although I would be proud to have done so), but one of us took a cherry bomb, pulled out the fuse, carefully emptied all the gunpowder, and glued the fuse back in place. Four of us quietly plastered our bodies against the phone booth door, which opened outward. The fuse was lit; the cherry bomb was dropped through the open slot. Bruce freaked out. He pounded against the door. It would not budge. Oh, you should have heard the screaming, crying, and whining that filled the phone booth and echoed down the hall. It was a most gratifying moment. I am sure Monica was impressed with her honey's courage. The fuse burned down to the end. Nothing happened. The screaming diminished to a whimper, followed by the best retort Bruce could muster, "You bastards!"

Another event, which should not go unmentioned, was the Wheelchair Race. I have no idea where it came from, whether a Brother needed one for an injury, or if the boys rolled some poor invalid and stole it. I lean toward the latter. Regardless, there was a wheelchair sitting around the frat house. Not to waste an opportunity for new fun, we enlisted the device for the first and last Wheelchair Race down the first-floor hall of the house. A brave volunteer sat in the chair while one Brother pushed the chair as fast as possible down the hall, turned around and raced back to the starting point. The rider, Barry Norman (aka Barely Normal), and I, the pusher, set a record time for the race. Part of the reason for our fast time was we forgot to stop at the end. The chair punched a big hole in the wall at the end of the hall. We got first prize, which consisted of the opportunity to learn the art of drywall repair.

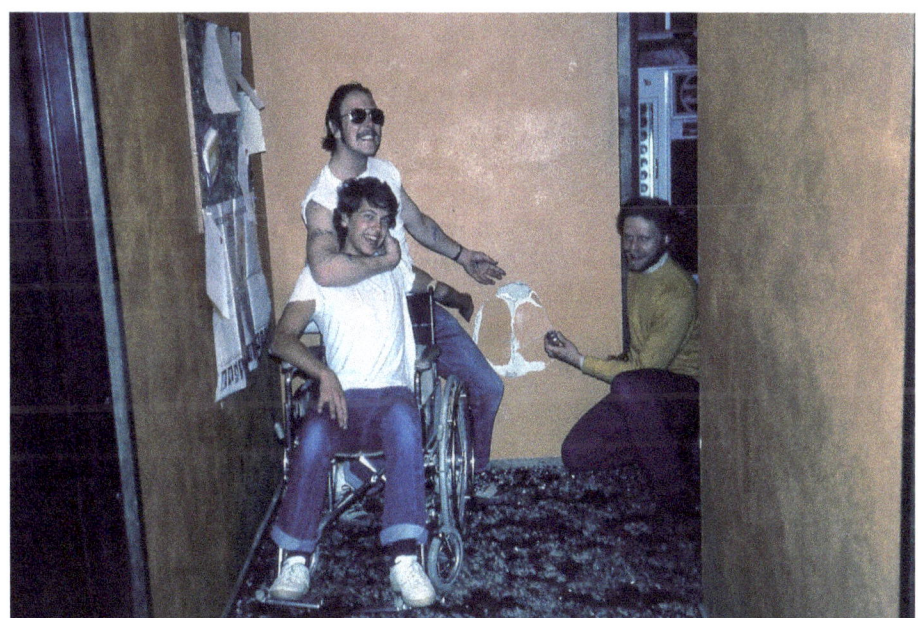

Wheelchair Race champions – Barely Normal & the author, Eric Eckelberg as timer.

Being students at a school called Mines, we were naturally attracted to holes in the ground. There was no lack of them around Central City, a mining ghost town in the Front Range, west of Golden. One of our favorite activities was exploring the old mine workings, a dangerous undertaking which involved hazards such as being trapped by cave-ins, crushed by falling rocks, falling down open shafts, and breathing poisonous air – all of which could easily be mitigated with sufficient quantities of beer. Armed with a keg of Coors (our version of a mine canary), the intrepid mine explorers of CSM wormed our way past caved-in portals to explore underground workings not seen for decades. It took another few decades before the foolishness of those days sunk into my pea-brain.

Central City was founded in 1859 when gold was discovered in Gregory Gulch. The following year, the population exploded to over 10,000 prospectors, miners, and camp followers. The boom turned to bust. In the 1970s, a mere 250 people lived there. Central City survived as a weekend and summer tourist destination, its historic buildings offering a glimpse into the past, the boomtown mining days of the 1860s. The bigger attraction for us was the smattering of rustic old bars: The Tollhouse, Gold Coin, and Teller House to name a few – ones that tended not to check IDs. Thus, Central City was the chosen site for the annual Fart Lighting Contest. Yes, Virginia, with proper training you can indeed light your farts, if you have them. Therein lies the challenge. It takes serious training to build up the quantity of methane required to rip a good combustible one. Contestants took the competition seriously. Each one employed a nutritionist, skilled in designing the most effective dietary routine to maximize flatulence. Red cabbage, hot sauce, and refried beans were favorites. A sure-fire winning regimen consisted of consuming an entire large combination pizza laced with crushed red pepper, flushing it down with copious quantities of beer the night before the contest, then showing up with a hangover and seriously upset stomach. Lactose intolerance was an added advantage.

Another benefit of holding the contest in Central City relates to the ideal gas law, which, as we budding young engineers-to-be learned in Thermodynamics class, is mathematically stated as $PV=nRT$. The amount of gas is "n" (measured in moles), which for the contest we can assume is constant as long as the contestants are not continually stuffing their faces with bean burritos. We also don't have to worry about "R", the universal gas constant, which, as the name implies, is a constant number (8.3144598 J K^{-1} mol^{-1}, in the unlikely event anyone is curious enough to want to know). Put in layman's terms, the law states that if the temperature (T) is constant, then pressure (P) and Volume (V) are inversely related; if one goes up the other must go down. Central City lies at an elevation of 8,510 feet, 2,835 feet higher than Golden. If you have ever taken a bag of potato chips, packaged at an elevation of 955 feet in Des Moines, up into the mountains, you know where I am going with this. Air inside the bag expands as the confining pressure at high altitude decreases; the bag looks like it is about to burst. The same thing happens when intestines filled with growing flatulence are rapidly transported to higher elevation. Simply put, that gas wants *out*! At about 9,500 feet elevation, halfway between Georgetown and Loveland Basin ski area, expansion increases to the point at which the control valve, the sphincter, fails and a rush of wind escapes. We dubbed this phenomenon, which we experienced every time we ascended Loveland Pass to go skiing after a night of partying, as the PDFs, Pressure Differential Farts. Central City was at the perfect elevation, where the ambient barometric pressure was low enough for the gas volume to increase, but the elevation was below the critical PDF level, where all the precious gases could be lost before the competition began.

On the day of the event, the contestants, judges, and spectators gathered at the Gold Coin bar. Competitors sat around a large table while the spectators, including dates of the guys who were intent on not having a second date, stood around the table. Rules were read: judges would award points for each fart lit, based on light and sound effects including volume, pitch, duration, and color – with extra points given for polychromatic displays. The contest began when the head judge announced, "Gentlemen, light 'em if you've got 'em!" When a contestant felt a good one coming on, he exclaimed, "Got one!" He leaned back in his chair, spread his legs, and held a lit match to the point on his jeans where the legs met. With any luck, good skill, and proper dietary preparation, a roar and flash of light would erupt from the point of ignition. Negative points were charged for emission of obnoxious odors (usually the result of incomplete combustion), ignition failure, and blotched shorts (judges could call a "shorts check" anytime they sensed a wet one being released, usually tipped off by a blown-out match and an "Ah shit!" from the contestant). Other than badly blotched shorts, the worst boo-boo was when a contestant exclaimed, "Got one!" then failed to produce anything. That was a disappointment for the crowd and an embarrassment for both the contestant and his trainer. Bonus points were awarded for exceptional displays of inflammatory flatulence including Tassle Whistles and Follow-up Flappers. The crowning glory was to rip a Triple Flutterblast, a spectacular display of rapid-fire farts accompanied by loud bursts of multi-colored flame, sort of like the grand finale at a Fourth of July fireworks show. As far as I remember, no one ever accomplished that lofty feat.

An entire year passed between Fart Lighting Contests, leading to all kinds of opportunities for boredom and subsequent displays of insanity. Imagine what could happen when a group of wild-ass Mines students, burned out from a week of sliding their K&E slide rulers back and forth and staring at differential equations, decide to blow off a little steam. Several of us were sitting around on a Saturday morning having a few breakfast beers. Soon, we were discussing ways to liven up the sleepy town of Golden on this crisp fall morning. The ingredients for the day's event started to fall together. Tom Palmer, alias Pax Dominos, the guy who had the pop bottle rocket blow up in his ear and was never quite right again, said he had some "vampire blood" left over from Halloween. Chuck Malachar announced he could borrow Tom Sander's shotgun and some blank shells (real blanks – not Alec Baldwin-style). Gary Lubers had his 1960 Volkswagen van. I had my camera. The plot was hatched; we would "murder" one of our fraternity brothers. Oh yeah, that should get people's attention! Pax volunteered to be the victim. He stood in front of Foss Drug Store, at the main intersection in town, with a mouth full of fake blood and a mixture of the fake blood and hamburger in his hands. Gary Lubers' Volkswagen van sped around the corner and slid to a stop in front of the store. The side doors flew open and Richard "Zombie" LaPrairie fired a round from the shotgun. Pax threw his head back and spit out his mouthful of fake blood, which landed on his face. He threw the hamburger in the air and collapsed in a heap on top of the bloody mess. Two gangsters, Richard and Chuck, jumped from the van, grabbed the bloody body, and threw it into the waiting get-away vehicle. In typical underpowered VW style, it chugged up the hill on 13th Street at a whopping 5 mph.

Engineers may be good at calculations, but are the antithesis of psychologists. They sometimes forget to enter certain variables into the equations, like human reactions. We planned the murder in great detail, but failed to consider what might happen next. The attendant of the Conoco gas station across the street from Foss heard the commotion, saw the body on the ground, and witnessed the VW speeding away from the crime scene. He jumped in his 4x4 pick-um-up truck with his loaded rifles hanging proudly in the gun rack across the back window with its NRA stickers and followed the fleeing culprits up the hill and across campus. Fraternity Row was located at the far end of campus, at the intersection of 19th Street and US Highway 6, the road leading up Clear Creek Canyon to the mountains. The intrepid attendant followed the van until it got near the intersection. He assumed the murderers were heading up the canyon to dispose of the body. He turned back to call the cops from the pay phone at the gas station.

Meanwhile, back at Foss Drug, a crowd was gathering to view the macabre scene of the murder. I was playing reporter and photographer. One of Golden's finest men in blue showed up a few minutes after the villains made their escape. I approached him as he stood next to the pile of vampire blood-soaked hamburger Pax left on the sidewalk.

"What happened here?" I inquired, obviously playing dumb, something that comes quite naturally to me.

"It's horrible. A man was shot here," the officer replied, glancing at the mess on the sidewalk, quickly looking away. His face was as green as a shamrock. I thought he was going to barf. I took a picture of the cop and the gore, which looked gross, but not exactly what one would expect to see at the scene of a shooting.

The perpetrators of the crime faked out their pursuer. Instead of driving up Clear Creek Canyon, they turned around at 19th Street and parked in the gravel parking lot shared by the Sig Ep and SAE fraternities. Richard carried Pax's limp vampire blood-soaked body into the fraternity house. The boys retired to Zombie's room, where they sat drinking beer and laughing about the prank.

The cops were not laughing.

An all-out alert was issued and the manhunt was on. It didn't take long before the fuzz spotted Gary's VW van. The helicopter dispatched to search for the get-away vehicle in the mountains was no longer needed and was called back. The parking lot began to fill with all kinds of cop cars delivering the Golden Police, Jefferson County Sheriff, Colorado State Patrol, detectives, inspectives, rejectives, and undercover cops in their unmarked cars. Police in a rainbow of uniforms were milling around the get-away vehicle, but nobody was doing anything – like looking into it or taking fingerprints. CSI this was not.

One brave soul in blue took action and came into the frat house. I met him in the downstairs hallway. He was frantic. "You've got to help us," he pleaded. "This guy has been hit bad. We need to get him to the hospital right away."

I was impressed that he was more concerned about helping the victim than in catching the bad guys. I took his portrait to commemorate the compassion of the moment.

We walked down the hall together, peering into one room after another. We came to Zombie's room, where the boys were quietly sipping their beers. The officer tried the door and asked, "Why is this door locked?"

I informed the cop, "The guys in that room went home for the weekend, so they locked their door." OK, so I lied.

"But this is the ladies' room," pointing to the sign on the door that did indeed say, "Ladies."

Oh boy, it must have been a long time since this genius went to college, if he ever did.

"This is a fraternity, not a sorority," I replied. "We don't have a Ladies Room. The guys put that sign on their door as a joke."

That seemed to satisfy the officer, who continued down the hall looking into empty rooms, then went out the back door empty-handed to join the meeting of the Fraternal Order of Police in the parking lot.

It is worth noting only a handful of the frat Brothers were in on the murder and knew what was going on. The rest of the guys stood around watching as the cops arrived, wondering why they were there. Unbeknownst to us pranksters, several of the Brothers, the ones affectionately known as "loadies", who spent a lot of time sitting in the parking lot smoking joints (weed was taboo in the frat house), decided the cops were there for a drug raid. Pandemonium broke out. A few of the loadies, Fat Alice and OD among them, rushed to the upstairs bathroom to flush their stashes of goodies down the toilet. Boy, were they PISSED when they found out what really was going on!

An ambulance arrived. The situation was getting serious. The police obviously believed the murder was real. The fuzz tracked down Gary. He surrendered, was led to a detective's car, read his rights, and put in the front passenger's seat.

I knocked on Zombie's door and talked to the boys, convincing them the situation was getting out of control; they needed to do something. A few minutes later, without finishing their beers, Richard and Pax walked up to the police, cans of Coors in their hands.

"Hi, officer, I pulled the trigger," Richard announced proudly.

"I'm the guy he shot," added Pax.

The cops were stunned. They immediately put Richard and Pax in the back of the police car with Gary in the front seat, but forgot to take their beers. Now, there was a rare photo-op – two guys sitting in the back of a cop car drinking beer. What ever happened to enforcement of open-container laws?

A lieutenant, who appeared to be more or less in charge of the fusterclick, approached the ambulance driver and asked if they received any calls while waiting to transport the gunshot victim to the hospital. The answer was, "No." We dodged a liability bullet on that one.

Richard, Pax, and Gary were hauled off to the cop shop. The parking lot emptied of police vehicles as the officers resumed their normal jobs pursuing robbers, muggers, rapists, real murderers, and little kids selling lemonade without a permit. The fraternity house was buzzing with news about the fake murder and the arrest of our beloved Brothers. We stood by waiting for their one phone call, thinking it might have been a good idea to send them off with a dime. Finally, the call came in; Richard was on the line. He told us they faced some serious charges and would need something like $4,500 bail – cash, or they would be stuck in the slammer.

Sig Eps do not abandon their Brothers to suffer the fancies of father-rapers and mother-stabbers in the county jail. We needed to raise the cash … fast. Everybody raided their beer savings and broke into their piggy banks. Even old Mrs. Mising, the House Mother, opened her purse and generously donated her bingo winnings. Foss Drug had a long line of Sig Eps cashing checks to raise the funds. We finally had the cash we needed.

Richard and Pax surrendering.   *Enjoying a beer in the detective's car.*

The phone rang. The call came in from the pay phone at the Coors Visitors' Center. Our beloved Brothers, the ones we were willing to give our last nickels and dimes, doubled down on the prank. They were released without posting bail. The only charges were discharging a firearm within the city limits and disturbing the peace. While we were busting our butts to raise money to free them, they were drinking beer at the brewery. Of course, we joined them there. But we thought maybe it would have been better to have left them to the devices of ole Bruno in the county jail for at least a little while.

In the end, the boys went free; all charges were dropped. It turned out the City Attorney was not just a Mines graduate; he was also a Sig Ep. He thought the whole prank was hilarious. It pays to have friends in high places.

The Ace Hi Tavern on Washington Street was our favorite watering hole. On many an evening, the halls of the frat house would echo with the inviting chorus, "Let's go the Ace, oh baby. Let's go to the Ace," (to the tune of "The Hop" by Danny and the Juniors). Like rats following the Pied Piper, a group of us marched downtown to finish our Fluid Mechanics or Thermodynamics homework over a mug or three of cold Coors. My good friend, Fred Heumann, and I closed down the Ace one Friday night and stumbled back to the fraternity house in search of additional beers to quench our parched throats. On the front lawn, we spied a little red wagon.

"Wow Fred, look at this, a Radio Flyer wagon, just like the one I had as a kid."

"Yeah, I had one, too. Those were a lot of fun."

"I remember being pulled around by my big sister. The first time I ran away from home, I loaded up my stuffed animals and my pop-up Winnie the Pooh book into my wagon and announced I was leaving, and not coming back. My mom called my bluff and said, 'Fine, have a nice trip.'"

"How far did you get?" Fred inquired.

"A couple of blocks. I forgot to take food."

"How old were you – five?"

"Heck no – 16!"

Fred chuckled, then pondered, "I wonder what it is doing here?"

"I have no clue, but I do have an idea."

We entered the house in search of beer. I knew there should be a cold six-pack in the refrigerator in my room. The door was locked, which could only mean one thing – Shorty, my roommate, was shacked up with his girlfriend from Greeley. I would spend another night on the living room sofa. I put my ear to the door, listening for the sloppy, sweaty, grunting noises that horny Mines students tend to make under such circumstances. All was quiet. The dirty deed had been done; the amorous couple was sleeping. I slipped my key into the lock, carefully opened the door; crept quietly across the floor to retrieve the six-pack, then grabbed my leather WWI pilot's hat from the shelf over my bed. As I was retreating, I noticed a bra and pair of pink panties with lace trim lying on the floor. *Hmmm,* I thought, *Kathy won't miss these for a few hours.*

Armed with suitable headgear and a fresh stock of Coors, Fred and I grabbed the little red wagon and pulled it across campus to Washington Avenue, the main drag in town. Here was a perfect hill to take the wagon out for a spin. At 2 AM, there was no traffic to dodge. Fantastic, we had the place to ourselves.

I donned my flyer's hat, adorned with Kathy's underwear. Fred wore his cowboy hat and had her bra around his neck. If we paid more attention to our structural engineering classes, we would have known the little wagon was not designed for the weight of two adults and six beers. Undaunted, off we went, beers in hand, flying down the hill until the wagon started wobbling and was impossible to control. We crashed, rolling over in the street. Hooting and hollering, we chased our beers as they rolled down the hill. Then we hauled the wagon up the hill and repeated the fun.

On the fourth run, a cop car came up behind us with his lights flashing. We thought about trying to outrun him – maybe head for the narrow alleys where a little red wagon could go, but a patrol car wouldn't fit. Golden's finest were not taking any chances early that Saturday morning. A second squad car sped around the intersection below us and headed straight for us. Damn, we were trapped in a pincher move. We dragged our feet to slow the wagon and pulled over to the left-hand curb. The officers stepped out and began the usual routine,

"Driver's license, registration, please."

"Sorry, ossifer, we don't haff the reg-tration."

The cops were very serious and stern faced. Some of our crashes happened in front of the D & N Motel (not that we had noticed). The manager called the police after the guests complained about the noise. The second cop drove down to the motel to ask if they wanted to press charges, disturbing the peace being the obvious misdemeanor. Drunk wagon riding and operating a little red wagon without insurance or registration were also possibilities.

The officer returned and said charges would not be pressed if we took our little red wagon and went home. Fred and I held a brief huddle to consider the terms of surrender. We wisely agreed to accept the offer. The fuzz lightened up. They were cracking up at our attire and our antics. As we started to leave, one officer suggested we try the hill on Maple Street, which went past the gym and the field house. Holy cow, that hill was way steep! I was not sure if the nice policeman was trying to help us or to kill us. We stuck to our word and went home with our wagon and our seriously shook-up beers.

Streaking was the "in" thing in the early 1970s. It was a right-of-passage to engage in at least one such venerable event. The cafeteria was hit early on, as were the library and the Ace Hi Tavern. In March came the announcement the University of Colorado in Boulder was planning what was billed as "The World's Largest Streak." That was something not to be missed – especially if it involved naked hippy chicks. We jumped into Duncan Lestina's 1965 Chevy van, dubbed the Purple Haze due to its lovely shade of purple paint; stopped at Golden Liquors to pick up a keg of Coors; then made the trek north to Boulder. We arrived at Libby Hall, the dormitory where the streak was supposed to originate, to find a large crowd milling around and a couple of guys on the steps using a bullhorn to encourage students to join the streak. We had only come for cocktail hour and to watch, but thought, *What the heck, let's go.* Duncan, Zombie and I jumped up on the steps, stripped down to the roar of the crowd, and walked into the dorm, carrying a tapped keg of beer and a supply of red Solo cups. The place was packed with potential streakers. Nothing was happening. We made our way upstairs, squeezing past pale, naked bodies and offering beers to any interested participants. Across the second floor, up to the third floor, then across the hall and down the stairs on the far end we went, sharing beers and ogling at the coed participants, who were sadly outnumbered by males (no worries – we were used to even worse odds).

We finally came to the end of the line at the main doors on the first floor of the dorm. By then it was way past the announced start time of the highly anticipated event. Eight thousand spectators anxiously waited outside, but nobody had the balls to go first – and the cutest ones had no balls at all, which was a bit of distraction. We almost lost Zombie on the second-floor landing when we passed a gorgeous redhead. The problem was the CU kids were afraid of being arrested or nabbed and expelled for lewd behavior or indecent exposure when they exited the dorm. Shucks, we didn't go to school there. We had no such concerns. The three of us stood at the exit doors and announced, "Let the streak begin!"

The big double doors swung open and three Miners in the buff carrying a keg of beer led the charge down the stairs and through a gauntlet of spectators. Camera flashes were everywhere. Several gals jumped in front of us to take pictures as we ran down the line. Oddly, I never got a flash in the face. Running with a keg of beer slows one's progress. After about 50 yards, we pulled over to the side to let the other 297 streakers run past. We spent the rest of the evening wandering around campus, drinking beer and chatting with people. Then we began the search for our abandoned clothes.

The next morning, the Rocky Mountain News carried an article on page 10 titled, "300 unrebuffed in mammoth CU streak" featuring a photo of none other than Richard's and Duncan's bare behinds (and the corner of my cowboy hat) with our keg of beer. Once again, the Miners aced out the CU students on their home turf. The article added,

> One coed, Patricia Mecurio, 21, was taken to Boulder Community Hospital by ambulance with chest pains and a rapid heartbeat. A hospital spokesman said her condition wasn't serious, but that her pulse was "being monitored."
>
> A Farrand Hall official at CU said Miss Mecurio could have been watching the streaking from her window in the dorm.

Behold the power of those Mines guys' behinds!

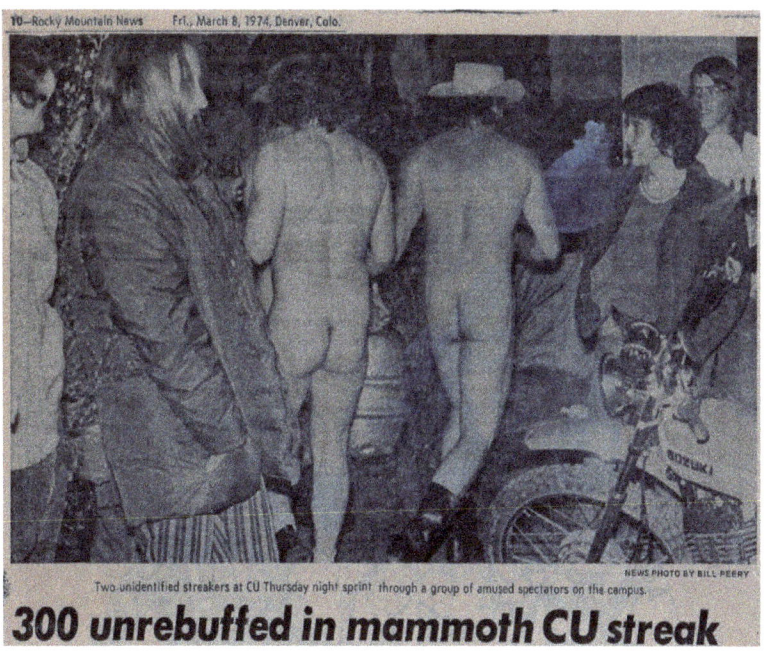

One Thursday evening, while celebrating some uneventful occasion – such as Thursday evening – we were introduced to a local yokel, Bob Coakley, at the Ace Hi Tavern. Bob owned a couple of small mines, the Blackjack Silver Mine and the Sun & Moon Gold Mine, in the mountains west of Boulder. Needless to say, with common interests of mining and drinking beer, we hit it off right away.

Coakley made us job offers we could not refuse. He needed help with geology, surveying, mining, and ore processing, all fields in which we had just enough expertise to be dangerous. In exchange for letting us hone our talents in his mines, he offered us the one thing that Mines students cannot resist – free beer. So, a group of us spent many a weekend up at the mines: surveying, mapping the surface and underground geology; learning to drill, blast, and muck rock. Sometimes we camped on the spot and sat around a bonfire telling stories and drinking beer into the wee hours. Other times, we drove back to Golden, stopping at every watering hole along the way for Mr. Coakley to make installment payments for our labors.

Working at Bob's mines made us feel like real Miners, fulfilling the rough and tumble image embodied in the school song, "The Mining Engineer":

Now here we have the mining man.
In either hand a gun.
He's not afraid of anything,
He's never known to run.
He dearly loves his whiskey,
And he dearly loves his beer,
He's a shootin', fightin'
Dynamitin' mining engineer

The Blackjack mine consisted of a horizontal drift, which went back about 200 feet. High-grade silver ore was mined in the 1930s from a few stopes above the drift. Near the end of the drift was a short crosscut with a winze, an internal shaft, which went down about 30 feet. The winze followed a rich ore shoot of silver mineralization with a host of rare silver minerals, including spectacular specimens of wire silver. We spent most of our time there, digging out specimens and working to sink the winze further. It was cramped down there and ventilation was pretty poor. Several of the guys almost got asphyxiated one Saturday morning. No, they didn't break into a pocket of bad air. Tom Orlin created one. Tom dined at Rosita's Restaurant in Denver late the previous night, consuming two orders of tortilla chips with hot sauce plus the house specialty, smokin'-hot jalapeno-laced Smothered Pablitos. The resulting explosion of noxious gases that spewed from his posterior was of record proportions. It rocked the whole mine, nearly causing a cave-in. The winze was quickly filled with a green jalapeno cloud that burned nostrils and lungs. Coughing and sputtering, three rather green Miners came running out of the adit in desperate need of fresh air. There was no way they were going back into the contaminated mine. It was eligible to be an EPA Superfund cleanup project.

The two big annual events at Mines were Homecoming and Engineers' Day. The celebrations featured several unique events, including a surveying contest, for which the team that completed a survey route with the smallest error received the Golden Plumb Bob award. Only true nerds entered that event. Real Miners competed in manly events such as hand steeling, in which one team member holds a piece of drill steel against a rock while his partner hits it as hard and fast as possible with a 4-pound sledgehammer to drill a hole into the rock (accuracy and trust are essential); the jackleg drilling contest, which involves wrestling a 120-pound air-powered drill to hammer a horizontal hole into a hard boulder; or the hand mucking contest requiring shoveling a ton of gravel into an ore cart as fast as possible. There were also the truly unique Cussing, Spitting, and Belching contest, the Push Cart Race, and the Push Ball contest. The latter event pitted Sigma Gamma Epsilon (the national earth sciences honorary society) against Theta Tau (the campus jocks). The opposing teams fought to push a massive 8-foot-diameter ball down the football field to score a goal. There were no rules; you could go after the ball or take out your opponents. Once the ball gained rolling momentum and started bouncing, it could crush people. Mayhem ensued. We Sig Gams almost always got the crap beat out of us, but we did have a few moments of glory.

This is gonna end in pain; the author (striped shirt) making a short breakaway run with the pushball.

All the honorary societies on campus were invited to enter a cart in the home-made Push Cart Race – one lap around the track at the football field. One such honorary was Press Club, the honorary of honoraries, whose cheer was:

> We're the best honorary.
> All the others suck.
> Press Club, Press Club,
> Rah, rah….. (insert "have a beer" – or something like that)

Press Club was originally formed as a group of literary types who worked on the Oredigger newspaper or the school yearbook. The club met every Thursday afternoon at Dud's Tavern to quaff a beer or two while discussing the finer points of such scintillating subjects as dangling participles, mangled metaphors, and proper punctuation. I actually worked as a reporter for the newspaper to earn my way into this select drinking society. Over the years, the bar for entry slowly lowered and eroded to the point the club became a drinking club with no further connection to its name or original purpose. Meetings, featuring kegs of 3.2 beer and lots of chugging contests, were held in the park by Clear Creek on Thursday afternoons.

An offshoot of Press Club was Zithers, the crème de la crème of beer drinkers on campus. Zithers consisted of seniors, who were 21 (or had good fake IDs) and could legally drink at the Ace Hi Tavern. Out of the blue one E-Day afternoon, while I was sipping a cold Coors and watching the mucking contest, I was notified that Zithers' entry into the homemade Push Cart Contest had not arrived on the track at the football field. The race was about to start in 20 minutes. Oh dear, there was a reason for that oversight. Perhaps, we forgot due to the stress of the many hours we spent intensely studying Statics, Mechanics, and Dynamics. Or maybe the lapse was due to our laser focus on drinking. Either way, we completely forgot about the race. As a Zither and an officer of Press Club, I felt a pang of responsibility. We could not suffer the embarrassment of failing to enter the contest, lest people think we were unimaginative tipplers instead of ingenious, although a tad tipsy engineers. I enlisted Gary Lubers with his VW van (of Pax Domino's murder fame). We raced across campus to the Sig Ep house in search of something to enter into the race. We looked for the little red wagon Fred and I rode down Washington Street. It suffered catastrophic failure in a later excursion and was useless. Searching the backyard, we found an old porcelain bathtub. *That is a start*, I thought, *but we need wheels*. We spied a possibility, seized the materials, loaded up, and drove to the football field.

Our last-minute entry was engineering simplicity in action. The bathtub sat upon three rounds of ponderosa pine cordwood. The chief engineer sat in the tub (along with an ample supply of cold beer) waiting for the starting gun to begin the 440-yard sprint. Two junior engineers pushed the tub forward and the rest of us passed the exiting round to the front of the line as the tub thumped along the course. We only made it about 10 yards before we gave up and sat around the tub drinking beer. We came in last place by a good 430 yards, but saved face with an ingenious last-minute entry, and got a lot of laughs for our engineering antics.

My first professional geology job came in the summer of 1971, between my 1st and 2nd senior years (some things are worth repeating?). My good friend and frat Brother, Larry Fischer, and I were part of a geological survey crew mapping and sampling gold and silver mineralization at the spectacular Red Mountains between Ouray and Silverton, Colorado. The Red Mountains lie along the margin of the massive Silverton caldera, the site of one of the most powerful volcanic eruptions in geological history. Twenty million years ago the area was much like Yellowstone is today, with hot springs bubbling and geysers erupting. Deep in the earth, hot geothermal waters cooked the rocks and left tiny grains of pyrite (fool's gold) disseminated throughout them. Gold and silver-bearing fluids circulated in fissures and eventually precipitated rich gold-and-silver veins. Uplift and erosion over millions of years exposed the altered rocks. Oxidation turned the pyrite into bright red hematite and yellow limonite, leading to the mountain's name. Quartz veins containing gold and silver stuck out like dogs' balls for prospectors to discover. Thus began the gold rush of the 1880s and development of the rich Red Mountain and Silverton mining districts. Old mines riddle the Red Mountains. Remnants of the ghost towns of Congress, Red Mountain, Guston, and Ironton are scattered along the lower slopes of the peaks.

Our job was to map the topography and geology of Red Mountain No. 2 (12,219') and Red Mountain No. 3 (12,890'), and collect rock-chip samples for gold, silver and multi-element analysis. Each day we drove the company Jeep to about the 10,000-foot elevation, then hiked up and over the mountains to do our plane-table mapping. Every day was an adventure: traversing unstable talus slopes and scary cliffs to get our samples; dodging lightning bolts during the daily afternoon thunderstorms; glissading down avalanche chutes when they were full of snow, or riding elevators of sliding rock after the snow melted; investigating old mines, mills, and ruins of the ghost towns; learning a great deal about geology. It was as good as a first professional job can get.

The Red Mountains in their iron-stained glory – our playground for the summer of '71.

That experience hooked me on exploration. Not because we got rich from our salary of $550/month, or the company's generous $5 per diem (which paid for three guys crammed into a room at the Ouray Hotel, and all the peanut butter we could eat), but because it was FUN. Although I was earning a degree in Geological Engineering (or at least attempting to), I decided the engineers could keep their slide rulers and trig tables. I was off to explore the mountains as a shootin', fightin', dynamitin' exploration geologist.

After finally graduating and narrowly managing to avoid an all-expense-paid tour of Southeast Asia, courtesy of Uncle Sam, I landed a job with Exxon Minerals doing uranium exploration in Colorado and Wyoming. I thought I had it made, until the boss told me if I wanted to continue working in the industry, I would need a Master's degree. Crap, that was not in my plans. So, it was back to CSM for another four years to earn a degree that should have taken two years, with summers and a few skipped semesters off working for Exxon in Colorado, Wyoming, Oregon, Arizona, New Mexico, Nevada, and Alaska to pay tuition and to support my skiing and beer-guzzling habits.

SECTION II

THE GREAT NORTH

Chapter 3

DOING STRANGE THINGS IN THE MIDNIGHT SUN

> Strange things have been done in the Midnight Sun
>   and the story books are full —
> But the strangest tale, concerns the male, magnificent walrus bull.
>
>   "Ode to an Oosik"   – Anonymous

    For more than 150 years, Alaska, the Land of Enchantment, with its spectacular scenery, fantastic wildlife, world-class fishing, and huge gold nuggets, has lured foolish young men (such as we were in 1972) from the lower 48 to fame, fortune, or ruin in the last frontier. While we were students at Mines, Fred (Ferd Berfel) Heumann, Randy (Black Bart) Palmer, Dave Scott (LD) and, I spent many a night at the Ace Hi Tavern, singing along to Johnny Horton's "North to Alaska" on the jukebox as we made plans to spend a summer prospecting for gold in the great northland. Ferd wanted to find enough gold to make a wedding ring for his girlfriend, Jane, alias "The Stick Woman." Black Bart wanted to go to Alaska for the fishing. He heard the fish up there were plentiful and big. But he hadn't heard they feed on swarms of bloodthirsty mosquitoes the size of bumblebees. The enormous size of the mosquitoes is due to the gallons of human blood the skeeters consume each summer. LD Scott and I wanted to go for the gold, fish, and most of all, adventure – but not the mosquitoes.

    Fred came to the Ace one night with a special announcement. He spent his hard-earned cash, not on gold mining supplies as planned, but for a wedding ring for The Stick Woman. He caved under pressure to get a job and get hitched immediately after graduation, before our expedition commenced. Randy and I were shocked and bummed out. We gave Fred a ration of shit for abandoning us. Several cold Coors drafts later, Fred claimed his decision not to go to Alaska was made for financial reasons. He foolishly said, "If I had $100, I would go to Alaska right now." I called his bluff, went to the bar, wrote a check for $100, and handed five twenties to Mr. Berfel. Not wanting to disappoint us twice in one night, Fred said, "OK, let's go." We loaded our carcasses and that of Mark Frauenhoff into Fred's CJ5 Jeep and headed north at 1:00 AM, with only $100 and the clothes on our backs. We made it halfway to Fort Collins before we realized our folly and turned back to Golden.

    Fred and Randy, never made it to pan gold and catch big fish in Alaska; it was many years before LD got there. I was extremely fortunate. Two years later, I was offered a summer job as Assistant Geologist working for Exxon Minerals in interior Alaska. My boss was Mel Klohn, a dedicated, hard-working geologist who served as both mentor and friend. We spent the summer searching the sedimentary basins of central Alaska for uranium, visiting old abandoned mining camps, catching lots of fish, and being in awe at the majesty of Alaska, the Last Great Frontier.

    I took a full-time job with Exxon Minerals in 1976 after earning my MSc degree. My assignment was to explore 663,000 square miles of Alaska for uranium – a task that obviously required some prioritization before starting fieldwork. I worked the field seasons of 1976-1981 with hard-working, fun-loving teams of geologists and helicopter crews in Southeast Alaska, the Seward Peninsula, the Alaska Range, and the Kuskokwim and Koyukuk regions of Interior Alaska.

Clockwise: Mel Klohn with a small mammoth tusk, Susitna River, 1974; the author on glacial ice at the south edge of the Alaska Range; Capps glacier & Alaska Range, 1974

Nome hotties.

Someone told us that to experience Alaska as an Alaskan; you must do four things: pee in the Yukon River, swim in the Bering Sea, shoot a bear, and make love to an Eskimo woman. We really wanted to become official Alaskans but were unsure about the third and fourth challenges, especially since the bears were cuter, smaller, and had more teeth than most of the Eskimo women we met. Mixing up the last two challenges might piss off the bears. We settled for focusing on the aquatic events, so we could at least qualify as half-ass Alaskans.

Peeing in the Yukon was easy. We hit both shores multiple times and got the middle a time or two from islands where we camped. Swimming in the Bering Sea proved to be more difficult – partly because it is frozen solid most of the time. We arrived in Nome, the only town of significance on the Seward Peninsula, on the Fourth of July weekend just in time to catch the big parade and join the festivities, which focused on drinking to excess, an activity in which we had much expertise. A chilly onshore wind was blowing. I was pleased to see the sea ice had broken up and was parked several miles offshore. That left plenty of room for a leisurely swim (skinny-dipping, of course) in the Bering Sea.

Training started at 4:00 PM with margaritas at the Nome Nugget bar. Margaritas were followed by beers, which were followed by shots of Jack. During the afternoon and evening debauchery, we slowly worked up the courage to complete the second half of the Alaskan aquatic biathlon. Around 8 PM, my besotted buddies, Bruce Cox and Mike Mancuso, and I decided we had infused enough antifreeze into our bodies and numbed our brains enough to attempt the swim. Equipped with towels from our hotel rooms and the foolish confidence inspired by whiskey, we left the bar, made our way over the jagged boulders of the breakwater, and walked down to the beach.

We stripped down and ran into the 38° F water. When we were thigh deep, we dove in and took a half dozen strokes forward. The frigid seawater took our breath away and instantly numbed our limbs, making them nearly useless. We immediately turned around and swam back. Three half-frozen idiots from the lower 48 emerged from the water and stood on the beach shivering and looking like members of the Blue Man Group on tour. It was only then we realized we were doing all this directly in front of the picture windows of the dining room of the Nugget Inn, where little old ladies on a group tour of the great Alaskan north were casually dining on reindeer steaks. It would have been a thrilling show for the ladies – perhaps the highlight of their entire Alaskan adventure, but what we offered were not just frightened turtles; we had seriously frightened *hatchling* turtles. We waved, wrapped our towels around our waists like sarongs, and hurried back to our rooms to take a hot shower and try to restore circulation to our frozen extremities, especially those little hatchlings.

Bruce Cox completing the Bering Sea plunge.

Nome owes its existence to the prospectors of the 1899 gold rush, who either struck it rich or went bust panning the gold-rich streams and beach sands. In the 1970s, town measured all of four blocks wide by ten blocks long. The main drag in town, which parallels the beach, is Front Street, a dirt road that alternates between a skating rink, a mud pit, and a dust-bowl disaster, depending on the prevailing weather. Front Street is lined by saloons, including my favorite, the Board of Trade. A few native craft stores are interspersed with the bars, selling a variety of local art items, principally soapstone and walrus ivory carvings. Among the various pieces of native art, we discovered walrus oosiks for sale. As the great bard of the north, Anonymous, wrote in the poem, "Ode to an Oosik":

"Oosik?" you say - and quite well you may,
    I'll explain if you keep it between us;
In the simplest truth, though rather uncouth
    Oosik, is, in fact, his penis!

Yes, indeed, oosik is the Eskimo name for what biologists call the *baculum* or *os penis* of the male walrus. The display cases of the native craft shops held several polished oosiks and a few carved like the ivory walrus tusks next to them. Some oosiks were larger than the tusks.

> Now the size alone of this walrus bone
> >would indeed arouse envious thinking -
> It is also a fact, documented and backed,
> >There is never a softening or shrinking

True enough, the oosik is technically a bone. Maybe that is where the term "boner" came from? Walruses are not the only critters to be blessed with such wonderful appendages, although they are the champions for size. Seals, raccoons, chimpanzees, gorillas, even rats have them. Somehow *Homo sapiens* got left out of the deal.

> This, then, is why the smile is so sly,
> >the walrus is rightfully proud
> Though the climate is frigid, the walrus is rigid,
> >Pray, why is not man so endowed?

Maybe the good Lord figured out that such an addition to his earthly image would make it difficult to wear tight designer jeans, let alone engage in sports like football, wrestling, or – heaven forbid – the hurdles.

As the little old ladies at the Nugget Inn could testify, we had absolutely *nothing* in common with the proud bull walrus after our dip in the Bering Sea.

Rich Rein and I were intrigued by walrus oosiks and vowed to find one before heading south at the end of the summer. The problem was we had no clue how to get one, other than starting with a walrus, preferably a well-behaved dead one, the only kind likely to give up his pride and joy without a fight. Eskimos hunt walruses on the pack ice during the winter, both for their meat and ivory. Sometimes a walrus is shot but falls off the ice and sinks into the sea before the hunters can recover it. Every winter, a few walruses perish from natural causes. Their bodies remain frozen in the ice, waiting to be released during spring break-up. So, an occasional dead walrus washes up on shore in the summer.

One July afternoon, we were flying down the west coast of the Seward Peninsula, returning from a day's work in the Kigluaik Mountains, near Teller. The pilot spotted a large object bobbing in the surf a few feet offshore. He circled over it. It was a walrus, a big, decidedly dead, extremely bloated, belly-up walrus. Stewart set the chopper down on the beach near the walrus. Brimming with excitement, I pulled out my big Buck knife. I waded heedlessly into the ice-cold water to get my very own oosik. But where to start?

Chopper in the mist;   *Rich Rein collecting rock-chip samples – Kigluaik Mts.*

As the bloated walrus rocked in the waves, I stuck the knife in the carcass above the butthole and tried to cut through the tough hide. I wasn't making much progress. Rich stood on the shore, offering much-needed advice. But it was the blind leading the blinder. Rich shouted out over the roar of the waves, which were splashing frigid water over my thighs, "Higher, cut higher." I took the knife and plunged it into the walrus about a foot higher than where I started.

Oh shit! The knife cut directly into the distended body cavity of the walrus, which was blown up like an oversized birthday balloon filled with putrid gas generated as the walrus slowly rotted. At that moment, a breaking wave struck the carcass, creating a hideous explosion of noxious gas and half dissolved innards that spewed from the fresh gash I cut. I stood there in thigh-deep water with stinking, slimy, rotten walrus guts splattered all over my pants and shirt. To add insult to the injury of not scoring an oosik, the pilot would not let me back into the helicopter until I had thoroughly washed all the guts off my clothes. Soaked and shivering, I finally was allowed back inside for the flight back to Nome. Stewart was kind enough to turn on the heater so that I didn't arrive in a severe state of hypothermia.

After that experience, we got smarter, but only a little. The plan was to seek the advice of experts, the local Eskimo hunters. Of course, the only ones we ever encountered were the drunks in the bars. Off we went to the Board of Trade Saloon in search of a qualified expert, one who had enough whiskey in his belly to tell outsiders his tribe's deepest, darkest secrets, but was still coherent enough to speak intelligibly. The latter proved difficult to find. Rich finally located a likely suspect, struck up a conversation, and bought him a drink. Then we popped the question,

"How do you get an oosik from a dead walrus?"

"Ya juss reach in dere, pull it out, and cut er off," was the curt, surly answer.

That is when it dawned on me. We somehow forgot our most basic biology lessons. What would you expect? We were geologists, not biologists. I had not dissected anything since Kermit, the leopard frog, in 10th grade. The walrus we were trying to remove an oosik from was a dang female. No amount of cutting in the right place would have produced the prize we sought since it was simply not there. Armed with our newfound knowledge, we resumed the hunt for a dead walrus, this time a verified bull. Sadly, we never came across another deceased walrus. I had to settle for buying an oosik at one of the local native craft shops. It is a beauty – a full 24 inches long with a gentle curve to it, like the neck of a swan. No doubt its former owner pleasured many a satisfied walrus cow with it. It now graces the mantle over my fireplace, where it serves as a conversation piece, a great ice breaker for any get-together, starting with a recitation of "Ode to An Oosik."

Nome in the 1970s had a population of about 2,500 permanent residents, the majority of whom were Inupiat Eskimos – the majority of whom were drunks. The majority of the minority of white people were also drunks. That left maybe 100 sober people in town. We based operations out of Nome for several weeks. We needed a vehicle to transport the field crew and gear to and from the airport, where our helicopter was based. Taxis were few and were generally unreliable. The taxi drivers made a living by shuttling drunks between the bars on Front Street: The Board of Trade, the Anchor, the Anvil, the Polar, the Bering Sea Saloon, and the Vitus Bering, even though they were only a couple of doors down from each other. There was no car rental agency in Nome. The few dirt roads heading out of Nome lead nowhere, so nobody ever needed to rent a car to drive them. We finally scored a vehicle from Riddel's Garage, which also served as the town's junkyard. Our chariot was an old yellow taxicab Mr. Riddel revived with a new battery and a rebuilt carburetor. It performed admirably for our purposes but came with an added challenge. Although it lacked the "Taxi" sign on the roof, the bar patrons recognized the yellow color and thought it was a taxi in service. They flagged us down anytime we drove down Front Street. They got royally pissed off when we didn't stop to give them a lift to the next bar. We probably could have made a living driving the drunks around town and might have gotten some helpful advice on oosik hunting in the bargain. But we didn't want to quit our daytime jobs to do it.

Our favorite hangout was the Board of Trade Saloon, which was seedy but less so than the other watering holes. On a busy Friday night, we were having a cleansing ale or two and checking to see if any eligible ladies were present. Few young white gals ever went to Nome, let alone stayed there. Hell, who could blame them? Most of the Eskimo ladies offered little of interest, being rather rotund as a result of annually packing on a few hundred pounds of excess adipose to survive the long winters, as well as lacking pearly whites. Occasionally, there was a cute young one, rarely a beautiful one. The jukebox was appropriately blaring the Eagles' hit "Desperado" as I desperately patrolled the dimly lit bar for young talent. To my utter amazement, perched atop a barstool at the end of the bar sat the most stunning, shapely, young Eskimo gal I ever laid eyes upon – her long, straight jet-black hair hanging down to the middle of her back, her high cheekbones and slightly slanted eyes beckoning in exotic seductiveness.

Luck could not have been more on my side. She was alone. Next to her was the only seat left at the bar, a high barstool on a tall pedestal with a black vinyl top. I sauntered over, sidled up to the bar, turned outwards, and sat down beside her, planning to strike up a lively conversation. Suddenly, the seat collapsed around me. I was bent in half with my arms and legs pointed up in the air. That was no barstool. It was an open garbage can with a black plastic liner. No wonder nobody was sitting on it. Perhaps they were smarter, soberer, or just more observant than I. So, there I sat, stuck IN it. It took two people to extricate my sorry ass. By the time I emerged from my plastic prison, the cute chick was long gone. She didn't bother to hang around and congratulate me for not spilling a drop of my beer throughout the ordeal.

Although we never scored a walrus oosik, we did have a gourmet walrus dinner. Not that we planned on it. Just for fun, Mike and I stopped by the local Eskimo grocery store while stocking up on camp supplies for 10-days of fieldwork. Inside a refrigerated case were several big walrus steaks. They were beyond ugly. We selected a particularly disgusting one and bought it as a joke to show the rest of the crew. As we checked out, I asked the Eskimo lady at the cash register,

"How do you cook walrus?"

"You boil it."

"For how long?"

The answer came back in a typical Eskimo deadpan, "Until it is done."

OK, it looks like that is all the detail we were going to get. We will go with boiling. We never really intended to eat the walrus meat. It was gross. The exterior was tough, leathery hide with wide-spaced, thick black hairs curving over it. Beneath the hide was a thick layer of white fat, walrus blubber, which transitioned below into gristle mixed with fat. The bottom layer was an incredibly tough mixture of pink meat and more gristle. Somehow, we screwed up the estimate of food we needed in camp. We ran out with two days left to finish the work. Trout from a high lake where we stopped for lunch filled the gap for the first night. The last night in camp, we were stuck with the walrus, a few cans of Olympia beer, and the remains of a bottle of Jack Daniels. Out of desperation to quell our hunger pangs, we dutifully boiled the

  walrus steak – until it was done. Then the challenge of actually consuming the boiled lump of lard ensued. If you could get over the smell, you were in for a helluva fight to chew the leathery meat. I think it takes more calories to chew the stuff than are gained from eating it. The real delicacy for the Eskimos is not the meat; it is the layer of blubber beneath the hide. We tried that. It covered the inside of our mouths with a slippery slime that wouldn't go away, no matter how much Jack Daniels we gargled.

Bruce Cox slicing walrus meat; Gene Franks trying to eat the blubber.

Working in the Land of the Midnight Sun does strange things to one's biological clock. In June and July, the sun only sort of went down at the end of the day. It rose just below the horizon, skimmed along for a while, then carved a giant arc across the sky. Finally, it dipped below the horizon, slightly south of where it rose. Twilight, with its soft light casting long shadows across the tundra, lasted for hours. We worked long days and often lost track of time. It was not unusual to arrive back in camp after 10 P.M., thinking it was late afternoon. The Eskimos, unburdened by modern chronometers, often go days without sleeping, then crash for a day or two.

Crossing the Arctic Circle in the Selawik Basin at 11:00 PM.

Alaska's Alexander Archipelago comprises more than 1,100 islands, which extend for 300 miles along the coast of southeast Alaska, from Ketchikan on the south to Glacier Bay on the north. It is a land of stark natural beauty: vertical-walled fiords, massive ice fields, tidewater glaciers, and timber-clad mountains soaring above the calm waters of the Inland Passage. Geologically, the archipelago consists of a potpourri of geological terranes: island arcs, deep ocean basins, and pieces of continental crust – exotic terranes, which over the past 220 million years slid thousands of miles northward along strike-slip faults similar to the San Andreas Fault and then were welded to the North American continent and to each other. The islands of today are the tops of submerged mountains, separated by deep channels carved by massive glaciers during the last Pleistocene glacial maximum from 30,000 to 16,000 years ago; then flooded when the glaciers started melting some 12,000 years ago, ultimately raising sea level more than 400 feet.

The outer islands of the archipelago: Prince of Wales, Chichagof, and Baranof Islands, protect the Inner Passage from the savage storms that spin out of the Gulf of Alaska. Shipping, fishing, logging, and tourism thrive in the protected waterways.

Southeast Alaska is almost as wet above the ocean as beneath its surface. Rain in one intensity or another is a nearly daily event from September through March. The rest of the year is only slightly less wet. Ketchikan, which receives an annual average of 160 inches of rain, has an average of only 42 clear days per year. The remaining 323 days have precipitation in one form or another. We lived in a world of rubber while working there: rubber boots, rubber rain pants, rubber raincoats, and rubber rain hats. Thus equipped, life in rainy Southeast Alaska goes on in just about any kind of weather.

Southeast Alaskan glacier shedding icebergs.

Sunny days in southeast Alaska are rare treasures, which are not wasted by the locals. I once flew into Juneau in February to visit the Alaska Division of Minerals. As the plane descended, I gazed in awe at the Juneau Icefield and the Mendenhall Glacier, just north of town, sparkling in fresh white snow under a clear blue sky. I was looking forward to my visit to catch up on mining activities and claim filings. The beautiful weather would be a bonus. Not exactly – the entire town shut down. There had not been a sunny day in three months. Schools, banks, government offices, even many private businesses were closed as if it were the Fourth of July or some other major holiday. And a holiday it was. Nearly everyone was out enjoying the sun and warmth. I had to postpone my visit to the Division of Mines, but I was far from upset. I joined the locals in the celebration, which naturally concluded with beers at the Red Dog Saloon.

Ketchikan, the southernmost town in Alaska, is a strip town about ¼ mile wide and four miles long plastered against the steep southwest coastline of Revillagigedo Island. There is little buildable land. Most of downtown is built on wood pilings extending into Tongass Narrows. Although Ketchikan lies on the mainland, it lacks road access and can only be reached by boat, floatplane, or ferry. Even flying to Ketchikan requires a boat ride since there is no room for an airport on Revillagigedo Island. A new airport on Gravina Island, across the Tongass Narrows, greeted us when we arrived in the early summer of 1976. The burned-out shell of an airliner lying on the south edge of the runway also greeted us. It was encouraging to know they got that mishap out of the way early, and pilots had by now learned how to land correctly on the new runway. We took the ferry across the Narrows, marveling at the bald eagles that cruised overhead and a pod of orcas that passed alongside the boat.

Our home away from home when we were not in camp was the Hilltop Motel. The Hilltop was cozy, and the staff (particularly a couple of young gals) were friendly, exceptionally friendly with Bruce and Dave, the lucky bastards. The other guys and I had to settle for evenings at the Shamrock Bar, the local strip joint down on Stedman Street next to Creek Street, which was the former location of Ketchikan's long lost (and sadly missed) red-light district, which hosted 20 brothels at its peak. There is no room for buildings along the steep side of the creek. They are built on stilts over the water. Creek Street's claim to fame is that it was the only place in the world where salmon and fisherman went upstream at the same time to do the same thing.

The Shamrock was a dark, dingy, narrow building with a small stage across the room from the long bar. Exotic dancers provided nightly entertainment for the patrons. The gals were exotic, all right. Mostly rejects from Seattle's raunchiest clubs, they came in all shapes and sizes, except shapely and attractive. Looks didn't matter to the horny fishermen and loggers who frequented the place. After a month or more at sea or in a logging camp, anything in (or out of) a skirt was fine by them. We went there to have a good laugh – not at the girls' expense – we were laughing at the antics of the inebriated patrons. The ladies may not have been beauties, but they tried their best to entertain the drunks with their somewhat limited talents. We cheered for them and passed a few tips their way, although we were secretly thinking, *Put your clothes back on*!

Another of our hangouts was the Focsle Bar, named for a contraction of "forecastle," the front of ships where crew quarters are located. The back of the bar featured a replica forecastle. We also hung out at the only bar I know that doubled as a post office. Behind the bar were post office-type slots where fishermen, who frequented the bar whenever they were in port, received their mail. The Rainbird Bar, Frontier Saloon, and Marina View were also fun watering holes. Bars closed at 5:00 AM to sweep the floors of trash and passed-out patrons. The drunks didn't need to be on the street for long since the bars re-opened at 7:00 AM.

All the bars had a brass ship's bell hanging over the center of the bar. Local tradition held that ringing the bell meant buying a round of drinks for everyone in the bar. Needless to say, I kept my distance from the bells. Fishing must have been lucrative. Several times when the fishing boats were in port, I had as many as four drinks in front of me but I hadn't bought a single one of them, nor did I know the people who did. Gosh, I miss those good old days.

Ketchikan has an enormous tidal range; at 22 feet it is enough to strand boats anchored in the wrong place when the tide goes out, leaving them perched precariously on rocks, awaiting the next high tide to set them free. During the summer, ferries full of tourists dock in Ketchikan and unload their passengers, principally little old ladies, who have outlived their husbands and are enjoying that expensive cruise their husbands wouldn't take. They spend the afternoon shopping in the town's curio shops or touring the city's historical attractions such as Creek Street and Totem Bight. One famous story involves a particularly observant little old lady. The cruise ship docked at high tide; the passengers disembarked down the ramp to the dock. When it was time to re-board the boat and continue the cruise, the tide had gone out by a good ten feet. The observant old bird refused to board the ship, saying, "I may be 85 years old, but I am not stupid. I walked down the boat ramp when we arrived, and I'll be damned if I am walking down it again to get on a boat that is sinking." It took a lot of effort by the ship's crew to convince the old gal that the boat was not sinking and was indeed safe.

Coastal Southeast Alaska lacks mosquitoes. I have no idea why, although it could be because it rains so much, they drown. I love it, but Mother Nature decided the lack of mosquitoes was unfair to the rest of Alaskans, who are pestered relentlessly by the nasty bugs. So, she substituted "white socks" for mosquitoes. White socks are a version of black flies with a white stripe on their six-pack of legs, little devils that take a chunk out of you when they bite, leaving a swollen bleeding hole. Fortunately, the white sox season is short.

Stone Rock Bay on a rare sunny day.

Stone Rock Bay, on the far southern tip of Prince of Wales Island – the southernmost point in Southeast Alaska – was our home for a large part of three field seasons. We built a seasonal work camp consisting of sturdy canvas tents set upon plywood platforms. Sleeping tents, a big mess tent, and an office tent were strung out along a trail that led to the beach. The tents had propane heaters for warmth and to dry our soaking wet clothes at the end of the day.

Communication with the outside world – to the extent we had any – was by single-side-band radio, patched into the phone system through an operator in Ketchikan. We gathered around the radio for evening entertainment, much like families did in the Golden Age of Radio in the 1930s and 1940s, before television was invented. The radio frequency was shared with several logging and fishing camps spread across the island near villages such as Hydaburg, Kasaan, and Klawock. Instead of listening to *Inner Sanctum Mystery* or the latest episode of *Captain Midnight* as our grandparents did, we listened to real-life soap operas. Loggers and fishermen took turns calling their wives and girlfriends back home to discuss the details of their everyday lives: finances, kid's softball games, diaper changes, and best of all – their sex lives. Inevitably, some exceedingly randy guy, who had been in the bush far too long and really, really missed his gal, forgot he was on an open radio channel and got carried away, sharing with everyone on the island all the details of how he was going to ravage his sweetheart as soon as he got in the door at home. It was like phone sex for groups of voyeurs. We voyeurs loved it.

Since it was always cool and damp, or more likely raining, we held off on bathing until we finished our field hitch, which was usually ten days but could be as long as two weeks. At the end of the anxiously anticipated last day of fieldwork, Tyee Airlines' De Havilland Beaver floatplane landed in the bay. It taxied over to the beach in front of camp, where a troupe of dirty, stinky geologists anxiously waited to be whisked back to civilization. On our first break, we headed straight to the upstairs lounge of the Hilltop Motel to wet our whistles. After a few drinks, it was time to check into our rooms, shower, change into clean clothes, and head downtown for dinner and partying. When I emerged from the shower, a horrible stench about bowled me over. *Oh, my god! What foul-smelling ogre entered the room while I was showering?* I traced the offending odor to a steaming pile of dirty clothes lying on the floor, the same clothes I had been wearing at the bar and for the previous ten days. Funny, but I never noticed the smell before then. I wondered if the other patrons of the bar noticed it. Thus, I was introduced to the phenomenon of olfactory fatigue, the ability of the sensory cells in your nasal cavity to become desensitized to a particular smell over time. Then new odors, like the bear sneaking up behind you, or the stinking geologist sitting next to you at the bar, can be detected. We had not noticed, but we had been getting powerfully gamey over the past week or so. I am sure our friends at the bar and anyone downwind noticed and wished they had been desensitized to the malodorous vapors that enveloped us. Hoping to keep the few friends we had made, we adopted a new protocol of showering and changing clothes *before* going to the bar.

One of Southeast Alaska's treats for outdoorsmen is devil's club, *Oplopanax horridus,* an extremely nasty plant, hence the *horridus* part of its name. Devil's club grows up to 10 feet tall, with its stalk and leaves covered in tiny sharp spines. The brittle spines have a tendency to break off in your skin, where they fester for days, causing itching and discomfort. The plants like to hide within the tangle of fallen trees that cover the forest floor. Walking through the rain forest is difficult due to fallen hemlock and giant Sitka Spruce trees. When an eight-foot diameter tree trunk blocks your path, you need to go around it, which can take you 200 feet off course. Often, before reaching the end of the fallen giant, another tree trunk, lying perpendicular to the first, may block your path, taking you another 100 to 200 feet in the wrong direction. The worst places were enormous "blowdowns," impassable tangles of trees stacked one on top of another like a giant game of Pick-up-Sticks, created by williwaws, intense local blasts of wind that tear down the mountains to the sea. The only way to get across such an obstacle is to find a place to climb on top of the log pile and follow from one log to the next until you reach the other side. Leafy undergrowth, including our friend devil's club, fills in between the stacked logs, obscuring the height of the pile above the ground. More than once, I slipped off a moss-covered log and crashed eight or ten feet to the forest floor below, which was mercifully covered with springy undergrowth, soft moss, forgiving mud, and dang devil's club. I then had to pull out the spines of devil's club that pierced my body and search desperately for a way back up onto the pile to continue my journey.

The camp manager and logistics coordinator at Stone Rock was Mike Mancuso, a heavy-set cheery fellow with a full black beard who had no background in geology, logistics, or even camping. He was a drama student. He was also the nephew of the company's VP, which explained his employment despite his lack of qualifications. Mike must have been an outstanding student since he never lacked drama. He spent a lot of time wandering the woods around camp while we were out on our geological traverses.

I was looking for something (probably a beer bottle opener or some other essential tool) in the mess tent when I discovered a zip lock bag on the counter with slimy mushrooms in it. I asked Mike what that was. "Magic mushrooms," was the answer. Apparently, "Agriculture 201– fun with fungus – taking a trip without leaving home," is a popular class among liberal arts students where Mike attended university. He told me his psilocybin mushrooms needed to ferment for a few days before gaining their full magical powers. Two days later, I arrived at the mess tent early in the morning to find Mike sitting there blurry-eyed, scratched, dirt smudged on his face and moss in his hair – basically a burned-out shell of his former existence. Mike had a date with his mushrooms the previous evening. He spent the night running around naked in the chilly rain forest, accompanied by Alice, the Mad Hatter, and a few other hallucinogenic friends. Poor Mike, when he was just small.

We zipped up the mess tent tight every night to keep out marauding critters. That didn't stop the damn mice. Soon the place was overrun with them. They were getting into the food and were leaving little nuggets as gifts for us everywhere they went. We caught a glimpse of a small tawny brown critter poking around the mess tent on several occasions. It would quickly vanish when we approached. We named it Spook. One evening while we were having after-dinner drinks and discussing work plans in the mess tent with the front door open, who should appear, but Mr. Spook. Now that we had a good look at him, we knew what he was – a pine marten. Martens love to eat voles and mice. We decided to conduct an experiment – leave the tent open at night so Spook could hunt the mice in the tent. Within a week, the mouse problem was solved, and Spook was tame enough to come into the tent while we were there. Spook was our pest-control hero. We held an election for Mayor of Stone Rock City, a mostly ceremonial position but prestigious, nevertheless. Spook won by a landslide, with Mike a distant second. That didn't bother Mike, who decided Spook could have the title and all the mice he wanted. Mike was busy searching the woods for more mushrooms, so he and Alice could be tall again.

Alaska Drilling Company, which oddly was based in Texas, won the bid to drill at Stone Rock during our second year working there. They might have regretted that. Not that the drilling conditions were difficult; shit just happened. The drillers brought two drill rigs, a big one that needed to be airlifted between sites by a helicopter and a small one, which the drillers could disassemble and carry through the woods in pieces.

The drillers weren't inept. In fact, they set a record for the fastest time to drill the first ten feet of a hole. Unfortunately, the record didn't make the Guinness Book of Records because they used an unconventional drilling method. It might also have been disqualified because the drilling was in the wrong place. The drill rig was being airlifted to a new site and was hooked to a long line, about 250 feet of cable, to allow the helicopter to lift it clear of the tall trees. Up and up went the rig; then the chopper flew over toward the next site, about a mile away. The helicopter arrived at the drill site, but the drill rig didn't. About halfway there, the hook released and sent the drill rig plummeting 500 feet to the ground. The driller and I hiked through the woods to look for the rig. We found the mangled top three feet of it sticking out of a patch of marshy muskeg; the rest was far below ground. The drill was a total loss. It turned out the hook of the helicopter was faulty and failed mid-flight, letting the load fall. Lord knows how many times Mike and I stood under the chopper, hooking up and unhooking loads. I was grateful the hook failed when it did, and nobody was hurt.

That was not the end of Alaska Drilling's woes. The drillers finished the last drill hole at the end of October, which is getting pretty late in stormy Southeast Alaska. A ridge of high pressure, the North Pacific High, which occupies the Gulf of Alaska in the summer, breaks down in September and is replaced by the Aleutian Low. The low-pressure system spins off nearly daily storms, which increase in intensity as winter approaches. When drilling finally finished, my assistant geologist, Dave Bickford, and I broke camp and carried everything to the edge of the forest. We placed it upon the tangle of driftwood logs marking the high-tide/storm-surge line. From there, we could move it by helicopter onto a boat for the journey back to Ketchikan. The *Alaskan Salvor*, an old tugboat converted into a salvage boat, steamed to Stone Rock Bay for the job. Temsco's Hughes 500 helicopter arrived in mid-afternoon. It lifted the sling loads to the boat and dropped them into its hold.

Slinging loads from a barge to the log pile.

Meanwhile, the drillers were screwing around and were taking their sweet time moving the man-portable drill rig and supplies to the beach. The weather was deteriorating rapidly. The chopper pilot was worried about flying back to Ketchikan safely. With darkness and heavy rain approaching, he announced he had to leave. The drillers had only moved half of their equipment onto the boat. The rest was sitting on the logs, about 100 feet back from the usual high tide line and a good 10 feet above it. Two drillers flew back with the pilot. Dave, the third driller, and I joined the two-man crew of the boat.

A nor'wester hits Stone Rock Bay.

We planned to sail the tug around the corner to McClean Arm, a much more protected bay, and spend the night there. The helicopter was to return in the morning and finish moving the drillers' gear.

The *Alaskan Salvor* sailed out of Stone Rock Bay into the open ocean in heavy seas and high winds. McClean Arm was a mere four miles away, yet that was far enough for Dave and me to turn a lovely shade of green down in the ship's hold. We slipped past the narrow entrance to the long inlet, expecting to find calm waters. But the approaching storm was no longer approaching; it had arrived.

The captain navigated the boat a couple of miles down the bay. He anchored for the night on the leeward side of the arm, where there was less wind. An hour later, the wind and waves pulled us off anchor and sent us adrift. The captain and first (and only) mate were professional deep-sea divers. Taking no chances, the captain moved the tugboat back to the leeward side of the arm, where he dove down and set anchors at the bow and stern by hand. We weathered the storm safely overnight in the dank, musty lower quarters of the old tugboat.

The next morning, the storm had not abated. It had intensified. We spent the day and another night rocking in the waves and gale-force winds.

On the second morning, the wind and rain died down enough for the captain to decide to seize the opportunity to sail to Ketchikan. He dove down and released the anchors. We steamed across the arm and headed east to the inlet. Just as we entered McClean Arm's narrow entrance, the engine quit. The captain and first mate rushed to the engine room and frantically tried to restart the engine. Dave and I stood on the deck anxiously watching as the waves pushed us toward the rock wall of the entrance. We wondered if we could scale the cliff after taking a swim in the frigid sea or how long we could survive in a raging storm with nothing but the soaked clothes on our back. When the boat was a mere 30 feet from the rocks, a reassuring roar came from the engine room – saved in the nick of time. The *Alaskan Salvor* sailed to Ketchikan in moderate seas with ominous clouds following it.

The Alaskan Salvor at Ketchikan Harbor – post-storm.

We escaped in the eye of the storm. Shortly after we docked in Ketchikan, the tempest returned and raged for another three days. Fortunately, we were in the safety of Ketchikan and its cozy bars, a far better place to weather a nasty storm than the dingy hold of a tugboat. The weather cleared on the fourth day. The head driller and I flew back to Stone Rock Bay. As we turned the corner to enter the bay, a disastrous scene greeted us. It looked like a bomb had exploded. Propane bottles and cardboard boxes containing various supplies were floating in the bay. The gravel beach was gone, as were most of the stacked driftwood logs. A few boxes of the drillers' equipment sat on the last remnants of the driftwood pile at the edge of a 10-foot cliff, which was not there five days ago. The drill rig and most of the equipment were underwater in the bay. Nothing was worth salvaging since the intense waves had ground everything up in the coarse gravel, like running ore through a ball mill.

The rugged coastline of Prince of Wales Island was hazardous, but the brutal sea was unforgiving. We could access the best rock exposures only during low tide when they were temporarily above sea level. We were constantly working the intertidal zone on very steep, extremely slippery rocks covered in seaweed and kelp. Lots of falls ensued. Fortunately, there were no serious injuries to our team. However, the contract survey crew that preceded us in the early spring had a bad experience, a very bad one. The survey crew was working at McClean Arm and camped at Stone Rock Bay. At the end of the day, three surveyors took a shortcut across the headland through the tangled woods to return to camp. One young surveyor decided it would be easier and would involve less devil's club if he followed the rocky coast. When he didn't arrive at camp, the crew went looking for him. They found him in the water, holding onto a log about 200 feet offshore. The strongest swimmer tied a rope around his waist and tried swimming to him. His limbs seized up after a few minutes in the frigid water. His buddies had to haul him back in with the rope. They stood on the shore and helplessly watched as their friend slipped off the log and disappeared. Divers found his body a week later in 80 feet of water.

With one notable exception, our medical issues were limited to daily scrapes, bruises, bug bites, devil's club spines that festered in our flesh, and the occasional case of bottle flu. On the second day of a 10-day hitch in camp, I came down with a severe toothache from an abscessed molar. The pain was so intense I couldn't eat, couldn't sleep, couldn't even think. It was like I transformed into a liberal Democrat or something. After three days of debilitating pain, I considered playing dentist and using my rock hammer and a chisel to knock out the offending molar. The floatplane to take us back to Ketchikan was not due to arrive for another eight days. I didn't think I could last that long. My last resort was to ask Mike for some of his magic mushrooms so I could get small or go on a binge with my buddy Jack Daniels for a week to blunt the pain. I was on the wagon at the time, something I used to do for a month each year to soften my liver a bit (I gave up on that stupid idea years ago). I picked the wrong month to give up drinking.

The drillers came to the rescue. No, they didn't drill out the tooth or knock it out for me. Instead, they recommended a home remedy, a poultice made from Bacardi 151 and Skol chewing tobacco. It sounded horrible (and it was), but I had nothing to lose. The driller took a shot glass, put a few pinches of Skol in it, and added enough rum 151 to make a paste. He instructed me to hold the paste against the abscess. It was pretty disgusting stuff for someone not used to smokeless tobacco products, or any tobacco for that matter. They also told me that under no circumstances should I swallow the stuff. I had a good idea of what would ensue after watching Randy Palmer while drinking Coors and chewing tobacco. He was using an empty beer can as a spittoon for the nasty juice. He accidentally switched the two cans and took a big gulp of the brown spittle. The gastric expulsion was nearly instantaneous.

The poultice had a slight numbing effect. I kept putting it on the infected tooth. I decided since I was getting alcohol through osmosis, I might as well break my abstinence. Pinches of the poultice alternated with shots of 151. After a couple hours of that procedure, I was thoroughly numbed. I stumbled down the trail to my tent and fell asleep for the first time in three days. I woke early in the morning and walked to the mess tent, only to realize halfway there that I no longer had a toothache. The driller's crazy remedy worked. I didn't see a dentist until I returned to Denver several months later. He didn't believe my story.

The Inland Passage, where water, land, and sky meet.

I partnered for the 1979-1981 field seasons with the one and only Dr. John Carden. John and I shared responsibility for designing, managing, and implementing the exploration program. We also shared camp meals, a tent, and a comfy camp commode.

John received his Bachelor of Science degree from Kent State University. He was there at the fateful moment when the Ohio National Guard opened fire on a crowd of students protesting the Vietnam War, killing four of them. The outrage over the massacre led to a turning point in public opinion against the stupid war. I think the trauma of that experience drove John to drink, not as your typical drunk, but as a professional beer drinker. No, he wasn't a brewery representative or a taster for Anheuser-Busch. John was a professional beer chugger, an undefeated one at that. John's unique physique is an essential asset in his line of work. He has a short, wide neck. Actually, I think his head sits directly on his shoulders. His mouth stretches all the way across his broad face, producing a smile that would make the Cheshire Cat or Joe E. Brown proud. He has been known to eat a Big Mac in one bite. On top of that, John has no gag reflex. Whatever hits his gullet keeps going. These genetic attributes make John a natural beer chugger. Once a beer glass hit his lips, the beer is gone. There is nothing to inhibit the flow of suds on the short descent to his stomach.

John attended graduate school at the University of Alaska in Fairbanks at the perfect time – when the great Alaska Pipeline was under construction. Texans and Oakies flooded into Fairbanks, wads of cash from the high wages they earned as equipment operators, welders, and pipefitters burning holes in their pockets. Whenever John needed some spending money, he visited the bars in downtown Fairbanks. He would strike up a conversation with some pipeline workers (easily identified by their cowboy hats, huge belt buckles, drawls, and boisterous babbling). John explained he was a student at the U. of A., working on his Ph.D. in geology. After some chitchat, John would inquire, "Do you chug beer?" The macho Texans sized up John with his scholarly coke-bottle glasses and concluded they had an easy kill with this nerd

Not so. In classic pool shark fashion, John played them like a fiddle. He held back on the first chug, barely beating the competition. Belching and feigning digestive discomfort, he challenged his quarry to a second chug with a $10 bet on the line. Once again, John taunted the pipeline boys, beating them by a hair. John raised the bet to $100, which was gladly accepted by the unsuspecting Texans. On the following chug, John's empty glass hit the bar shortly after the competitor's glass reached his lips. Leaving the losers in shock, John cruised down to the next bar in search of new victims with more money than brains to repeat the feat.

Dr. Carden was most dangerous when he had serious competition. Fortunately, that was not often. Besides John's already impressive chugging talents, he has fast reflexes. If it looked like a chugging contest was going to be close, John blasted his beer glass down at Mach 10 speed. His near-sightedness didn't help, either. On occasion, he slammed his glass into the bar, sending glass shards flying in all directions. He was less than politely invited to leave a few bars when that happened.

Alaskan mosquitoes are legendary for their size. They have been designated as the unofficial state bird and are rumored to carry off sled dogs and small children. They are also famous for their ferocity and tenacity, capable of stampeding herds of caribou and driving geologists crazy – and for their sheer abundance, thick as molasses and able to blacken the sky. The size of Alaskan skeeters may be exaggerated, but I can testify the latter claims are fact rather than fiction.

The Kuskokwim Mountains lie between the Yukon and Kuskokwim Rivers in the southwest part of the state – the middle of mosquito country. We worked there and in the Kokrines Hills and Nulato Hills north of the Yukon River in the summers of 1976 and 1979-1981. The region is largely uninhabited, with only a scattering of small Athabascan fishing and trapping villages along the Yukon: Nulato, Koyukuk, Galena, Ruby, and Kokrines; nothing along the Kuskokwim, save McGrath with a whopping population of 350 sourdoughs. This part of Alaska is characterized by hills and rounded pine-forested mountains, separated by broad flat valleys, centered on lazy meandering streams winding through oxbow lakes and muskeg. Although not nearly as swampy as the flat Yukon-Kuskokwim Delta in the lower reaches of the rivers, which is more water than land, there are a lot of wetlands in the Kuskokwim country, certainly enough for the incubation of quadrillions of bloodthirsty mosquitoes.

The mosquitoes that plagued us were not large enough to carry off dogs. But they were pretty damn big, about half an inch long plus a proboscis about half the length of the body. With a poker that big, they could easily penetrate two layers of clothing to reach the coveted arteries and veins that flow beneath. Despite summer temperatures often in the 80s, we wore long underwear and heavy flannel or wool shirts. That combination was usually, but not always, enough to prevent the bloodthirsty devils from reaching their target.

While I was bent over, looking at an outcrop of greywacke laced with quartz veinlets, John came up behind me and said, "Don't move."

"Why?" I nervously inquired, thinking that maybe a bear was about to eat me unless I stood still and blended in with the rocks.

"I'm going for the record." Whack! A big hand came down on my back. The official count began:18 smashed mosquitoes were stuck to John's hand; 24 carcasses were collected from my shirt, for a total of 42. Forty-two big-ass mosquitos killed in one swat! That beat the previous record by four skeeters. It could have been more, but several of them escaped between John's spread fingers.

For fun and delightfully cruel revenge, we liked to catch mosquitos, being careful not to squash them. They were big enough so that it was easy to grab them by the wings. We used the small scissors on our Swiss Army knives to cut off the mosquito's proboscis. Thus modified, we let them loose to fly around – unable to feed or suck blood, doomed to a slow death by starvation. Even more fun was to trim off half of one wing so that the little bitches flew in ever-decreasing spirals until they flew up their own assholes. Ah, revenge is sweet!

One of the more grueling customs of working in the Alaska bush is the Saturday night bath. The challenge lies in getting naked and bathing quickly enough to complete the mission before losing more than a quart of blood to the clouds of mosquitoes that constantly envelop you. The bathing process goes like this: walk to the river's edge with soap and shampoo in hand; strip quickly at the river's edge; jump into the river and wash up, making sure no part of your body below your nose is ever above water. When finished, run to the shore. Grab your towel and sprint to the tent, swinging the towel wildly to avoid being covered by the squadrons of mosquitoes that patiently waited for you to emerge from the water. Upon reaching the tent, dive in and zip up the screen door; then hunt down and smash the mosquitoes that followed you into the tent. And you wonder why we bathed only once a week? The sourdoughs gave up on bathing altogether. I don't blame them.

After our Saturday night bath, we built a smoky fire. Then, we went to the stinky pile of clothes we abandoned at the edge of the river. Armed with a willow stick, about three feet long, we retrieved our gross, grungy, stinky underwear. After wearing them for a week of hiking, sweating, and farting, there was no way they could possibly survive a second or third week. Thus ensued the "Weekly Burning of the Shorts". With our sacrificial underwear skewered on the end of the willow stick, and with a small dose of jet fuel (a double dose for particularly crusty shorts after really sweaty weeks, especially if beans were on the dinner menu), we roasted them like marshmallows, but without the ensuing consumption of tasty melted goo.

The Weekly Burning of the Shorts, Anvik River, June 1980.

Logistics are a huge part of conducting exploration in remote Alaska. When we first worked a new region, we usually had a logistics assistant precede us to make lodging arrangements, find jet fuel sources, and arrange for fixed-wing transport of fuel and our camp gear. After that, we were on our own to keep the program running smoothly. Doing so required a lot of time planning meals, obtaining food and supplies, and estimating helicopter fuel consumption and placement of fuel caches. Complicating those efforts was the need for a backup plan for nearly every day of work. It never seemed to fail that when the pilot asked where we were going as we boarded the helicopter early in the morning, I pointed to the darkest, most threatening clouds. Then it was time to execute Plan B.

Bush pilots and their planes were crucial to the success of our work in interior Alaska. We dealt with a number of legendary Alaskan pilots: Lloyd "Stinky" Hardy of Seward Peninsula Flying Service, Norm Yaeger of Galena Flying Service, and Martin Olson of Olson Flying Service, to name a few. Where decent airstrips were near our work areas, we hired Fairchild C119 Flying Boxcars to deliver up to 40 barrels of fuel. In other areas, we relied on Cessna 207 Skywagons to deliver three barrels of fuel or Cessna 185s to fly two barrels of fuel to short, primitive landing strips in the bush. From there, we used the helicopter to sling two barrels of fuel at a time to stock remote fuel caches.

For work along the Yukon River, barges brought our fuel to ports at small villages. When we worked on the Seward Peninsula, Martin Olson, who owned Yutana Barge Lines, used a WWII LCT landing craft to transport fuel from Arctic Lighterage in Nome to Golovin, a tiny fishing village on a spit in Golovin Lagoon on the southeast end of the peninsula. Martin was the grandson of Ollie Olson, a Saami reindeer herder from Lapland who came to Alaska in 1898 and married an Eskimo lady. Martin and his wife, Maggie, were two of the nicest people we met in Alaska. Martin also ran Olson Flying Service with his sons, who flew our gear and fuel to bush airstrips at Omilak, Aggie Creek, Elim, and others.

An old aviation maxim says, "There are old pilots, and there are bold pilots, but there are no old, bold pilots." That is especially true for bush pilots in Alaska, who must deal with mountainous terrain, extreme winds, blizzards, rough or nonexistent landing strips, and a general lack of backup support if something goes wrong in the middle of the Alaskan bush. The FAA has estimated the risks of flying in Alaska are five times what they are in the lower 48. When we first worked with Martin Olson in 1976, one of his sons had perished in a fiery crash in his Cessna 170 just a few miles north of Golovin. A year or two later, a second son died in a plane crash. On a dreary November day in 1980, Martin was flying a heavy load of reindeer meat from Shismaref to Golovin in his Cessna 206. He climbed too high to get above the cloud cover and became hypoxic and disoriented. By the time he descended below 10,000 feet and regained his wits, he was low on fuel and facing an oncoming snowstorm. Martin and his cargo never made it to Golovin. In the spring of 1988, Norm Yaeger and his chief pilot were ferrying a Beachcraft Baron from Galena to Anchorage. Their flight plan took them over the Alaska Range through Rainy Pass at an elevation of 3,500 feet. Caught in a snowstorm and high winds, they attempted to fly IFR but drifted off course and flew straight into a mountain. Another legend lost.

Deep in the Kokrine Hills, about 30 miles northeast of the Yukon River village of Ruby, lies an Alaskan treasure, Melozi Hot Springs. At least 20 individual hot springs pour their steaming waters into Hot Springs Creek, a tributary of the Melozitna River, which flows southwestward to the Yukon River. The hot springs in and of themselves are worth visiting for a good soak. But what greeted our field crew in August 1976 was truly exceptional – Melozi Hot Springs Resort, the lifetime dream-come-true of Len and Pat Veerhusen. In 1966, the Veerhusens obtained a 20-year lease from the BLM to develop a resort at the hot springs. What a beautiful place they built! The central point was a large log lodge with a huge set of moose antlers hanging over the entrance. The lodge featured a high-ceiling living room with a towering rock fireplace, a spacious dining room, and a country kitchen. Several rough-hewn log cabins, a smokehouse, and a greenhouse surrounded the lodge. The cabins were plumbed with hot water from the 130° F springs, which kept the rooms warm. Even the toilets used water from the hot springs, which created an odd sweaty situation if one took too long to perform their daily "business."

Melozi was a welcome respite from roughing it in our camps. In the evenings, we relaxed in the lodge reading (my favorites were Alaska Magazine and Mother Earth News, both of which made me yearn to build my own wilderness retreat, but it was not to be) or listening to Pat's stories of building the resort and the many hardships and adventures she and Len shared there. Sadly, Len passed away seven months before we arrived at Melozi; I never had the pleasure of meeting him. From Pat's stories, it was apparent that he was an amazing man, a true Alaskan legend. Shortly after obtaining a lease on the land, Len walked a D-8 cat over 30-some miles of reindeer trails to construct a landing strip so that supplies could be flown to the site. He and Pat fixed up an old reindeer herder cabin to live in while he built a sawmill. Using local materials: trees that he cut down, rocks, sand, and gravel from the river, Len built the resort. Nails were so precious that Len carefully straightened any bent ones to reuse them. Pat told us of frightening bear encounters, chasing moose out of her beloved garden, and the time Len broke his leg while working alone on a bulldozer two miles from camp. He set his badly broken leg and survived the frigid night by stuffing his overalls with dried grass for insulation. Pat radioed for help in the morning. A helicopter from the Air Force base in Galena found him and airlifted him to the hospital.

Each morning we woke early from our cozy beds, then followed the enticing aroma of sizzling bacon to the lodge. Pat's delicious sourdough pancakes greeted us there – a welcome improvement over cold cereal served in camp. Dinners included fresh vegetables from Pat's well-tended garden, often accompanied by moose meat or by fresh fish we caught on the way back to camp. Grayling were in abundance in the cold, clear streams. Northern pike ruled the sloughs and backwaters of the Yukon and its tributaries. Monster sheefish, the "tarpon of the north," a silvery variety of whitefish, cruised the deep waters at the confluence of the Melozitna River and the Yukon and were more than willing to hit a well-placed lure and give a great tussle before either breaking our 10-pound-test line or yielding to the frying pan.

The crew and I returned in 1977 to spend a week at Melozi. We once again thoroughly enjoyed our stay. Two summers later, we stopped by Melozi to do a few days of reconnaissance around Wolf Mountain. I was looking forward to seeing Pat again. Our Hughes 500 helicopter buzzed the lodge, made a hard banking turn, and slowly descended to a small clearing near the lodge, kicking up a cloud of dust as we landed. Once the dust settled, we exited the chopper and were greeted not by Pat but by Roslyn, a cute young blonde who was running the lodge with her boyfriend, Tom. I was dismayed to learn Pat had succumbed to cancer over the winter. Our stay was pleasant but was not the same without Pat.

Roslyn and Tom operated the Melozi Lodge for several more years until the Bureau of Land Mismanagement, in its infinite wisdom, raised the annual lease payments in a successful attempt to force them out. With the lodge abandoned and in disrepair, nature has been slowly reclaiming the buildings and their contents. That was not good enough for the BLM. A 2012 BLM announcement read in part:

July 13, 2012

City of Ruby,
P.O. Box 90
Ruby, Alaska 99768

To Whom It May Concern:

This August the Bureau of Land Management (BLM), Central Yukon Field Office plans to begin a phased cleanup of a former lease site located at Melozi Hot Springs near Ruby, Alaska.

After the death of Leonard Veerhusen and his wife, several different lessees took over Melozi Hot Springs Lodge, with the last being the Melozi Hot Springs Incorporated. The latest lease was not renewed and expired on May 27, 1996.

Today the site includes numerous buildings and other improvements, two bulldozers, solid waste and hazardous materials. Most of the improvements have not been maintained and have fallen into disrepair, causing public safety issues. Several buildings have collapsed, others will likely do so without further maintenance, and the airstrip is unusable due to erosion and re-vegetation.

With costs increasing every year and public safety at stake, the BLM has obtained funds to begin clean-up of the site.

Isn't it wonderful how the Federal government deliberately creates a problem, then claims "public safety" (in a place few people can access) as an excuse to spend public funds on an unnecessary mission to destroy a wonderful piece of history? Len and Pat would roll over in their graves.

Stew Hill, chopper pilot (left), and the author at Melozi Hot Springs Lodge with a stringer of northern pike and sheefish, caught during a couple stops on the flight back from a day's fieldwork – dinner for all and plenty for the smokehouse. The plaque on the left reads:

IN LOVING MEMORY OF
LEONARD B. VEERHUSEN
1913   1976
A DREAMER WHO HAD THE COURAGE
TO MAKE HIS DREAMS COME TRUE

The great Alaskan geo-adventure came to an abrupt end in September 1981, following an epic battle between the forces of lift and gravity (hint, gravity won). More on that later. In seven field seasons working in the Land of the Midnight Sun, I managed to make a good dent in the goals that Fred, Randy, LD Scott, and I set at the Ace High Tavern. The stories Randy heard were true. Fish were both abundant and large, as were the bloodthirsty mosquitos they ate. We fished to our hearts' content in the crystal-clear, cold streams, catching arctic grayling with their speckled dorsal fins held aloft like ship's sails. I remember catching at least 20 grayling from one small pool in the stream next to our camp. We speculated the grayling have short memories; we might have caught the same fish over and over again. Whenever we had a choice between riding with the pilot back to refuel the chopper or being dropped off at a stream or high mountain lake, we opted for the latter. At an unnamed lake high in the Kuskokwim Mountains, John Carden and I spent an hour catching nice trout on 50% of our casts. It doesn't get much better than that.

John Carden with high lake trout.

A gorgeous arctic grayling.

Author with a Kuskokwim northern pike; and a flying chum salmon.

We cheated by flying down rivers during the spawning season until the clear, calm waters turned black from salmon fighting side by side to make their way upstream to spawn. We landed on the nearest gravel bar and did our best to temp the salmon or the Dolly Varden trout that accompanied them to take a lure. From our camp at Stone Rock Bay, we launched our rubber dingy in the evening. We spent an hour or two sipping Jack Daniels while we bounced our bait off the bottom of the bay in search of rock cod and the occasional small halibut ("chickens"– the tastiest fish ever), occasionally coming up with a starfish or sea urchin instead. Backwaters of the major rivers such as the Yukon, Koyukuk, and Kuskokwim were the lair of toothed monsters, the northern pike, one of my favorite fish to catch, but my least favorite when it comes to removing a lure from its toothy mouth. And I will never forget the fight of 15-pound sheefish in the deep waters of the Yukon.

Fred probably did the right thing buying a wedding ring for his gal. Unfortunately, despite many hours standing in freezing water panning for gold, I accumulated only a few flakes and one rice-grain-size gold nugget from my efforts – certainly not enough for a wedding ring, even for a stick woman.

As for adventure, there was a lot of it: going where few men had gone; flying over some of the most spectacular mountain scenery in the world; walking on moving rivers of ice; exploring old abandoned homesteads and mining camps; surviving pounding storms, bars, and bears; then fighting (and losing) that infamous battle with gravity.

Although I made substantial progress on the list of things one must do in Alaska, I left still looking for that bear (or was it an Eskimo lady?) to make love to so I can become a true Alaskan.

Horny geologist – still looking for that special bear – I mean lady.

Chapter 4

IN SEARCH OF SMOKES

"Me no Katmai. Life better than money."
— Chief of Katmai Village to Dr. Robert Griggs,
leader of the 1915 expedition to Katmai Volcano

The Valley of Ten Thousand Smokes — Robert Griggs

The villagers of the fishing village of Kodiak looked to the west in the early evening of June 6, 1912, to see a rapidly approaching dark cloud accompanied by spectacular lightning and deafening booms of thunder, phenomena unheard of in that part of Alaska. Gray volcanic ash began to fall as the menacing cloud descended upon the bewildered and frightened residents. Within an hour, the ash fall was so heavy, darkness enveloped the town. Ash continued to fall intermittently over the next three days while numerous earthquakes shook the town. Pandemonium broke out. The eruption buried the town of Kodiak under more than a foot of heavy ash, crushing buildings. Unstable accumulations of ash on steep slopes avalanched, further destroying buildings. Meanwhile, explosions were heard as far away as Juneau and Fairbanks. Ash fell over a broad area, extending as far south as Seattle. Thus, the world was introduced to the second-largest volcanic eruption in recorded history, the eruption of Mount Katmai.

While we were making preparations for the 1980 field season in Alaska, my field partner, John Carden, introduced me to a fascinating book, *The Valley of Ten Thousand Smokes*, published by the National Geographic Society in 1922. The book details the 1915-1919 scientific expeditions to Mount Katmai and the Valley of Ten Thousand Smokes, led by the book's author, Dr. Robert F. Griggs. The expedition discovered one of the true natural wonders of the world, a steaming volcanic wonderland. Although the valley may sound like a great playground for the Marlboro Man, it has nothing to do with smoke. The "smokes" were fumaroles, white steam vents that rose hundreds of feet in the air like smoke from giant chimneys. In addition to documenting many scientific discoveries, Dr. Griggs related spellbinding stories of the arduous and often dangerous conditions the explorers encountered. John and I discussed making a trip to Katmai at the end of our fieldwork.

Oh boy, an adventure in the Alaskan wilderness with volcanoes, glaciers, and cool rocks thrown in! I was hooked. We had to go to Katmai.

Mount Katmai lies at the east end of the Alaskan Peninsula, inboard from the Aleutian trench, where the dense Pacific tectonic plate dives under and buckles the lighter North American Plate, creating an arc of mountains and active volcanoes – the Ring of Fire. The eruption that buried the town of Kodiak in ash was traced to the chain of volcanoes forming the backbone of the peninsula, more than 100 miles west of Kodiak. The actual source remained a mystery for several years until the Griggs expedition discovered the Valley of Ten Thousand Smokes in 1916.

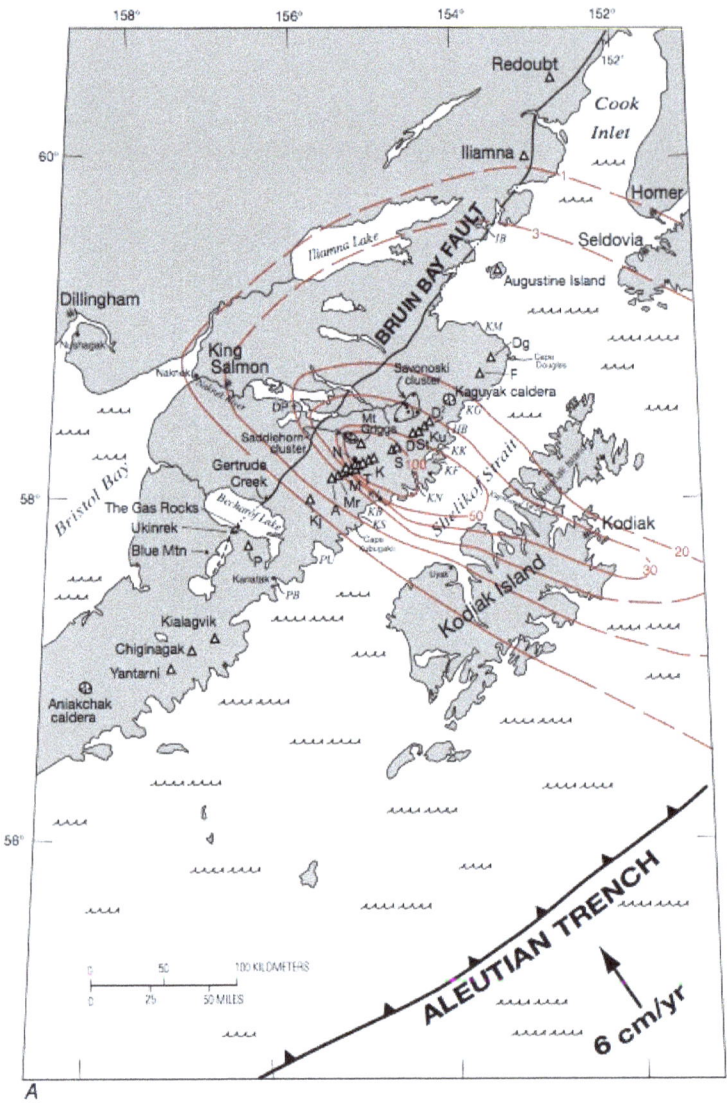

Thickness of ashfall (cm) from the 1912 eruption (USGS PP 1791).

Recent studies by the US Geological Survey determined that during the three days when ash fell on Kodiak, three major pulses of explosive eruptions at Katmai hurled roughly 6.7 cubic miles of ash, pumice, and rocks high into the air. For those challenged by the size of cubic miles, that equals 147 billion cubic feet. Of that amount, about four cubic miles of hot ash and pumice (ignimbrite) flowed into the valley of the Nukak River, obliterating the valley and all life in it. The water in the river and surrounding marshes in the valley turned to steam beneath the hot layers of ignimbrite. The steam followed fissures to the surface and shot skyward as fumaroles to form the Ten Thousand Smokes. Mount Katmai lost the top 2,500 feet of its summit during the eruptions and developed a caldera three miles wide and more than 3,000 feet deep.

We finished our fieldwork in mid-August. After five weeks of living in tents, battling mosquitoes, and traipsing across rough mountainous terrain, we should have elected to return home to civilization for some rest and relaxation. But the lure of Katmai was too great.

Before we left Denver, along with our standard field gear, we packed ice axes, climbing harnesses, crampons, ice screws, a good climbing rope, and ski goggles for the Katmai adventure. John brought his favorite indestructible mountaineering tent and his copy of Griggs' book, which we read enthusiastically in camp. We carefully laid out our plan to spend a week hiking in the Valley of Ten Thousand Smokes at the conclusion of our fieldwork in Southeast Alaska and the Kuskokwim Mountains. Intrigued by Dr. Griggs' accounts, John and I were eager to visit some of the fantastic places he described. As geologists, we wanted to study the volcanoes, their recent deposits, and Novarupta – the source of the eruption.

Being adventurous young bucks, we yearned to experience some of the challenges and adventures of the Griggs expedition, especially to scale Mount Katmai to gaze down into the deep blue waters of the crater lake, 3,000 feet below.

Whoever came up with the old saying, "Be careful what you wish for," was wise indeed.

The Griggs expedition approached Mount Katmai from the south, landing at Katmai Bay and following the Katmai River north. Their journey was difficult due to steep terrain, unconsolidated ash, mud, quicksand, nearly impossible river crossings, and bad weather – really bad weather. Griggs hoped to hire natives to help porter supplies. The few who returned to the devastated village of Katmai refused, believing the mountain to be cursed. The first expedition failed to even reach Katmai Pass. It was not until near the end of the second summer that the expedition viewed the Valley of Ten Thousand Smokes on the way down to Katmai Pass after successfully climbing what was left of Mount Katmai.

Our expedition would be easier, accessing the valley from the north and taking advantage of modern transportation to drop us at the very edge of the Valley of Ten Thousand Smokes, something that would have saved Griggs one-and-a-half seasons of hard work. We flew from Anchorage to King Salmon, then on to Brooks Camp in Katmai National Monument. Peninsula Airways' Grumman Goose, a 1940s eight-passenger amphibious plane, delivered us to the calm waters of Naknek Lake, next to the park headquarters. Despite its age and the oil leaking from the engine on my side of the plane, making my view blurry, the plane was both air- and seaworthy. I assumed the pilot had a clearer view from the cockpit and a few quarts of oil for the trip back.

The plan was to camp overnight at Naknek Lake and catch the daily park service van to the base of the Valley of Ten Smokes in the morning. That left us with half a day to kill. Canoes were available for rent at Brooks Lodge. Rather than paddling around the placid lake, as most tourists do, John and I decided to paddle up the Brooks River, portaging around the long set of rapids below Brooks Falls, where we watched bright red sockeye salmon jumping the falls on their migration to the place of their birth in Brooks Lake and beyond. After getting our fill of leaping salmon, we shot the rapids and floated down the river.

Near the river's mouth, sitting in the quiet water along the far shore, was a big brown bear poking around in the weeds for a snack, perhaps a salmon that floated down the river after spawning and dying. I was sitting in the front of the canoe; John was in the back. I put down my paddle and grabbed my 35 mm camera to take pictures of the bear. For several minutes, my eye was glued to the camera's viewfinder while

I took shots of the bear using a zoom lens. When I put down my camera for a second, I about shit. We had drifted across the river. My lens was zoomed out, not in. The bear was only 20 feet in front of the canoe – a distance it could cross in a flash. Plus, it was a good six feet closer to me than to John, a key factor in my decision to turn to John and say quietly but sternly, "Paddle backward, NOW!"

We spent the evening setting up camp, cooking dinner, and walking along the shore of the lake. A bleached moose skull with a giant full rack lay along the beach a short distance from camp, a reminder of the brutal Alaskan winters. As we walked along, we entertained ourselves by throwing pebbles and cobbles of white pumice into the lake and watching them float, bobbing in the gentle waves like a miniature flotilla – not normal rocks at all.

In the morning, the park van whisked a group of tourists and us along 23 miles of gravel road through a dense forest of spruce, poplar, and birch to the Three Forks Overlook at the north end of the Valley of Ten Thousand Smokes. The Valley kicked its smoking habit years before. Below us lay the Valley of No Smokes, a three-mile-wide barren plain surrounded by high snow-capped peaks, with not a smoke to be seen. The spectacular fumaroles that once filled the valley with steam lasted only about a decade before they began to decrease in intensity and eventually died. Fortunately, the amazing volcanic features from the eruption remain little changed since their birth in 1912.

The Valley of Ten Thousand Smokes from Three Forks overlook.

The overlook is about as far as most tourists go. It was just the beginning of our adventure. We bid adieu to the van driver after reminding him to pick us up five days later, shouldered our 90-pound packs, and marched off. Low clouds hung over the mountains. To the left, we could see the base of Griggs Volcano. To the southeast lay the volcanic backbone of the Aleutian Range. Katmai Pass was dead ahead with Martin and Mageak Volcanoes on the west side of the pass and Trident and Katmai Volcanoes to the east. We were in a volcanic wonderland, a veritable geologist's paradise

There are no official trails in the Valley of Ten Thousand Smokes, only general "suggested routes." A lot of the suggestions have to do with avoiding deep chasms, cliffs, and rivers that can't be crossed. Our route took us southeast along the barren ash-and pumice-covered wasteland that was once the verdant valley of the now buried Nukak River. The going was easy on a firm pavement of pumiceous sand and gravel. We followed the west side of the River Lethe, a deep narrow gorge cut into the poorly consolidated ash. The river was born about two days after the eruption and cut half of its current 100-foot depth by the time the Griggs expedition reached it four years later. Several miles upstream, we waded across the frigid glacier-fed river after it ceased coursing through a vertical-walled chasm.

Walking along the rim of Lethe Gorge.   *John fording the glacier-fed River Lethe – brrrr!*

As we hiked merrily along, the cloud cover lowered. We paid little attention to the weather. After all, we were prepared for anything. Dr. Griggs' words should have been heeded, "The Alaskan Peninsula is notorious as a storm-breeder, and before the eruption, Katmai Pass had a reputation for bad weather not to be matched elsewhere on the American Continent."

About 11 miles into the hike, we stopped for the day. We set camp on a flat spot on the flank of Baked Mountain. Our campsite was two or three miles north of Katmai Pass, not the best place to be. We would be there for only one night. What could possibly go wrong in that time? We purposely chose a site near where the Griggs expedition had camped in 1919. This was not just one of the many campsites for the expedition – it was the most notorious one.

According to Griggs:

> All went well at Baked Mountain Camp as long as the weather remained good, but when the storms struck we encountered a fury that no tent could endure. The configuration of the mountains is such that Katmai Pass is a double-ended funnel, through which the wind sucks with terrific violence.

The guidebook to Katmai, *Exploring Katmai National Monument and the Valley of Ten Thousand Smokes,* adds a stern warning:

> The narrowing of the mountain walls which flank the pass produces a natural venturi, accelerating wind velocities up to speeds in excess of 100 miles per hour (160 km/h). Winds of such velocity are more than sufficient to blow a hiker off his feet and carry him along the surface of the ground. Avoid this area if storms threaten.

Too late, we were already there.

In the middle of the night, I dreamt we were camped on the Southern Pacific railroad tracks, and a big locomotive was bearing down on us. It was not entirely a dream. We awoke to a deafening roar, quickly followed by a massive blast of wind that rocked the tent. Then it was quiet again. Ten minutes later, another roar came from the south, accompanied by a bigger blast of wind. This continued throughout the night, with each gust getting stronger. About the time the third blast of wind hit, my mind turned to recollections of what the Griggs expedition experienced:

> By the time it was fairly dark the storm had increased to considerable proportions. It came in great intermittent gusts, with intervals of quiet between. We could hear them coming over the mountains long before they reached us, their terrible roar as they tore their way down through the Pass for a while more frightful than the blow when it struck. Quaking with dread, we lay huddled together beneath the tent awaiting the crash.

Oh, my God, we were actually reliving the experience of the Griggs expedition. This was a bit more than we bargained for. Another quote came to mind, this one from the back cover of the Katmai guidebook:

> In 1912, Katmai was rocked by one of the most violent volcanic eruptions in modern times; today it is a tranquil land of untouched natural beauty.

Tranquil, my ass! The authors of that book sure as hell didn't camp here.

The tent stakes were not holding in the soft ground. Wind gusts started to lift the tent off the ground. Anticipating stronger winds and not wanting to fly down the valley in a jumble of nylon and broken tent poles, John and I donned our ski goggles to protect our eyes from the blowing ash and went outside to secure the tent before the next hurricane-force blast. We drove our ice axes into the ground to serve as heavy-duty tent stakes and tied our heavy backpacks to the tent corners to help hold the tent down.

The blasts of winds continued, coming more frequently and stronger each time. Once again, the Griggs expedition prepared us for what was to come:

> The pumice came beating against it with the noise of a hailstorm on a tin roof; but these stones were sharp at every corner, not round and smooth like hail. The impact was so heavy that our flesh would not tolerate the pain if, trying to keep the tent in place, we placed our bare hands or arms against the wall.

Heeding the warning from six decades prior, we placed our foam sleeping pads against the back of the tent and leaned against them. That way, we hoped to support the tent without being beaten to a pulp by flying pumice. The storm that hit the Griggs' expedition ended up destroying their entire camp, shredding each canvas tent to pieces. They had to retreat to a lower camp, which had been spared the destructive winds. We feared for the worse since we had no backup camp for retreat. According to the Park Rangers, we were the only people in the entire valley – perhaps other people knew something that we didn't – like maybe the weather forecast. Between each powerful wind gust, John and I exited the tent and hurriedly re-staked it in hopes of surviving the night.

It didn't work.

The wind gusts started flattening the tent, despite our valiant efforts to hold it up. About three o'clock in the morning, an intense gust of wind snapped the poles of John's tried-and-true, indestructible Alaskan tent. Just like the Griggs team, we spent the rest of the night huddled beneath our collapsed tent. The storm continued through the early morning; then the winds suddenly subsided.

The author inspecting the broken tent between blasts of wind, dust, and flying pumice.

Shortly after the winds died down, the sky cleared, and we were back to a beautiful sunny day. John's tent was destroyed. Having read the accounts of the Katmai expedition, we were prepared. I carried a spare tent in case such a disaster struck. Calamity was avoided. We ate a quick breakfast, broke the broken camp, and hiked further up the valley to visit Novarupta, the source vent for the colossal eruption of ash and pumice in 1912 and home to the last of the smokes.

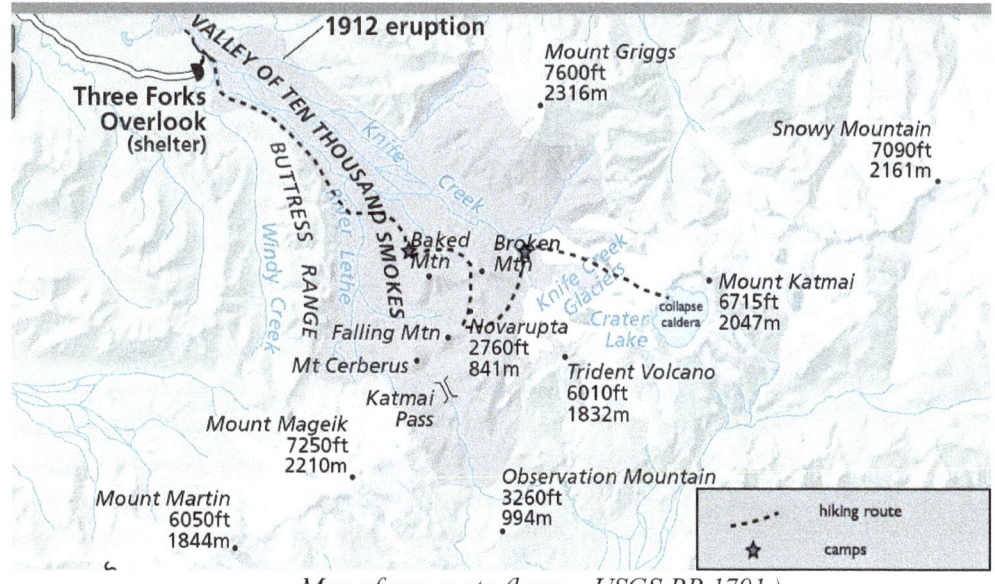

Map of our route (base – USGS PP 1791.)

61

We left camp, watching white clouds build behind the high peaks of the Aleutian Range. As we hiked up the valley, the clouds suddenly broke over Katmai Pass and poured down into the valley. Previous experience told me such meteorological events usually portend big storms. I hoped that was not the case, especially since we had just experienced a horrific windstorm and thought we were due for a break.

Ominous clouds flowing over Katmai Pass.

Hiking to Novarupta along the side of Baked Mountain.

Large breadcrust bomb, dome of Novarutpa (the rubble area) in background; John scrambling up the rubble of Novarupta – in search of smokes

We spent the morning clambering around the unstable rubble of the rhyolite dome that filled the eruptive vent of Novarupta. We found a few wispy fumaroles, the last weak remnants of the original 10,000 smokes, to warm our hands over. John and I studied the dome, the cone of ejecta around it including large volcanic bombs, and the unique black-and-white banded pumice blown out of the vent, the product of mixing two very different magmas, a rare phenomenon first reported from here. Despite being topographically unimpressive, the shallow basin surrounding Novarupta was the source of nearly all the ash and pumice that erupted over three days in 1912. Six miles to the east, the giant caldera of Mount Katmai – three miles wide and over 3,000 feet deep – was not caused by the mountain blowing its top but by collapse after the magma under Mount Katmai drained to feed the eruption at Novarupta.

A misty rain started while we were at Novarupta. We donned our rain gear. Soon more clouds broke over the pass, ugly black ones this time. In a few minutes, heavy rain driven by high southerly winds engulfed us. Another passage from Dr. Griggs came to mind:

> During the season of 1917 there was rain almost every day of our stay – not the gentle mist familiar to dwellers of Southeastern Alaska, but real rain in big drops, driven before the gusty winds that penetrated everything, until our tent roof leaked like a basket.

Oh no, did we have to relive this part of the expedition, too?

We had several more miles to hike to our destination, Knife Creek, at the base of Knife Creek Glaciers. We put our heads down and marched steadfastly into the storm. Meanwhile, the rain was busy marching right through our clothes.

Reaching the slope overlooking Knife Creek, we breathed a sigh of relief. A good site for a camp was nestled in the narrow canyon, protected from the worst of the winds by the front of the glacier. We scrambled down the steep slope, waded across Knife Creek, and arrived at the site. We immediately dropped our heavy packs. I untied the straps securing my sleeping bag to the backpack and held a 2-pound-2-ounce down bag that weighed about five pounds. The relentless wind drove the rain right through a supposedly waterproof stuff bag.

"Don't worry, we can share my sleeping bag," John said confidently.

"We are good friends, but not that good. Besides, what makes you think your bag will be dry?"

"It's in a waterproof bag inside my backpack."

John took out his sleeping bag and found it equally soaked. Here we were in the middle of the Alaskan wilderness, soaked to the bone and with wet sleeping bags. This was truly a predicament. Although we were out of the worst of the wind, it was still windy with the temperature at 45 degrees and dropping, perfect conditions for hypothermia. We quickly set up the reserve tent, started our little butane stove, and made hot soup. We stuffed the wet sleeping bags in the overhead webbing of the dome tent to dry. A few candles accompanied the stove to help heat the tent. There we sat in our wet clothes – chilled, but not too severely, because we wisely (for once, some wisdom in this misadventure) were wearing wool, which retains some body warmth even when wet. Wool is wonderful: just ask any sheep – or sheepherder. By about 2 AM, the sleeping bags had dried enough for us to crawl in and get some sleep.

Knife Creek camp (yes, there is ice under that ash).   *John at camp.*

The tempest refused to abate. We were pinned down by rain and wind for most of the next two days. We made a few brief excursions to investigate our surroundings during occasional lulls in the storm. Those breaks gave us a chance to study the volcanic deposits and glacial features, such as kettle holes left when blocks of glacial ice trapped in outwash gravels melted, leaving gaping holes. We climbed up the slope across from the glacier to scout our route up the face of the glacier and across the irregular maze of ash-covered glacial ice to Mount Katmai.

The storm let up considerably on the third day but was far from over. Time was running out for our planned climb to the caldera rim. We needed to be picked up at Three Forks Overlook the following afternoon and were scheduled to fly back to Anchorage the next day. It was now or never. We decided to go for it, despite the heavy clouds hanging over the glacier, hiding our destination from view. We grabbed our ice axes and the climbing rope, strapped on our crampons, and ascended the steep front of Knife Creek Glacier.

At the top of the ice wall, we entered a maze of heavily gullied, hummocky, ash-covered ice strewn with unstable boulders that rolled down from the surrounding mountains. The Knife Creek glaciers were beheaded by the collapse of the summit of Mount Katmai and lost the source of most of their snow and ice. The lower parts were starved of new snow and melted under a blanket of ash, creating the rough, pockmarked surface. The next mile or so would be challenging up-and-down travel over loose footing.

John on lower Knife Creek Glacier.

We made it across the badlands and found easier travel on the snow-covered active portion of the glacier. Soon we entered a fog. Visibility dropped to less than 100 feet. Despite the glacier being beheaded, the upper parts were once again alive, with new snow and ice covering the ash. We navigated by compass bearing, occasionally deviating from our course to work our way around crevasses, being ever aware of the dangers of collapsing snow bridges. Where necessary, we set ice screws and belayed each other across sketchy sections.

Before long, the fog turned to snow. We walked into a blizzard – in mid-August. Undaunted, we pushed on. Navigating by compass and keeping a close watch on our altimeter, we cautiously approached the rim of the caldera. Eventually, the slope dropped off. A check of the altimeter told us we must be at the rim. Not wanting to push our luck, knowing that the rim is unstable under normal conditions, much worse with a fresh cornice of snow, we declared the climb a success. Sadly, the magnificent view we came for was nothing but white.

John inspecting small kettleholes in glacial outwash gravels.

John crossing Knife Creek: "Although narrow spots do exist where it may be possible to jump across rivers, jumping is not recommended due to the risk of serious injury and/or death." – Hiking the Valley of Ten Thousand Smokes, Katmai NP brochure.

We planned to retrace our route down the glacier by following our footprints in the fresh snow. It was snowing hard. Our tracks were soon obliterated. Once again, we reverted to following a compass bearing. We finally exited the snowstorm and the fog below it, and followed our tracks across the rocky badlands back to our campsite.

We broke camp and packed up to hike back to Novarupta, cutting a few miles off the long return hike. The next morning, we awoke to blue skies with scattered clouds, the weather we had been waiting for. It came a day or two late, but it made for pleasant conditions on the long hike back to the overlook.

We climbed the hill at the end of the valley to reach Three Forks Overlook, where we planned to meet the park van. Instead of the van, we found a note saying it wasn't coming. Windy Creek flooded with the heavy rains. The road was washed out at the ford. So, we shouldered our packs again and hiked the additional five miles to the new pickup point. We came across two geologists studying the record of alternating soil and ash layers in a road-cut about a mile before Windy Creek. We chatted and told them about our adventure. They were amazed. After working in Katmai for more than 10 years, they had never seen such an intense storm as this one. They were glad they were staying at Brooks Camp during the storm, unlike us.

Resuming the hike, we reflected upon our great timing for the adventure, coinciding with some of the worst summer weather in years. A mile down the road, we forded Windy Creek and thankfully found the van waiting for us on the other side.

On the placid shore of Lake Naknek, under brilliant blue skies, we celebrated the success of the trip with a sumptuous gourmet meal of rehydrated dehydrated beef stroganoff and applesauce. Although we failed to accomplish one of our main goals – seeing the caldera and its lake far below; we successfully reached the caldera rim; we saw lots of cool volcanic geology. We also got much more of a taste of what the early explorers experienced than we wanted. Dr. Griggs would be happy to know some things have not changed.

Moon over Naknek Lake; Mt. Mageik with a fresh coat of August snow.

Chapter 5

FEAR OF FLYING

Both optimists and pessimists contribute to the society. The optimist invents the aeroplane, the pessimist the parachute.
— George Bernard Shaw

That is a rather odd title for this story. I don't fear flying. In fact, I love flying. Flying is fascinating; it is exhilarating; it is fun. Of course, I am talking about actual flying, not about all that pre-flying crap: the hassle of getting to the airport three hours ahead of time; the long check-in lines; excessive charges for baggage and carry-ons – even charges for your seat; not to mention the hassle of being x-rayed, strip-searched and felt-up, and being treated like the Unabomber by TSA idiots.

I am not the type to sit at the gate waiting to board the plane while nervously worrying about turbulence, getting hijacked to Cuba, or being blown out of the sky by an underwear or brassiere bomber. Nope, I have infinite faith in the efficiency of TSA, the skill of the pilots who fly the planes, and the mechanics who service them. It is my fellow passengers that concern me. I spend my pre-boarding time sitting in the holding pen of the gate, people watching. I like to board early, so I can put my carry-on bag into the overhead compartment before the little old grannies cram their oversized shopping bags into it, rendering it useless to anyone else. I take my seat and wait, sweating profusely as I watch the other passengers walk down the aisle. *That one will do*, I say to myself, as a svelte, young blonde approaches. She takes seat 13 B, two rows in front of me, leaving me exposed and vulnerable. Then comes the behemoth, the 400-pound gorilla I spied at the gate. With fat rolls bouncing, he waddles down the narrow aisle, seatbelt extension in hand, squeezing his way past passengers in the aisle seats with a lot of "Sorry" introductions as he smashes them with his excess adipose. Row 11, 12, 13, 14…*Shit, he is still coming. Oh Lord, please make him keep going!*

I can even put up with the tiny seats, which seem to shrink annually, and recline a maximum of two degrees off vertical, and the in-flight meal of six mini-pretzels, as long as I get a window seat that is not above the wing. You can tell the geologist on any commercial flight. He is the one with his nose glued to the window, intently watching the landscape and geology of the world go by while his fellow passengers, oblivious to the fascinating natural world below, whittle away the time reading about the affairs of irrelevant celebrities in People magazine, watching Hollywood's latest movie that bombed at the box office (worse yet reruns of I Love Lucy) or trying to keep the obese guy, who ended up in row 18 *(thank God)*, from flowing over into their seat.

Hell no, I don't fear flying. What I do fear (other than sitting next to those obese passengers) are those sudden unscheduled stops in a flight; you know, the ones that are preceded by a lot of screaming, crying, praying, and a "Mayday" or two from the pilot. Most people call them crashes. Pilots prefer to split hairs and to categorize them either as "crashes" or "hard landings" – the distinction being you can walk away from the latter. Since I am alive and writing this, I can accurately state that I have never been in an aircraft crash. Hard landings are an entirely different matter.

A quick review of aviation history and basic aeronautical principles is helpful before continuing with this topic. Man has been fascinated with flight for thousands of years. No doubt the Neanderthals looked up at eagles soaring effortlessly overhead and longed to join them in flight. The Greeks were the first to give it a shot (the mythological ones, that is) when Daedalus and his son, Icarus, attached feathers to their arms with wax and took off flying. It didn't work out too well for Icarus, who should have flown lower, or maybe used Super Glue instead of easily melted wax. In the 15th century, Leonardo da Vinci took a break from painting foxy ladies with impish smiles and designed several flying machines. Leo wisely never built any of the contraptions or tried to fly them. Over the next few hundred years, several crazies attempted to duplicate the flight of Daedalus by attaching wings to their arms and jumping off bridges. Icarus could relate to the results. In the early 1800s, Sir George Cayley experimented with gliders. The first successful unpowered manned flight was made in 1848 in one of Cayley's gliders. A skinny young English lad was the pilot. He should be famous, but Cayley got all the credit; some things never change. It wasn't until 1903 that the Wright Brothers built and flew a powered, controlled (well, sort of) flying machine that flew a spectacular distance of 852 feet. That doesn't sound like much today, but modern aviation was born that day in Kitty Hawk.

The flight fancies of the Greeks, the designs of da Vinci, Cayley's gliders, and the Wright Brothers' groundbreaking flight, as well as everything from Cessnas to gigantic Boeing 777s, share one thing in common. In fact, all of nature's flying creatures: eagles, bats, butterflies, hummingbirds, and bees, all have them. Yes, I am talking about wings, those marvelous appendages that create lift and allow for control of flight. Anything that attempts to fly without them is stuck with the glide path of a flying squirrel or a flat rock.

Along came an emigrant engineer, Igor Sikorsky, who had a better idea. Roughly translated from his native Russian, it goes like this, "We don't need no stinkin' wings!" So, in 1942, Igor developed the first commercially produced helicopter. Sikorsky's invention was revolutionary. It could do things that no other aircraft could do, like take off and land vertically. It could carry people and cargo to remote places lacking airports or landing strips. It could even star in movies in which intoxicated doctors shout, "Incoming!" or bandana-clad soldiers hang out the doors and blast anything that moves, while loud rock music blares from big speakers, scaring the shit out of the local peasants.

For geologists and other explorers, helicopters are a true blessing. They take us to faraway inaccessible places. They allow geophysical surveys over rough terrain to gather data we can use to interpret the deep structure of the earth in search of oil & gas, coal, gold, and silver. They carry our gear, ferry our rock samples, and deliver much-needed groceries and Jack Daniels to our remote fly camps. Yes, Igor's invention is a wonderful thing.

I started working with helicopters in the early 1970s doing airborne geophysical surveys in the lower 48: Colorado, Wyoming, and Arizona. My job was to design the surveys, serve as navigator to fly them, and follow up any anomalies on the ground. It was a fun way to see some spectacular landscapes and geology.

Working for years in Alaska, where helicopters are essential for transportation as well as for conducting airborne surveys, left me with a different attitude, more of a love-hate relationship with the contraptions. The hate side may be due in part to over two thousand hours of flying time spent listening to the high-pitched whine of the transmission that sat next to my head. I attribute my deafness to that, but Led Zeppelin and the Rolling Stones may share some of the blame. My dislike of helicopters is also a function of their one major design flaw – they ain't got no wings, the things that allow every other flying creature or machine to get airborne and stay there, at least for a while.

While attending graduate school at CSM in 1974, I was offered a dream job, spending the summer working in Alaska for Exxon Minerals Company. I hired on as an Assistant Geologist, helping Mel Klohn evaluate the uranium potential of the Tertiary sedimentary basins of central Alaska: the huge Susitna Basin, extending from Anchorage to the south edge of the Alaska Range; the Matanuska Basin; the Eagle Basin; and the Healy Basin in the Alaska Range. I got to see an amazing amount of the state. Mel was a great mentor, who loved teaching younger geologists, and had endless stories to tell. He also had a work ethic that fit me well: work hard and play hard. Keith, a soft, somewhat overweight fellow, a grad student at some lesser university down south, joined us in Anchorage.

Mel contracted a Bell 206A Jet Ranger helicopter from ERA Helicopters to fly us around the state. Our pilot was Chet Simmons, a little wiry ex-Vietnam pilot with shoulder-length hair, a big black mustache, and a chunk of his left bicep missing from some war-related mishap. Like most pilots who served in Vietnam, Chet was just a tad crazy. You didn't mess with Chet. He always had a pistol and a huge Bowie knife strapped to his hip. While we did our ground surveys, looking for anomalous radioactivity with our scintillometers (fancy Geiger counters), Chet blasted away at whatever target he could find. He was deadly accurate, all the while shooting from the hip, quick-draw style.

Quick-draw Chet with the formerly flying salmon.

We stopped for lunch along the lower reaches of the Chulitna River in the Susitna Basin one fine summer's day. King salmon were running up the river. There was a good size one, about a 20-pounder, lazily swimming in a deep quiet pool, about 50 yards long. The salmon was resting before resuming its long struggle up to the headwaters to spawn and die. Keith wanted to catch the salmon for dinner. He threw in a lure. The big fish was not hungry, but after several casts, it got annoyed and attacked the lure. Keith struggled to land the brute. It wasn't long before the 10-pound test line broke – so much for dinner. The big king was free, but it was not happy. It swam rapidly to the left end of the pool, did a 180-degree turn, and came swimming back toward us at salmon warp speed. Chet was standing next to me, hands at his side, his pistol holstered. As the angry fish approached, Chet said, "It's going to jump." That it did, followed by a "bang." When it hit the water, it was stone cold dead. We managed to snag the fish and bring it to shore. It had a big hole in its back, about six inches behind its head. Chet looked at it disappointedly and lamented, "Damn, I was aiming for its head." No, you don't mess with quick-draw Chet.

Chet liked to screw with us. While working near Eagle in east-central Alaska, we set out to investigate a uranium prospect at Calico Bluff, high above the Yukon River. There was just enough room to park a helicopter between the top of the 800-foot-high cliff and the pine forest that encroached upon it. Chet landed the helicopter at the lip of the cliff with the tail hanging over the edge. When it was time to leave, a sane pilot would have slowly lifted off, turned 90 degrees, and flown off level. That was far too tame for Chet; he was no sane pilot. He lifted the chopper, backed up about 30 feet, and dropped backward over the cliff. Looking up at blue sky, we thought we were all going to die. Chet rolled the chopper 180 degrees to the left, and we screamed down the cliff face toward the Yukon. Less than 100 feet above the river, he pulled up hard. As we rocketed back up, the G forces sent my esophagus to meet my sphincter, with my partially digested lunch scrunched somewhere in between. Damn crazy 'nam pilots!

A river-bar landing pad along the Little Susitna River.

The Jet Ranger was underpowered for the load of three geologists and our equipment. It gets pretty warm during the summer in interior Alaska, warm enough to increase the density altitude to the point the helicopter lacked enough lift to make a vertical takeoff in the heat of the afternoon. Many times, when Chet landed in a small clearing in the cool morning, we returned in the afternoon to see him busily cutting down small trees to make a runway for an airplane-style takeoff. Chet was pretty good at estimating just how much he needed to cut down for the escape. More than once, a treetop passed between the skids as we barely cleared the stunted taiga forest.

In early July, we based out of McKinley Village, at the entrance to McKinley National Park. On our second day there, the weather turned inclement. We flew north, trying to reach the Healy area, but were turned back by a low cloud ceiling, very low – actually, at ground level. After having lunch at the lodge, Mel decided we should fly south to Cantwell to refuel and see if the weather was better down there. Chet did the usual airplane-style takeoff, using the road in front of the lodge as a runway. When we were at full throttle and speeding forward about 25 feet off the ground, there was a loud "whack" and a jolt. Chet turned the helicopter around and landed to see what happened. He forgot to check for utility lines, like the telephone wire, that until a few minutes ago, stretched across the road to the lodge. The broken telephone line lay in the street. Black marks on the Jet Ranger indicated the line hit at the middle of the bubble, slid up, and caught on the cowling just below the main rotor, where it snapped. If it had gone higher, it would have tangled in the rotors, and you would not be reading this story. Shaken by the incident, we canceled the trip to Cantwell and spent the rest of the afternoon at the Village Pub, the lodge's cocktail lounge.

Those black lines should not be there.   McKinley Village, July 1974.

At the end of the 1976 field season, my boss, Art Pansze, flew to Alaska to review the results of the summer's efforts. We toured around the Seward Peninsula and interior Alaska, visiting prospects and enjoying the glorious autumn weather and lack of mosquitoes accompanying the hard frosts of fall. Our last stop on the tour was a visit to a uranium occurrence near Juneau. The day started as a typical September day in the rainy panhandle of southeast Alaska. The weather forecast called for early morning fog and light rain followed by lowering clouds with scattered rain showers, turning to steady rain in the afternoon with periods of heavy rain and sideways rain, then decreasing to moderate rain for a while before coming down like a bull moose pissing on a flat rock, finishing by raining cats & dogs in the evening. In a part of the world that sees as much as 200 inches of rain per year, weathermen need to be creative with their reports. People want details. It doesn't work just to report, "It will rain all day today like it did yesterday and will do tomorrow."

We flew up the Lynn Canal for about 40 miles to William Henry Bay, a short southwest-trending bay on the west side of the canal. Stewart Hill, our pilot for the summer, was at the controls of the Bell 206 Jet Ranger, which was equipped with pontoons in case a water landing was required. Our destination was about four miles up the bay, on the ridge forming the north side of the bay, at an elevation of 1,800 feet. When we arrived, the cloud level was at about 1,600 feet. The pilot landed the chopper among grassy hummocks just below cloud level. Art and I hiked up the ridge to see the prospect. The fog turned to rain. We were wearing our impervious Helly Hansen rain gear, so it did not bother us. When we finished our work, we walked back down the ridge in the fog. We walked for what seemed too long and became a bit bewildered as to our exact location – "lost" is not in a geologist's vocabulary. Finally admitting our predicament, we radioed the pilot and asked him to start the helicopter so we could figure out where the hell it was. Sure enough, the cloud cover had lowered, and we walked right past it in the fog about a hundred yards up the ridge.

Back at the chopper, we sat for a while as the fog and rain swirled around us. Stewart checked the weather. The ceiling was supposed to be at 1,400 feet. Art and I walked down the steep south side of the ridge for at least 300 feet but did not get below cloud level. We hiked back up and once again had to ask Stew to start the engine so we could find the helicopter nestled between hummocks in the pea-soup fog.

We sat in the relative warmth of the helicopter, waiting for the cloud cover to lift. It was not lifting; it was lowering. Worse yet, the wind picked up and was flipping direction faster than your average politician changes campaign positions. Finally, Stew decided since the helicopter was pointed facing out over the bay, he could hover down the slope a few feet above ground level until we got out of the fog; then fly out over the bay. He had performed this hovering maneuver many times that summer up in the foggy Seward Peninsula. It was always a scary undertaking, but in that part of Alaska, everything is above timberline. There are no trees or high bushes to worry about, only moss and lichen. Plus, whenever we hovered like that, there was very little wind.

Stew cranked up the turbine, and we started hovering down the slope. The constantly shifting wind was buffeting the helicopter, making the tail swing from left to right. Stew was rightly worried about snagging the tail rotor on a bush, which would have spelled disaster. He pushed the cyclic forward, and we headed out toward the bay – enveloped in fog – a world of white with no reference points.

Just what happened next is unclear. Perhaps Stew got vertigo in the fog, or maybe the headwind suddenly changed to a tailwind. The cause is less important than the effect. The helicopter suddenly shuddered. Sitting in the front seat next to Stew, I immediately scanned the gauges and saw the airspeed was a whopping zero. Not good! Helicopters do not stay airborne unless they are moving or are just a short distance above the ground (has to do with lift over that non-wing thing).

A second later, we were out of the fog, falling like a rock toward the steep green slope below.

If you ever feel the urge to be in a helicopter crash (or hard landing, if you prefer), there is no better pilot to be with than Stewart Hill. Stew is a skilled practitioner of the art. He was a Warrant Officer serving as a helicopter scout in Vietnam. That assignment relieved Stew of the need to wear a uniform or to salute. He was deemed dispensable. The Army did not want to waste a good uniform on such a soldier. Stew's mission was to fly low over places suspected of harboring the Viet Cong. The Army generously armed him with a pistol and put him in the oldest most beat-up Jet Ranger in the Army's junkyard. The idea was to draw fire so the Cobras, sleek, heavily armed helicopters following behind him, could blast the VC to smithereens. Stew was shot at daily. He never officially crashed but was shot down and made hard landings seven times. This was a talent very much in demand at the moment. Could he make it eight hard landings, or would this be an official crash?

The ground was rushing up to greet us at an alarming rate. To make matters worse, the helicopter was angled across the slope. This was definitely one of those *"Oh, shit!"* moments you dread. It seemed inevitable we would slam into the hillside and roll down it in blazing ball of flame. Just before impact, Stew turned the chopper so that it was facing straight uphill and was parallel to the steep slope. We hit hard. The pontoons popped but bounced us up in the air. The chopper was about to roll over backward, but Stew went into hard-landing action. He somehow turned it 180 degrees and pointed it straight downhill. This was more of an accomplishment than you may think. At this point, the helicopter was technically unflyable.

The Jet Ranger careened down the slope, like an out-of-control toboggan with a weed whacker on top, chopping down small bushes along the way. The lower bubble shattered; wet leaves blew in, plastering wet orange and red autumn foliage against the cabin and its occupants. Stew gave the chopper full power. The Jet Ranger picked up speed, but it refused to fly. It was like being on a wild amusement park ride that went horribly wrong and flew off its tracks. Several times we hit bumps and were airborne for a short distance, then slammed back down.

Ahead of us was a deep ravine, about 200 yards wide, with a vertical wall of black rock on the opposite side. I knew it was all over. We would either fall off the edge of the ravine and crash to the bottom or take a short flight and slam into the rock wall on the other side. Neither option was good.

Right at the edge of the ravine, the helicopter hit a big bump, which bounced it into the air. Suddenly we were flying! Stew banked the wounded aircraft to the left in front of the looming rock wall; we flew down the drainage toward William Henry Bay.

As we crossed the bay, Stew hit the radio microphone and announced to anyone who might be listening and maybe gave a shit, "Mayday, Mayday! Helicopter three-seven-four echo hotel, William Henry Bay, Mayday, Mayday!"

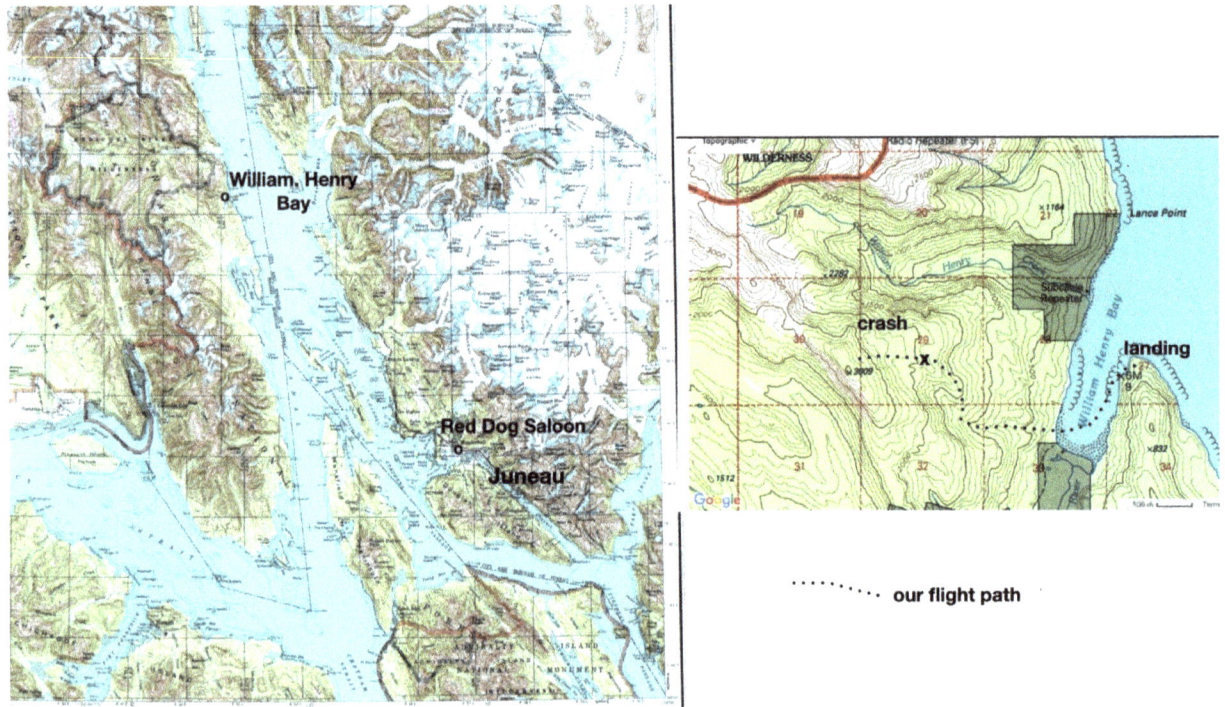

.... our flight path

The glacially scoured walls of the canals and fiords of Southeast Alaska tend to plunge straight into the ocean with little or no beach. William Henry Bay was no exception. There was no place to land the helicopter along the length of the bay. Near the mouth of the bay was a headland with a narrow rocky beach, which is exposed only at low tide. Could we make the four miles to get there? The helicopter was severely damaged and was making disconcerting noises. Would the tide be out enough to land on the beach, or would we be forced to land in the bay with our popped and useless pontoons, only to sink to visit Davy Jones in his locker?

Suddenly, I felt a knot in the middle of my back. I was convinced I received a spinal injury from the first impact. I tried to ignore it and focus on willing the helicopter to keep flying long enough to land on the rocky beach.

Stew took a beeline for the headland. We were in luck. The tide was out. He landed the chopper among the rocks on a short stretch of beach, to a great sigh of relief from all.

Art and I got out to kiss the ground and to inspect the damage. The pontoons were deflated, the skids were bent up along the sides of the helicopter, the lower bubble was shattered, and the tail boom was kinked and bent at an awkward angle. A couple of bushes caught in the skids accompanied us on the trip, bearing testament to the wild ride. The impact cracked the big crystal in our airborne spectrometer – an expensive loss, but we didn't care. Amazingly, the only injury was a chipped tooth Art suffered in the initial impact. The pain in my back went away. Apparently, it resulted from the aftermath of a major adrenaline rush.

A Coast Guard helicopter crew received the Mayday call just as they were landing at the Juneau airport. They immediately flew to the rescue. Meanwhile, Stew was on the radio talking to ERA's mechanic in Juneau to salvage what was left of the chopper. The Coast Guard arrived first. Art and I gladly accepted a ride back to Juneau. After notifying Exxon's headquarters of the situation, we moseyed over to the Red Dog Saloon to calm our nerves over a few (make that several) cold brews.

Art Pansze with N374EH beached at William Henry Bay.

Bill, ERA's mechanic, flew out in Coastal Helicopters' Hughes 500D, accompanied by a mechanic from Coastal Helicopters. Bill surveyed the damage to faithful ole N374EH. His conclusion: it was basically fucked. This part of Alaska has a 15-foot tide range, which equates to the tide rising or falling by over one foot per hour. The tide was coming in fast. Stewart wanted to fly the helicopter to a grassy spot above the high tide line that he found about 100 yards away. The conversation went like this:

"We are going to have to lift the helicopter over there using the 500D before the tide comes in," Bill declared.

"How about if I fly it over there. It would be a lot easier."

"No, you can't."

"Why not?"

"Because, legally as a mechanic, I cannot sign off on that. Besides, from the damage I have seen, it cannot fly."

"But I flew it here. I can fly it a little more," explained Stewart.

"Beats the hell out of me how you got it here. This helicopter cannot fly!" End of conversation.

The mechanics won the argument but may have regretted it. They tried to lift the wreckage with the 500D, but it was too heavy. The solution was to dismantle the helicopter – remove the engine and transmission, and lift them and the airframe separately. By the time the mechanics finished, they were in thigh-deep frigid seawater.

By then, Art and I were thigh-deep into ice-cold beers.

Thus, ended the 1976 field season.

For the 1977 field season, we upgraded to a larger, more powerful helicopter, a 5-passenger Aerospatiale Alouette III. This model of helicopter has a huge bubble and large plexiglass doors, which open forward. With the doors open, it bears a strong resemblance to a certain Disney character with huge ears. Thus, we dubbed our ride "Dumbo." The pilot was Gene Franks, an affable fellow in his 40s, who always had a joke ready. Gene had big ears and was the perfect match for Dumbo. He was the safest and sanest pilot I have ever flown with. Come to think of it – Gene was the ONLY sane pilot we ever had. Unlike the crazy ex-Vietnam pilots, Gene didn't learn to fly in the military. He received his Army training as a mechanic; then, he learned to fly as a civilian. Gene was also a stickler for safety. On the very first day, he informed us, "There will be no one-skid landings." For the uninitiated, one-skid landings are made where there is not enough room to land the helicopter safely. The pilot places the end of one skid on a steep slope, a big rock, or the edge of a cliff and hovers while the passenger exits by standing on the skid, then stepping off. It is a risky but all too often necessary maneuver.

Typical helicopter landing spot on the rugged coast of Prince of Wales Island.

Gene was true to his word. For the first couple of days, we didn't do any one-skid landings. As it turned out, we did very few landings at all. It was impossible to make a normal landing in nearly all the places we needed to access along the rugged coastline of Southeast Alaska. Within a few days, a frustrated Gene gave up on his prohibition. He went one better. Instead of one-skid landings, we were doing "no-skid" landings. For those, Gene hovered the helicopter above the ground. We climbed out onto the skids, dangled from them, and dropped off. My favorite no-skid landing was on a ridge above Stone Rock Bay. Our daily routine was for Gene to fly each geologist to a peak or high ridge and drop him off. We hiked down, doing our geological thing along the way; then we got picked up at a predetermined landing spot on the coast. Gene flew me to the ridge top and circled, looking for a landing spot. There was none. Thick brush covered the entire ridge. It was time for another no-skid drop-off. I opened the door, stepped out onto the skid, crouched down, closed the door, and held onto the skid. Then I lowered myself and hung from the skid, ready to drop into the brush a few feet below me. I expected the brush to be the usual four or five feet high. I gave Gene the signal that I was dropping off so he would be ready to counter the weight shift. I let go. Boy, was I surprised! I fell and fell, ricocheting off springy green branches until I hit the ground with a splat. Looking up, I saw Gene leaning out his window and laughing his ass off. That made his day, but it was a rough start for mine. Scratched and bruised from the 10-foot drop, I began my descent to the sea.

The start of another day's trek to the sea.

Later that summer, we finished work in the interior and were returning to southeast Alaska for further work there. The rest of the team took a commercial flight from Anchorage to Ketchikan. I elected to accompany Gene as he flew Dumbo down the coast. It would be a two-day flight past some of the most spectacular scenery in the world – one of those once-in-a-lifetime experiences I could not miss.

We flew east up Turnagain Arm, over the Chugach Range to the Gulf of Alaska. Clear blue skies greeted us, making for perfect flying and sightseeing. Our route hugged the coast past the spectacular lofty peaks of the St. Elias Range, including 18,008-foot Mount Saint Elias; past enormous glaciers calving icebergs into the ocean at their terminuses; over colonies of seals floating on icebergs; even over a pair of giant blue whales beginning their long annual migration south to spend the winter vacationing in Baja.

On the second day, we were flying along smoothly over the islands south of Wrangell when Gene asked,

"Do you have your shoes off?"

"No, my boots are on. Why?"

"I smell something."

"Thanks, but it is not me."

Gene looked over his shoulder, then immediately put the helicopter into a steep dive and announced, "When we land, don't wait for me to shut down. Get out, run like hell, and take cover."

Gene put Dumbo down in a clearing of muskeg. I jumped out, ran about 50 yards over the soggy ground, and hid behind a big hummock. Gene followed only a few steps behind.

Puffs of white smoke issued from the cabin as we watched the rotor blades slowly stop spinning.

"Gene, what the hell is going on?"

"The battery overheated and is melting down."

"Is that so bad? Why did we have to run?"

"It sits next to the fuel bladder."

"Oh!" That was all the explanation I needed.

After a few tense minutes of expecting to hear a loud boom and see flames shooting into the air, the smoke stopped. We cautiously approached the helicopter to assess the situation. Poor Dumbo looked sad, listing as it sank into the soft marshy ground. There was no damage other than a melted battery.

Gene asked, "Where are we?"

Oops, I had been playing tourist and not paying particular attention to my navigational duties. "On Etolin Island, I think. But I am not sure where on the island we have landed." Etolin is a big island.

The dead battery meant we were going nowhere. It also meant the radio did not work. We were stuck. Gene wanted to turn on the Emergency Location Transmitter (ELT), which would have announced a plane crash to any and all aircraft – including commercial airliners, resulting in the launch of a major search and rescue operation. I convinced him to hold off on using the ELT while I hiked to the coast to see if I could better pinpoint our location and maybe find a cabin or fishing boat to get some assistance. About a mile down the coast, a long, narrow bay jutted eastward into the island. Just offshore was a small recreational fishing boat. I called to the people on board and informed them of our situation, carefully explaining that it was not an emergency and asking if they could radio Temsco Helicopters in Ketchikan to pick us up.

Gene and I waited for the distinctive bumblebee buzz of Temsco's Hughes 500. What we got was the loud "whop, whop, whop" of a huge Sikorsky S62. The Coast Guard was coming to the rescue, uninvited. They overheard the radio call to Temsco and misinterpreted our emergency landing as a crash. Gene thanked them and sent them on their way since Temsco was also en route. When the Temsco helicopter arrived, Kenny Eichner, the owner of Temsco, greeted us. Kenny was also under the impression that we had crashed and was concerned enough to come to the rescue personally. Maybe Gene should have used the ELT after all.

Gene Franks and Dumbo after the meltdown, Etolin Island, August 1977.

Much of our work revolved around flying radiometric surveys across the rugged terrain of southeast and interior Alaska. Typically, such surveys are contracted to geophysical companies, which conduct the surveys, usually by flying along parallel lines across a predetermined area. Weeks or months later (if at all), geologists were supposed to check any anomalies in the field. Our approach was to fly the surveys ourselves, with one geologist navigating and the other watching the analog printout from the spectrometer. Whenever we flew over "hot spots," we broke off the survey, turned back, and landed as close as possible to the spot. That way, we were able to evaluate the anomalies immediately.

Such work is both demanding and dangerous. The danger is rooted in fundamental physics: the intensity of radiation decreases with the inverse square of the distance from the source to the detection instrument, the former being the ground and the latter the spectrometer in the helicopter. The bottom line of this equation is that altitude is critical for a good survey. Optimal ground clearance is 300 feet and should not deviate by more than 50 feet. That is easy to do if you are in Kansas. However, it is nearly impossible if you are flying over mountainous terrain in Alaska. There were many times when the radar altimeter indicated ground clearance of less than 50 feet as we flew up steep mountainsides. I worried it might become zero. On the downhill sides, the pilot struggled to force the chopper down to a clearance of 500 feet. To avoid that, we did a lot of "contour flying," which involved circling around mountains, much like following contour lines on a topographic map. That solved the altitude issue but presented a few difficulties of its own.

The real problem with flying such airborne surveys is the low altitude required for the flight – 300 feet – gives little or no time for a pilot to react should something go wrong when flying over rough terrain at 80 mph. If there were a mechanical failure, all we would have would be a wing and a prayer. Wait, helicopters don't have wings. We would be down to just the prayer.

Late into the 1980 field season, our pilot, John, was getting a little nutty and was making stupid mistakes. For starters, we ran out of fuel on the way to Poorman, a tiny placer mining camp south of the Yukon River, where we were camped. It was bad enough to be forced to walk through the bush the last few miles. But rather than setting the chopper down when the fuel warning light came on, John flew the helicopter until we were running on fumes – not a safe thing to do. Worse was when he dropped us off on a ridge to do some geological work while he flew to our fuel cache along the Yukon to refuel. We stupidly forgot to grab our radios when we exited the helicopter. What should have taken an hour turned into nearly four worrisome hours. We wondered if the pilot had crashed or was pissed off (as he usually was) and simply abandoned us. After a couple of hours, John and I were planning a 40-mile hike to the Yukon when we heard the helicopter flying along a ridge several miles in the distance. Then it was gone. The prick was lost. An hour later, we heard the helicopter again. After flying over about every ridge in the area, he finally got close. We flashed him with a signal mirror. He headed our way. The lying, flying sack of poop would not admit to being lost and made up some BS story about problems with the fuel barrel pump to explain why he was so late returning. The next time we refueled, we counted the fuel barrels. There was an extra empty one. He had flown so long over God knows where that he had to refuel a second time.

Near the end of the field season, we were refueling at Cantwell when John met a fellow pilot. He learned his colleague was making more money than he was. On the ride back, he threw a hissy fit, jerking his hands (which were on the critical controls like the cyclic and collective), making the helicopter bounce wildly in all three dimensions. The final straw happened when we were flying to Lake Minchumina. I was in the middle seat, next to the pilot. Geologist John was on the right. Out of the blue, the pilot turned to me and said,

"Here, take this," as he handed the cyclic to me.

"You're kidding."

"No, I have to get something out of the back seat. Just hold it steady."

I grabbed the stick unsteadily with my left hand and held it in a death grip, fearing that any false movement, maybe a sneeze or sudden passage of gas, could cause my hand to jerk and send us plummeting to the ground in an uncontrollable spin.

The pilot climbed into the back seat and rooted around for whatever was so important that he would give control of the helicopter to an unqualified person, a lowly geologist at that, who had never taken a single flying lesson. If I had, the instructor certainly would not have started me flying left-handed.

I did my best to hold the stick steady. After a few minutes, we were slowly losing altitude. John popped his head up from the backseat and provided the following advice, "When the trees start to get big, pull back on the stick."

The trees were indeed looming larger. I cautiously pulled back on the cyclic. We slowly regained altitude; the trees returned to their smaller, less threatening size.

During the 1981 field season, John Carden and I flew from Anchorage to Fairbanks, where we stayed overnight before mobilizing for fieldwork in the interior. Following a nice dinner (basically our Last Supper before being stuck with camp cooking), which was preceded by cocktails, washed down with copious quantities of red wine, and finished with snifters of brandy, we retired to our hotel room. We sat there, discussing our plans. John reached into the nightstand drawer, only to find Gideon's Bible. Gideon checked out and left it, no doubt, to help with our revival. John flipped through the book and read a few verses.

"John, you ever read the Bible?
"No, how about you?"
"Nope. I went to parochial school as a kid, and we had Bible studies, but I never actually read the good book."

John handed the Bible to me. As I thumbed through it, a light bulb went off in my head. The only reading material we brought along was technical geological literature. We needed something more inspirational to get us through the summer. We decided to take the Bible with us, assuming Gideon wouldn't mind. The goal was for each of us to read at least one entire chapter. John decided on Chronicles – not that he was really interested in biblical history – it was all the begetting that intrigued him. I picked Ezekiel simply because it sounded like a cool name, and I once knew a dog named Zeke, who was likely a close relative.

In addition to our large office/cook tent and our sleeping tents, our camp included a small high-peaked tent with a hole in the floor. That was our portable outhouse. A screened tent was essential for a privy in the Alaskan bush. We weren't concerned so much about privacy as about protection. Without the tent, the voracious mosquitoes attacked your private parts as soon as you dropped your drawers. In the middle of performing one's daily duty is not a good time to multitask by serving as a blood donor for hordes of savage mosquitoes. With mosquitoes plastered against the tent screen but getting no further, it was possible to have a few relaxed personal moments in the little tent. There was even time to do some reading. Sitting on a box next to the toilet seat was Gideon's Bible. Each morning we ambled over to the little tent to take care of business, sat down, and read a few pages from our designated chapter. Then we tore out those pages (to keep track of where to start reading the following morning), and used them for – well, you know. Forgive me, Gideon! We forgot to stock enough TP.

The last leg of the summer's work was at the Titna project. The project owed its name to the Titna River, a horribly twisted stream flowing westward to the Nowitna River, a tributary of the Yukon River. John Carden and I found uranium anomalies there while doing airborne surveys in 1980. We returned the next summer to spend a week or two doing detailed groundwork: geological mapping, radiometric surveying, and rock and soil sampling.

The Titna project centered on two hills, having about 800 feet of relief and rising slightly above timberline. The hills are separated by a two-mile-wide valley filled with a nearly impenetrable tangle of alder, willow, and underbrush. Each hill had a roughly north-south orientation. The smaller one lay east of the larger one. For map reference, logic would suggest that the hills be named "West Titna" and "East Titna." But John, being the joker he is, called them "Left Titna" and "Right Titna." He used this nomenclature during a presentation for Exxon Minerals' management and was quick to point out, "The Left Titna is larger than the Right Titna." The presentation generated a lot of laughs, even from the females present. Back in those good old days, people had good senses of humor. Nobody got their titties in a wringer or sued for sexual harassment over a simple attempt at levity.

We originally planned to camp at Lake Minchumina, using the helicopter for the 40-mile commute. However, John and I were burned out from too many hours of dangerous flying. We decided to set up a base camp at the lake but stay in a fly camp at Titna. We could then send the helicopter back to Ketchikan and have a peaceful break from flying for a week or two. Bill Kennedy, a field assistant who helped with logistics, came out to stay at the Minchumina camp. Bill served as our communications link via FM radio. He would also arrange for a helicopter to come from Palmer, near Anchorage, to pick us up when we finished our work.

A week later, we finished our work at Left Titna, broke camp, and hiked the two miles over to Right Titna through the tangle of trees and undergrowth.

We spent the next three days working at Right Titna. When we finished our work, we packed up our camp gear and samples for the helicopter to move to Left Titna. We hiked back to Left Titna and called

Bill at Minchumina to have the helicopter transport us and our gear to Lake Minchumina the following afternoon.

Back at Left Titna, we arranged the equipment and samples into properly sized loads for the helicopter to transport. The chopper, a Hughes 500D from Rocky Mountain Helicopters, call letters N8363F, showed up at the scheduled time, much to our delight. We put the loads into a net that was slung under the helicopter. It took three trips to ferry the equipment to the airstrip at Lake Minchumina. The last trip was to be for John, me, and our precious rock samples.

John hooking up the last sling load at Left Titna camp.

John and I were excited. Our work was finished; this would be our last helicopter ride of the year. After ten days of hiking, sweating, and drenching ourselves in mosquito repellent, we were looking forward to a hot shower, a tasty meal, and a soft bed. We were flying back to Fairbanks via a fixed-wing plane in the afternoon. The next day, we planned to fly to Anchorage and take a couple of days of R&R ice climbing on the glaciers near Anchorage before flying home to Denver.

Robert Greenhaw, the pilot, was an old hand at the business, old enough that he flew in Korea instead of Vietnam. The flight to Lake Minchumina was going splendidly. It was a gorgeous blue-sky fall day. Beneath us, the blueberry bushes and willows were showing off their bright orange and red fall colors. We passed over a herd of caribou grazing on the tundra grass, fattening up for the oncoming winter. Snow-capped Mount Denali stood out in bold relief in the distance. Cold beers and comfy beds were waiting for us in Fairbanks. Life was good.

That changed when there was a loud "pop" from the engine, accompanied by an annoying beeping in our headphones. A big red "Engine Out" light was flashing on the instrument panel.

This was not good. Engines come in very handy when flying over the Alaska wilderness, especially when the nearest landing strip is a good 15 miles away. That fatal flaw in Igor Sikorsky's flying contraption, the lack of wings, was about to bite us. An airplane's wings will continue to provide lift and keep the plane aloft on a shallow glide path long after the engines stop operating. There have been some spectacular examples of this, such as US Airways flight 1549, which lost power while climbing out of LaGuardia Airport, when a gaggle of geese committed suicide in the engines. No problem, the pilots glided the jetliner to a perfect landing in the Hudson River. Similarly, Canada Air Flight 143 from Ottawa to Edmonton ran out of fuel over Manitoba halfway through the flight (thanks to a mistake in converting English to metric units — oopsie!). The Boeing 767 became a rather heavy glider; yet, the pilots made a nearly perfect dead-stick landing on the runway of an abandoned Air Force Base about 15 miles away.

A helicopter's rotor blades actually can behave much like a wing, creating lift and allowing control of forward speed and direction – as long as the rotors are powered and the chopper is moving along briskly. When a helicopter's engine quits, Aerodynamics 101 dictates it has the glide ratio of a flat rock. I looked down at the pine forest beneath us and despaired. We would need a long glide to clear the trees. I keyed my microphone,

"Is there a clearing you can make?"

"I think so," the pilot responded, in a not overly confident tone.

There was no time for a Mayday; Robert was busy trying to fly a flat rock.

This is when you are supposed to see your life flash before your eyes. Instead, I was thinking maybe we had gone overboard messing with Gideon's good book. Were we going down because our blasphemy had pissed off Howard? I hadn't talked to him since my Catechism classes back in 7th grade, but I remembered the prayer well:

"Our Lord in Heaven, Howard be thy name…"

Good ole Howard and I were on a first-name basis back then. I attended parochial school from 2nd to 6th grade before transferring to a real school where I could finally learn something useful. I escaped the daily wrath of the nuns, but the Church kept a grip on us defectors. On the first Monday of each month, we Catholic boys were let out of school early to attend catechism class and confession at the Catholic Church across town. I remember well pulling back the curtain on the little confession booth, kneeling down in the dark and announcing, "Bless me, Father, for I have sinned," then rattling off how many times I cursed, or lied, or swiped an apple off the neighbor's tree. After that, I got to the good stuff. I confessed my friend, Clayton, and I raided his dad's stash of Playboys and lusted after images of sexy naked women. Father Donavon was all ears. Apparently, the rectory library was seriously deficient in its selection of essential manly periodicals. He told me, "Son, that is a grave sin; please tell me more, so I can decide your penance." More than happy to oblige, I described the centerfold in as much detail as a 12-year-old could muster, after which our session concluded with my penance: "Say five Hail Mary's, six Our Fathers and get back to me next month with a report on what is in the November issue." OK, I confess – I made up part of that, but I am sure the good Father was thinking along those lines and would have loved to join us on our next Playboy raid. My buddies sitting in the pews waiting to clear their consciences of the past month's sins and save their wretched souls must have thought I was one heck of a sinner since I took so long in the confessional.

It had been a long time since we last spoke. I hoped ole Howie still remembered me as that good church-going boy who kept Father Donavon up to date on the Girls of Dartmouth and other important issues of the day.

Despite their lack of wings, helicopters do have one saving grace. Given the right conditions and a skilled pilot, the engine can fail, and the helicopter can fall out of the sky, yet make a safe soft landing. The process is called "autorotation." It involves disengaging the main rotor from the engine and dropping the pitch of the rotor blades, letting the air out of the blades. The blades spin rapidly as air passes through them while the helicopter descends at a near vertical angle. Then, just above the ground, the pilot flares the nose to reduce forward speed and "pulls pitch," using the energy of the rapidly spinning blades to bring air back in, creating an air cushion under the chopper. A skilled pilot can make an autorotation landing, pick the helicopter back up, spin it 180 degrees, and land again – all without engine power. Having watched pilots practice this maneuver successfully, I was not as worried as I should have been during our descent.

When the engine quit, we were at optimal speed and altitude – 100 knots and 700 feet – to perform an autorotation. The problem was there were trees below us – lots of trees – small scruffy pines typical of the taiga. Helicopters and trees do not mix well. Even small trees win.

We made it over the trees to the clearing. But we paid the price. As we approached the ground, we carried too much airspeed – both forward and downward – from our long glide. I watched as the pilot yanked up on the collective to pull the pitch, expecting the helicopter to slow down and land softly. Not! It barely slowed at all. We hit the tundra hard, very hard. The long glide necessary to make it to the clearing required slowly bleeding the pitch out of the main rotor. There was nothing left to provide that much-needed cushion at the end of our flight.

The bubble shattered in front of our eyes, vanishing into a million pieces. The chopper bounced, tore off the left skid, landed a second time, and rolled onto its left side. Instantly, the spinning rotor blades hit the ground, flipping the chopper right and left violently. The right door flew open and was chopped in half by the blades. John, who was sitting next to the door, unbuckled his seatbelt and started to climb out. I grabbed him and said, "No, you don't." Stepping into the shrapnel of disintegrating rotor blades would not do much to improve John's good looks.

The rotor blades reduced themselves to stubs and finally stopped spinning. The helicopter was on its left side. I was lying on top of the pilot, and John was on top of me. Robert was yelling for us to get off him. John and I stood up and stepped out through what had been the bubble of the helicopter.

We helped the pilot out of the helicopter. Although the engine failed, the turbine was still exhausting red-hot gas when the broken beast came to rest. The dry tundra grass behind the exhaust was in

flames. John is pretty much blind without his coke-bottle glasses. They flew off his head when we hit the ground and were nowhere to be found, rendering him pretty much helpless. The pilot had a gash where his thumb joined his hand. His thumb was at an awkward angle. Worse yet, he complained of back pain. In a few minutes, he was unable to walk, then unable to stand. I felt blood running down the right side of my neck. Looking at John, I noticed a matching stream of blood coming down the left side of his neck. Either we had banged heads awfully hard during the violent death throes of the helicopter, or something in the back of the chopper had become a projectile and passed neatly between our heads, cutting both of us as it passed. It could have been worse. Rather than passing between our noggins, it could have taken our heads off.

I searched for the fire extinguisher. It was not in its mount by the pilot's seat where it should be. I finally located it about 20 feet in front of the wreck. By then, the fire had spread further downwind into the tinder-dry tundra moss and grass. More importantly, the fiberglass of the helicopter was burning. I pulled the pin and aimed the little fire extinguisher at the flames – too little, too late. The extinguisher ran out long before it had much effect on the burning fuselage.

The fire spreads; note damage from rotor blades to the passenger door (broken white frame in front of the skid) – the door John was trying to exit through.

    Plan B: retrieve items of value and anything useful for survival from the wreckage before the fire consumed them. I found a big plastic tarp and grabbed that. We could use it as a shelter. I also rescued our scintillometers. Not because we planned on doing any radiometric surveys, because they were expensive. Plus, we could disassemble them and use the metal bases as water containers. I looked for the pilot's pistol, which he said was under his seat. I could not find it. Before I left the burning wreckage, I did something bizarre. I took the keys from the ignition. Don't ask me why. Perhaps I was worried someone would find the helicopter, start it up, and fly away with it? Hey, cut me some slack; I had just suffered a hard blow to the head.

    The injured pilot lay down on the tarp. The fire was spreading rapidly downwind. Oddly enough, it was also progressing slowly upwind. We had to drag Robert further upwind several times as the fire advanced our way.

Our savior was Bill, who was at the Minchumina airstrip waiting for us to arrive. Bill knew how long it took for the first three round trips. He became worried when we were seriously overdue. Eventually, a small plane landed at the airstrip. The pilot flew in from Fairbanks to go fishing.

"Did you see anything unusual on the flight in?" Bill inquired.

"Nah, just a small fire burning north of here."

Bill pulled out a map and asked the pilot to show him where the fire was. It was directly on our flight path. Bill about shit a brick. He explained the situation to the pilot, who flew him to where he spotted the fire. That provided our first ray of hope. The little Cessna 180 came out of the south, circled, and flew low over us, tipping its wings in recognition. Our smoke signal worked. A rescue would be on the way, after all. What we didn't know at the time was that Bill had seen John and me standing but had not seen the pilot lying on the tarp. He assumed that there were only two survivors. He was freaking out.

Alaskan redneck.

Ironically, the BLM operates a firefighting base at Lake Minchumina each summer to combat the many lightning-caused fires that plague the area. The fire season ended a week prior. The base had been decommissioned and moved to Fairbanks. Helicopters, water tankers, medics, and firefighting crews had been only 15 miles away. Now, help had to come from 150 miles away.

Our smoke signal.

It wasn't long before a Cessna 207 flew high overhead. It circled over us a couple of times, then left. That was the spotter plane; we had been spotted. About 30 minutes later, a CASA C-202, a twin-engine turboprop boxcar, arrived. It also circled and dropped streamers to determine the wind direction. On the next pass, parachutes containing gear were dropped. On the third pass, five smoke jumpers parachuted – angels from the heavens. Thank you, Howard!

The BLM firefighters landed within 200 feet of us. They methodically gathered up their parachutes and calmly walked over to where we (well, two out of three of us) were standing. The thought of fighting the fire was not even on their minds. Each of the jumpers was a qualified medic as well as a firefighter. They did a quick triage and bandaged the pilot's cut and broken thumb. By then, our split heads had stopped bleeding. They checked us out and allowed that we would live, but we would need several stitches. I was not sure because ever since we landed, I could barely hear a thing. I was convinced I had a severe head injury that had destroyed my auditory nerves. Even worse, I thought precious gray matter was oozing from my head wound, mixing with the blood that dripped down my neck. Then I remembered I put in earplugs to dampen the high pitch screaming of the helicopter engine. I took them out and instantly restored my hearing.

The fire fighters' nonchalant attitude toward the fire, which was now burning vigorously in the trees we managed to clear, was soon explained. A bomber flew over, banked, and dropped a load of fire retardant on the advancing flames. A second pass pretty much extinguished the fire. The smoke jumpers' work had just begun. They explained they needed to stay on-site for a whole week, putting out hot spots. Tundra fires can burn under the surface and erupt in a new location several days after the fire appears to be out. Our parachuting angels were not at all upset. They would be paid 24 hours per day, racking up some significant overtime.

BLM smoke jumpers attending to Robert, the injured pilot.

I am not a big fan of government agencies. I find them to be overly bureaucratic and inefficient – think DMV! In contrast, the BLM response to our dilemma and the resulting fire was a marvel to witness. These were true professionals at work.

When the smoke cleared, there was not much left of the 500D. The smoldering remains could fit in the bed of a pickup truck. The mangled tail boom lay 50 feet in front of the wreckage. The tail rotor was nowhere to be found. John's eyeglasses had disappeared. The pilot's scorched pistol lay in the burned grass in front of the chopper.

Charred tundra.

John with the melted remains of the helicopter.

Noranda Exploration, a competitor of ours, was working on a project west of Lake Minchumina. Their helicopter pilot heard the radio communications regarding our plight. There is honor among pirates, thieves, and geologists. Noranda generously sent their helicopter to our aid. The medics made a stretcher out of sapling trees, put our pilot on it, and strapped him to one of the helicopter skids. John and I rode in the helicopter for the flight to the Minchumina airstrip. Noranda's helicopter was a Hughes 500D, the same model that lay in a melted heap in the tundra. I turned to the pilot and asked, half-jokingly, "When was your last 100-hour inspection?" He did not see the humor in that.

The cargo plane that dropped the smoke jumpers was waiting at the Minchumina airstrip to transport us to Fairbanks. A couple of hours later, two grubby, stinky, bloody geologists arrived at the ER of Fairbanks Memorial Hospital. The doctor inspected our head wounds and informed us we would each need several stitches. He also said we would not be allowed to wash our hair for a week. What? We had not bathed in ten days. My hair was dirty, greasy, and matted with sweat, mosquito repellent, and blood, but thankfully no brain matter. Another week of bad-hair days would be untenable. We convinced the doctor to allow us to shower and wash our hair before getting the stitches.

John complained of a very sore neck and was admitted for observation overnight. I was released and went with Bill to check in at the motel. I called my boss in Denver to notify him of the incident.

"Hi, Ken, I have some good news."
"Oh, great. What is it?"
"We finished our work at Titna a few days early."
"Fantastic."
"There is also some bad news."
"Uh oh, what is that?"
"We lost all our rock samples."
"How did that happen?"
"They burned up with the helicopter when it crashed."
Stunned silence.

The motel clerk handed me a package when I checked in. I asked my friend, Don Suttie, to send some ice screws for our planned glacier climb. I opened the package to find Don included a bottle of Mogen David wine – MD 20/20 – Mad Dog, as it is fondly known among wine connoisseurs. With a big wallop from its 20% alcohol content, it is truly rotgut wine – the kind skid-row bums drink from paper bags. I am sure Don sent it as a joke as a reminder of the annual fraternity Christmas parties at which everyone exchanged gifts of the cheapest, most awful wine possible: Thunderbird, Ripple, Bali Hai, and Mad Dog. *What the heck?* I took a sip. It was indeed rotgut but tasted awfully good after several weeks in the bush and a traumatic ordeal to overcome. Gene Franks, who now managed the Alaska Division of Rocky Mountain Helicopters, was on his way from Palmer to interview the pilot and meet me. I had a couple of hours to kill. I spent the time in deep conversation with my new best friend, good ole Mogen.

About seven o'clock, Gene and Bill knocked on my door. We chatted for a moment, then went to a Mexican restaurant to discuss the events of the day over margaritas, tequila shooters, and enchiladas. I recall employing my very best slurred speech to recount what happened. I gave the rescued helicopter keys to Gene, who gave me a rather puzzled look. "No, I have no idea why I took them!" Things got a little fuzzy after that, but I am sure a good time was had by all.

The following morning, I awoke with a splitting headache, undoubtedly the result of the whack I took to the head rather than any misbehavior on my part. I stumbled into the bathroom. As I passed the mirror over the sink, I was shocked to see crusted blood covering the side of my head. Uh oh, I had no idea what transpired after dinner, but it sure looked like I hit my head on something and split my stitches wide open. *Great,* I thought, *another visit to the hospital seamstress for more stitches*. I splashed water on my face to clean up the coagulated blood. Some of it dripped to the corner of my mouth. Hmmm, quite spicy blood – kind of reminds me of enchilada sauce.

Over a late breakfast of scrambled eggs, bacon, sourdough toast, strong coffee, and a good dose of aspirin, Bill filled in the gaps of my memory. In the middle of dinner, in the middle of our conversation, in the middle of a sentence, in the middle of a polysyllabic word, my head suddenly slumped forward, yielding to the overwhelming force of gravity. Fortunately, there was a soft cushion to absorb the impact – my plate of enchiladas. I spent the rest of the dinner in a peaceful but somewhat soggy coma, free of thoughts of wingless aircraft falling out of the sky. Bill and Gene finished their dinner while doing their best to pretend they did not know me. They were kind enough to carry my limp corpse back to the motel and tuck me – enchilada sauce and all, into bed.

John was released from the hospital the next day, wearing a brace for his sprained neck. Robert, the pilot, sustained a broken thumb and a broken back. Fortunately, the latter was compressional fractures of two vertebrae. He would recover fully. Sadly, our glacier climbing expedition was canceled. We flew back to Denver to deal with paperwork and the investigation into the accident

A month later, I received a package from Palmer, Alaska, home of the Alaska Division of Rocky Mountain Helicopters. Inside was a pair of keys on a white plastic key fob with "Hughes Helicopters" embossed in bold blue letters. Attached to the keys was a tag with this message, "Bob C., – Now you have your own set of keys to a $308,000 Hughes 500D. – Gene." How sweet, just what I always wanted.

I never finished reading Ezekiel. Gideon's gift was a victim of the conflagration that consumed the helicopter, our prized rock samples, and John's eyeglasses. I no longer tear pages out of Bibles and have become a big fan of Charmin. I also try to avoid helicopters like the plague and have flown in them only a handful of times. I might change my mind if they ever put wings on them. Until then, I am restricting their use to essential purposes – like heli-skiing. At least that is something worth dying for.

I haven't been back to Alaska since the fateful end of the 1981 field season. I would like to return someday, as a tourist. It would be wonderful to see the brilliance of massive glaciers as they transport their loads of gleaming white ice and dark bands of rock to the sea; to view the glory of bald eagles gliding effortlessly overhead in search of their salmon dinner; to witness the migration of herds of caribou moving like waves across the verdant tundra; to observe Toklat grizzly bears foraging for blueberries (from the safety of a tourist bus instead of a flimsy tent); to gaze upon the majesty of Denali, the Great One, one more time; to catch a salmon or grayling, maybe even a northern pike in Lake Minchumina. If I make it to Minchumina, I will bring my helicopter keys along and make the journey to visit old N8363F in her final resting place to reunite them. That must be why I rescued the keys from the burning wreckage.

Lake Minchumina and Denali, the Great One, waiting for my return.

SECTION III

AUSTRALASIA

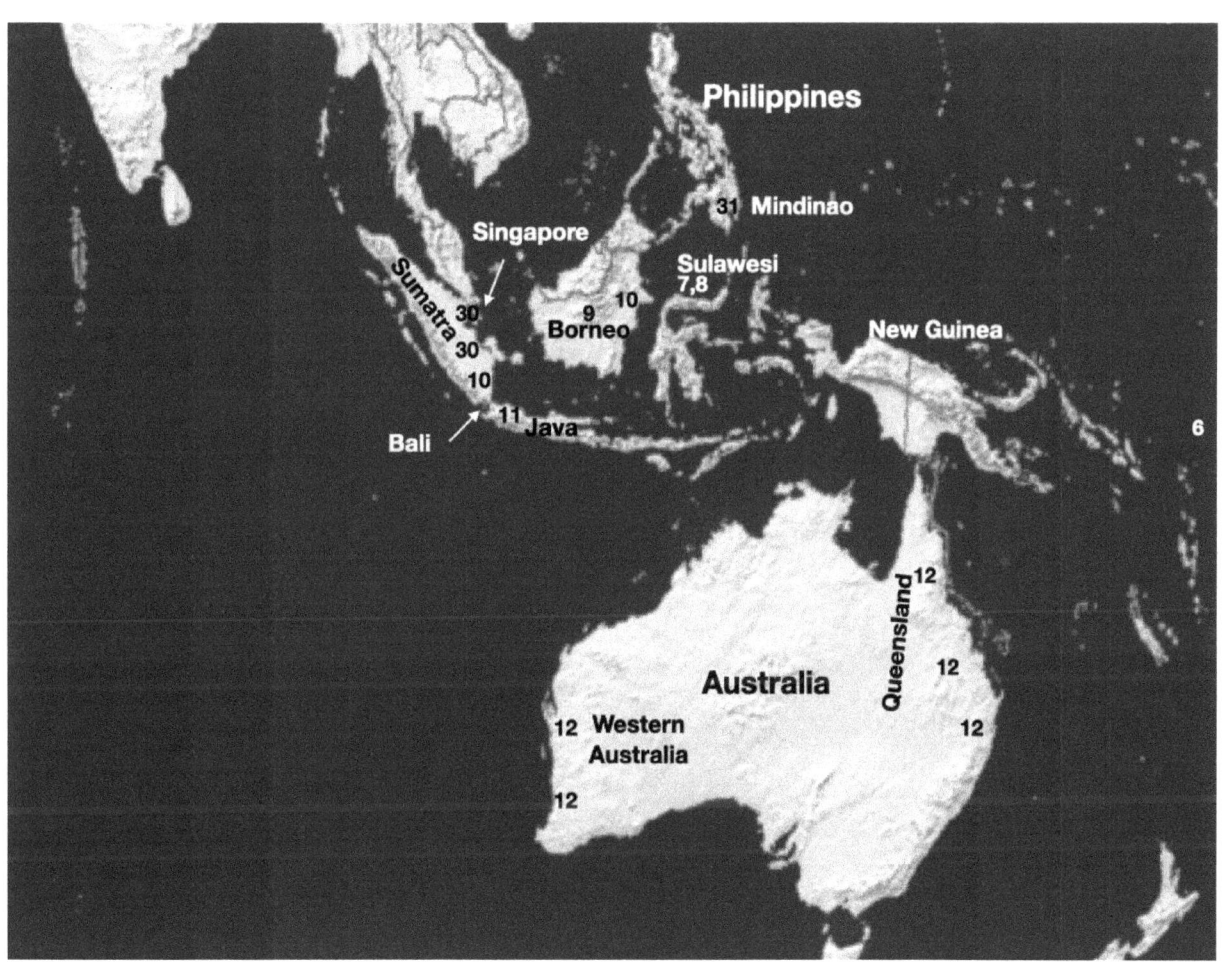

Index to chapters

Chapter 6

INTO THE SOUP POT

Getting into hot water in the Cannibal Isles.

Two burly miners, as black as the inky darkness of their workplace, wrestled a jackleg drill; then loaded, blasted, and mucked shots, while a third miner sprayed cold water on them to prevent them from having heatstroke. Such was the daily routine on the 15th level of the Emperor Gold Mine, which had intersected the scalding waters of a subterranean hot spring. The mine eventually shut down that drift and sealed off the steaming workings. Only fools would go back into that part of the mine.

Enter two fools.

My first foreign assignment came in 1981 when I was drafted to be part of a team evaluating the Emperor Gold Mine on the island of Viti Levu, Fiji, which Exxon Minerals Company was considering purchasing. Before heading overseas, I flew from Denver to the company's headquarters in New York City, the Exxon Building, 1251 Avenue of the Americas, to prepare for the assignment. I spent the evening walking around Manhattan, awed by the sites of famous landmarks: Radio City Music Hall, Carnegie Hall, Broadway Theatre, while relishing a classic hot dog with kraut and mustard from an iconic New York City hot dog stand. In the morning, I met Ken Cornelius, the geologist who was leading the effort. I received a briefing on the mission and got inoculations for various tropical diseases. The following morning, with a rather sore behind from being poked by several oversized knitting needles filled with vaccines, I boarded a 747 bound for Nadi, Fiji, with connections in Los Angeles, Honolulu, and Auckland.

After nearly 30 hours of nonstop travel, I walked up to the car-rental desk at the Nadi airport, where a rental Jeep awaited me. Tired, seriously jet-lagged, and somewhat disorientated, I opened the door and sat down in the driver's seat. *What the heck?* There was no steering wheel. This was long before the advent of autonomous vehicles. I was a bit bewildered. I looked to the right, and there it was. *Oh yeah, they drive on the left over here. Wonderful.* That was not the challenge I needed in my dazed state. I moved over one seat and commenced the next leg of my journey, driving the Kings Road along the northwest coast of Viti Levu to Tavua on the north side of the island, then inland to the village of Vatukoula and the mine site.

Having recently watched Brooke Shields frolicking on the beach in the blockbuster movie, *The Blue Lagoon*, I expected to be driving along an idyllic tropical coast, scanning panoramic views of pristine white sand beaches lined by swaying palm trees. If there were any scenic views, they were blocked by walls of 10-foot-tall stalks of stout grass. Vast fields of sugarcane, the backbone of the Fijian economy, lined both sides of the road for miles on end. Occasionally, I saw workers cutting the cane and loading it on rail cars bound for the sugar mills. Around a corner, a slender brown critter with a bushy tail dashed across the road. It was a mongoose, one of Rikki Tikki Tavi's cousins, off to do battle with the Fijian equivalents of the evil cobras, Nag and Nagaina, or maybe just to get a sugar fix.

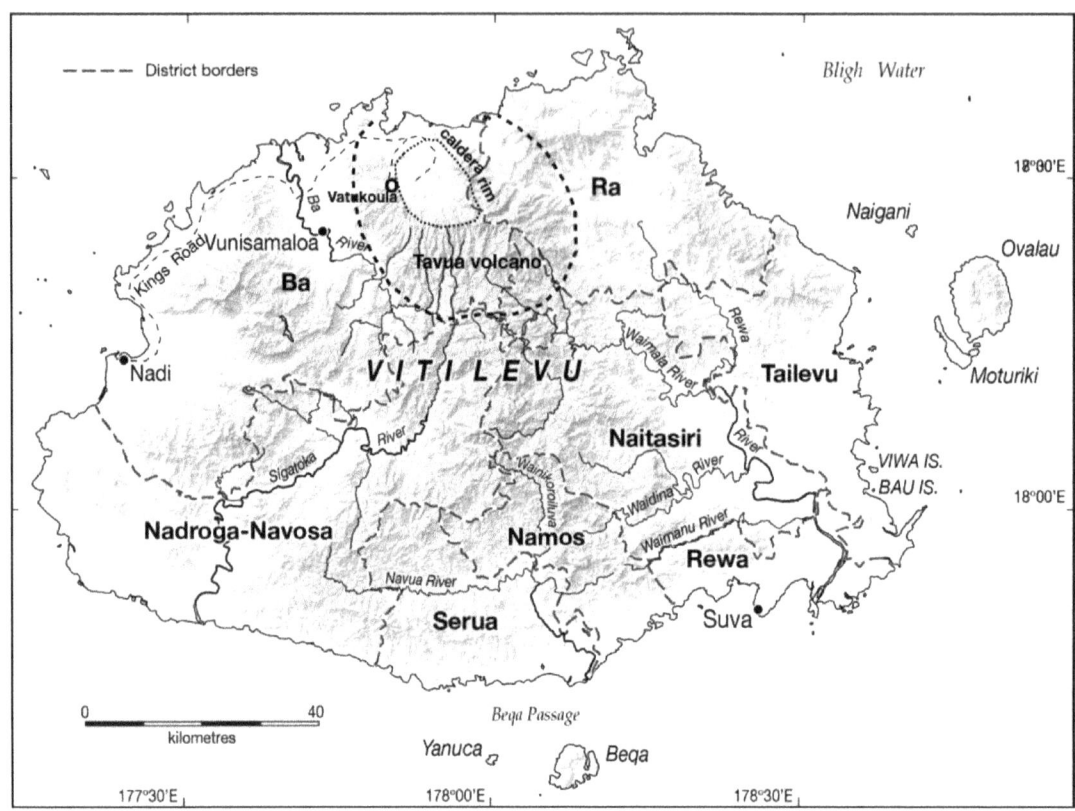

I arrived at the mine guesthouse without any serious incidents, despite a few temporary excursions down the wrong lane of the road. Ken Cornelius met me there. After a lunch of spicy curry, we reviewed maps before going to the mine office to meet the other team members: Sandy Sergiades, a Canadian mining engineer; Tony Greenish, a geochemist; and Walt O'Toole, a consulting geologist. Sandy's assignment was to study the mine reserves, the mining and processing methods, and the mining costs to place a value on the mine. Walt, Tony, and I were tasked with evaluating the exploration potential of the other 40 square miles of the Emperor Gold Company's land holdings.

The Emperor Mine lies in the Nakaudavadra Mountains, along the west rim of the Tavua Caldera, the collapsed top of a large volcano – basically a big hole in the ground. Grass-covered and locally forested mountains, clad in native hardwoods and Caribbean pines, ring the caldera walls, which rise as much as 2,500 feet above the low rolling hills of the caldera floor. Small streams drain into the basin, where they join the Nasivi River, which breaches the west wall of the caldera just north of the mine. There it turns north to the sea. The caldera would be our playground for the next two weeks.

At the mine office, Sandy and Tony were having difficulty getting the mine staff to produce the records they needed. The office workers were accustomed to "island time" – a light workload, loose deadlines, and lackadaisical management. Ken, the quintessential alpha male, jumped into action, giving orders left and right. The office was immediately humming with activity.

Working with the one-and-only Ken Cornelius was a trip. A brilliant geologist, he earned BS and MSc degrees in Geology at the University of Arizona, and a Ph.D. from the University of Queensland, Australia as a Fulbright scholar. In addition to being smart, Ken was also physically imposing. Stocky and muscular beyond his 54 years, he sported a glossy baldhead and one gold front tooth – likely the result of some past boxing match. He was the living embodiment of Mr. Clean, the bald mascot for the product of that appellation. Ken had a propensity to engage in crazy physical challenges. He met his wife-to-be in a tiny outback town in Australia where she was a schoolteacher. Hard to miss, he was the guy doing a handstand on the handlebars of his bicycle as he rode down the dirt main street of town. While visiting a prospect high in the Chilean Andes, Ken grew tired and saddle-sore riding a donkey. On the 10-mile-long descent from 15,000 feet, he challenged the donkey riders to a race to the bottom – and won. Arriving in a remote village in Mexico, he tore off his shirt and challenged the village blacksmith, the toughest hombre in town, to a bare-knuckle boxing match. I think that one was a draw – minus one front tooth.

We met a group of young Fijian men cutting sugar cane during one of our field excursions. They split open some pieces of cane with their machetes and offered us a taste. It was overly sweet and definitely not something your dentist would recommend eating regularly. Like a kid energized on too much sugar, Ken challenged the Fijians to a handstand pushup contest. He took off his signature beret, revealing his shiny crown, sprang onto his hands, and cranked off ten quick pushups in vertical takeoff mode. The Fijians were amazed by those antics. The next several minutes were filled with riotous laughter as the natives tried unsuccessfully to get into a handstand position, let alone do pushups. They either ended up stuck with their heads on the ground and their butts high in the air or came close to attaining a handstand only to roll over onto their backs or fall over on their sides, laughing.

Ken's next move was to challenge the group to a footrace. Ken and the chosen runner – young, tall, lean, barefoot – lined up at the starting line. At the countdown from three, they were off and running down the road alongside the cane field. As expected, Ken was left in the dust from the start. He knew he didn't have a chance. That was part of his plan – to display his prowess, followed by showing humility in defeat. That crazy approach won Ken the respect and cooperation of local peoples wherever he went in the world. I learned a special lesson that day.

Placing bets on the footrace.

Walt, Tony, and I visited gold prospects scattered around the margin of the caldera. Tiny villages, with unpronounceable names such as Matanangata, Ndavotu, and Waikumbukumbu, comprising clusters of just a few *bures*, traditional Fijian houses, were nestled in verdant valleys or perched along ridge tops. *Bures* are rectangular one-room houses with high, steep, thatched straw roofs overhanging vertical walls of woven coconut fronds. The resulting form is a shaggy rectangular mushroom, one of the more iconic images of the South Pacific islands.

Bure in the village of Waikumbukumbu

On a hill above Mbasala Creek, while Walt and I were in search of long-lost outcrops, we walked past a one-room tin shack next to a small field of corn. A young Fijian lady, wearing a threadbare formerly white, flower-patterned dress, and her daughter, whose fuzzy afro hairdo made her head look three times its size, came out to greet us. A black pot hung over a low fire along the side of the hut. After some small talk, the lady said she was making hot milk with honey and invited us to have some. Walt turned to me with a grimace, indicating he wanted no part of this tea party. I didn't want to insult the lady by turning down her offer. I accepted, being of strong constitution and weak mind, and hoping any deadly bacteria had been killed in the heating process. The milk was sweet and delicious. I was sad that we had nothing to give these poor people in exchange for sharing what little they had. The lady and her daughter were thrilled just to have met two strange men from a foreign land.

Gold was discovered in the Vatukoula area in 1930, and the Emperor Mine commenced production in 1935. We worked with historical geological and geochemical data generated in the 1930s and 1940s. Trying to follow up on that work proved challenging since much of the land had been cultivated after the original exploration was conducted. Key outcrops had been plowed under, leaving no trace of the mineralization. Teams of oxen pulling plows continued the destruction, much to our dismay.

Walt with the Fijian ladies.

While reviewing underground mine maps with the mine geologist, we noted a drift deep in the mine that crossed the caldera wall. Ken inquired whether it would be possible to enter that working to inspect the fault breccia at the contact. The mine geologist hesitated, saying the drift had been closed off for years. Ken pressed him further, asking if it could be reopened. "Yes," was the answer, "But that working was flooded when it tapped into a hot spring." When an attempt was made to pump the hot water out of the mine, a hot spring at Waikatakata, three miles away, stopped flowing. The flow of hot water in the mine was eventually stopped, and the hot-spring flow returned to normal. The working was abandoned with 110° F water partially filling it.

The mine geologist thought it was not a good idea to enter the old drift. He did his best to convince us not to go there. Ken is not easily dissuaded. Using his unique version of personable yet forceful persuasion, he gained permission to go underground to the 15th level of the mine. The mine geologist must have thought we were crazy.

He was right.

Ken and I donned our hardhats, steel-toed boots, and mine lamps and proceeded to the Cayzer shaft. The mine geologist wisely declined to join us. We entered the cage with a Fijian miner, who rang the proper number of bells to lower us 1,800 feet down the shaft. We stepped out of the cage on the 15th level. The miner unlocked the bulkhead door sealing the drift and opened the door. Bidding us good luck, he stepped back into the cage and rang the bells to be lifted to the surface. We were on our own.

We entered the drift, wading into ankle-deep bathtub-temperature water. As we proceeded down the working, the water became hotter. We were in a giant, dark hot tub. The breccia contact was 1,500 feet away….a long, hot 1,500 feet. Ankle-deep water turned to knee-deep water, which turned to waist-deep water. I began to question the wisdom of our mission as we sloshed along in the hot water, breathing stale, humid air that was half water. Our mine lamps projected diffuse beams of light penetrating the misty darkness, reminding me of scary monster movies moments before the Creature from the Black Lagoon or the Son of Rodan erupts from the black water. All that was missing was the foreboding creepy music.

We occasionally stopped to bang on the basalt ribs of the drift with our rock hammers to look for alteration as we approached the contact. We finally reached our destination and chipped off a couple of samples while standing chest-deep in 110° F water. By then, both Ken and I were feeling a little light-headed. We threw the samples in my pack and wasted no time turning back. About halfway to the shaft, I became slightly delirious and convinced myself we had been set up by the natives. They ran out of missionaries to boil and decided geologists would be just as tasty. They sent us down to precook in the hot waters of Level 15. After all, these were the Cannibal Isles.

About 100 yards from the shaft, I saw a light shining on the black basalt ribs in front of us. It was moving towards us. *"Oh, no,"* I thought. *"Now they are coming with the herbs and spices to add to the broth."* I snapped out of my delirium when the miner greeted us and said the mine manager, who was concerned about our safety, sent him to check on us. We walked to the shaft together. The miner rang the hoist man to send the cage back down to us. Ken and I sat down while we anxiously waited for the cage, sweating profusely and wondering if we would pass out before it arrived.

Back at the surface, I took a deep breath with a new appreciation for the fresh cool air, sweet with the fragrance of sugar cane and tropical flowers. We spent the rest of the day looking at outcrops in the mine area. I drank gallons of water and immediately sweated them back out. We spent too long in the hot water and raised our core temperatures to a dangerous level, somewhere between the melting points of aluminum and steel, I believe.

But we were safely out of the cooking pot and would not be served as dinner after all.

Chapter 7

TARZAN OF SULAWESI

Aahuaaa uaaa uaaaaaa!

As a young boy, I often daydreamed of being Tarzan, swinging through the trees on jungle vines, carrying on cerebral conversations with chimpanzees, and chasing that hot babe, Jane, around our treetop house. Denied that, I yearned to be on a tropical expedition, hacking through the jungle with machetes to find fantastic natural wonders and mysterious ancient civilizations, just like Tarzan did.

Alas, for years, I was stuck in the good ole U.S.A. and had little opportunity to play Tarzan beyond my childhood fantasies. That changed in 1994 when I got more than I bargained for in jungle expeditions. In the process, I learned what the Tarzan movies didn't show you: the daily downpours, accompanying floods, bottomless mud, annoying and potentially deadly biting insects, tropical diseases, flesh-shredding thorny vegetation, jungle rot, and sweltering heat and humidity. I don't recall Tarzan and Jane ever breaking into a sweat. I sure as hell did.

In the fall of 1994, Newmont closed its Reno Exploration office. After a nasty dust-up with the HR Department (the Inhumane Resources Department, as it was affectionately known), I ended up resigning and taking a severance instead of relocating to Elko to work at the company's mines. That left me unemployed, but not for long. Three weeks later, I was on a plane bound for Indonesia to work as a consultant for, of all companies – Newmont, specifically the company's Indonesian subsidiary, Newmont Minahasa Raya.

Before traveling overseas, I made the obligatory trip to my doctor to learn about the myriad of tropical maladies that might plague a geologist playing Tarzan in the Indonesian jungle. I got shots for hepatitis, tetanus, typhoid fever, Japanese encephalitis, and cholera. I received a supply of Ciprofloxacin to combat amoebic dysentery and Larium pills to ward off malaria, which was rampant in the area (roughly 65% of the population of Sulawesi being victims of the disease). I also learned about the hungry parasites and odd tropical diseases which may beset you, for which there are no vaccines or pills: lovely tongue twisters like schistosomiasis, lymphatic filariasis (elephantiasis), fasciaolopsiasis, meliodosis, and dengue fever, to name a few. What was I thinking when I signed up for this mission?

I flew to Singapore to arrange a work visa. Then, I caught a flight to Jakarta and on to Manado, at the north end of the island of Sulawesi, a mere one degree north of the equator. Newmont was in the predevelopment stage of the Mesel Gold Mine near the village of Ratatotok. At the end of each workday, we sat on the porch of the living quarters and watched giant dump trucks hauling overburden from the mine. As trucks coming down with a load passed empty ones returning, the two trucks stopped next to each other. After five to ten minutes, the trucks resumed their journeys. Needless to say, this was not an efficient way of mining. The Minahasan people are very sociable. It turned out the drivers stopped to chat because they were lonely. Solution – hire a person to ride with the driver and keep him company.

After a few days of orientation in the Mesel area, I was on my way to the Garini project, sweating profusely as I followed a group of porters up a crude jungle trail to the camp high above Ratatotok, accessible only on foot or by helicopter. I was to spend three weeks there learning the ropes of jungle exploration and filling in for the project manager, Alister, an Aussie geologist, when he went on break.

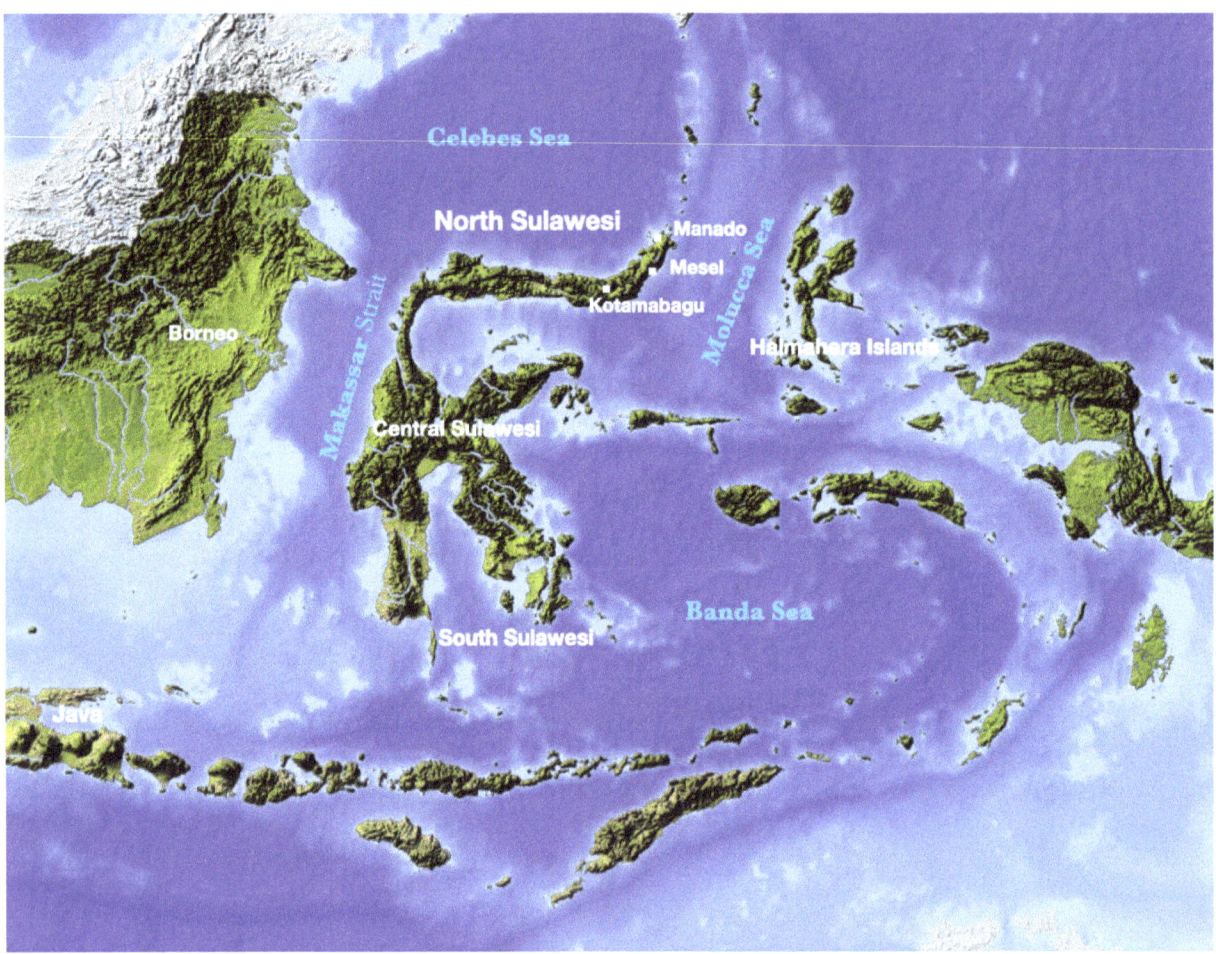

The second day in the field, I made a tactical error. We were walking down a steep muddy trail when I slipped and started to fall. Instinctively, I grabbed the nearest piece of vertical vegetation to catch my balance. That worked, but what I grabbed was a rattan vine covered in nasty thorns, one of which punctured my right hand just below the middle finger. Shortly after, my hand was so swollen I couldn't close it to make a fist. My hand was basically useless. A few days later, I was finally able to clench my fist, but when I did, my hand stayed in that position. I had to pry each finger loose in order to return my hand to a useful configuration. After a week, only the middle finger remained in the frozen position, a situation that made flipping the bird a rather tedious two-handed operation. It took a month for my hand to fully recover. Lesson learned – it is better to fall on one's ass than to suffer the consequences of blindly grabbing poisonous shrubbery.

One of my first duties was to visit the cliffs northeast of camp, where, a couple of months prior to my arrival, a young field assistant fell to his death off the side of a waterfalls. The locale was considered taboo – cursed, and nobody would go there. The area needed to be mapped, so I enlisted Terkelin, a young Indonesian geologist who spoke good English, to accompany me on a traverse through the cliffs. We proceeded with extreme caution and mapped the area without incident, other than some nasty welts from stinging nettles we encountered.

After a week in camp, I was getting used to jungle exploration and camp life; I even picked up some of the Indonesian language. I screwed up again when I accompanied a survey team on a reconnaissance mission south of camp. *Garini* is Indonesian for "in the clouds." "In the rain clouds" would be a better name. Heavy rain hit in mid-afternoon. We slid and fell our way down the steep mountain slope (being careful not to grab any green things). The last challenge was crossing Garini Creek to reach camp on the other side. The creek had rapidly morphed from a clear mountain stream into a muddy mess. Halfway across the stream, I caught my foot between two unseen boulders and twisted my ankle. I limped across the creek and into camp. That caused some excitement since part of the crew's job is ensuring the safety of the *bules* (a somewhat derogatory term for white guys). Alister left the day before. The only other person in camp who spoke decent English was Terkelin, who called on one of the local helpers, a traditional medicine man, to check my ankle. He poked and twisted for a while, then said something to Terkelin in Indonesian, which elicited a chuckle. I asked what was so funny. Terkelin replied: "He wanted to know if white people are put together the same as Indonesians." I got a kick out of that. I told Terkelin to reassure him that beneath the skin, we are really all the same. What followed was a rather painful massage of my sprained ankle with some magical jungle oil, probably extracted from python testicles or centipede brains.

It didn't take long for the sprained ankle incident to become a big deal with Newmont's management. Word reached the Mesel office via single-sideband radio that Mr. Bob suffered a *sakit kaki* (literally "sick leg" in the rather nonspecific Indonesian language), which was soon translated to the brass in Jakarta as a broken leg. Before the day was over, I was on the radio, explaining, "No, I don't need a helicopter to evacuate me, just a few days of rest for my ankle to recover."

In the evenings, the geologists sat around drafting tables on the office porch and compiled the day's data on maps. A generator provided lighting over the tables. The lights attracted a variety of insects, from tiny moths to giant rhinoceros beetles, footlong walking sticks, and monstrous praying mantises. *Panikis* (the Manadanese term for bats) flew over the drafting tables, then around the back of the office tent, doing endless loops over our heads as they picked off bugs for their dinner. One evening, there was a hatch of thousands of tiny bugs, which filled the air, choking us and landing on our maps, making drafting impossible. One of the local hires went off into the jungle. He came back a few minutes later with a small branch from a bush, which he tied over the drafting table. Within a few minutes, the little bugs were either gone or dead. No need to go the store to buy bug spray – as the locals say, "We have everything we need for free in the jungle."

To my surprise, the jungle was anything but tranquil. In the morning, the rain forest is filled with the sounds of insects chomping on wood, lizards scurrying through the forest litter, birds singing in the tree tops, monkeys chattering and screaming, and the loud honks and beating of giant wings as hornbills fly through the trees like pterodactyls. But that is only when the cicadas are quiet. As the day heats up, so do the cicadas. By mid-afternoon, the deafening buzz and ringing of a chorus of millions of the bugs fills the jungle, making conversation difficult. The sound of hordes of cicadas, the loudest insects in the world, has been documented to reach 90-100 decibels, equivalent to a truck or Harley Davidson roaring by.

By night, various frogs in the streams and forest serenaded us. A couple of frogs would start to trill; soon other frogs picked up the call and passed it along. The chorus increased in intensity as more frogs joined in. The crescendo would pass upstream until it faded, then return in a few minutes, sweeping over us like a "wave" at a football game as it worked its way back downstream. Pretty magical stuff!

After three weeks in Garini camp, I hiked (slowly, I admit) down the trail back to the Mesel camp. I spent the next few weeks assisting with various projects in the Minahasa Contract of Work, surrounding the mine. Then I began my primary assignment – prospecting and conducting reconnaissance exploration of the company's new concession, the Kotamobagu Contract of Work, which covered nearly 700,000 acres of rugged mountainous jungle to the west of Ratatotok.

Typical topography and jungle vegetation in the Kotamobagu Contract of Work.

The north part of the island of Sulawesi consists of a narrow, curved isthmus lying between the Celebes Sea on the north and the Molucca Sea on the south. The isthmus is composed of a series of coalescing volcanoes, forming the mountainous backbone of the island. Volcanoes become younger and more active northeastward along the range, with several active and temporarily dormant volcanoes at the northeast end of the island, including Ambang, Tongkoko, Mahawu, Klabat, Manado Tua, and Lokon, which erupted in both 2011 and 2012.

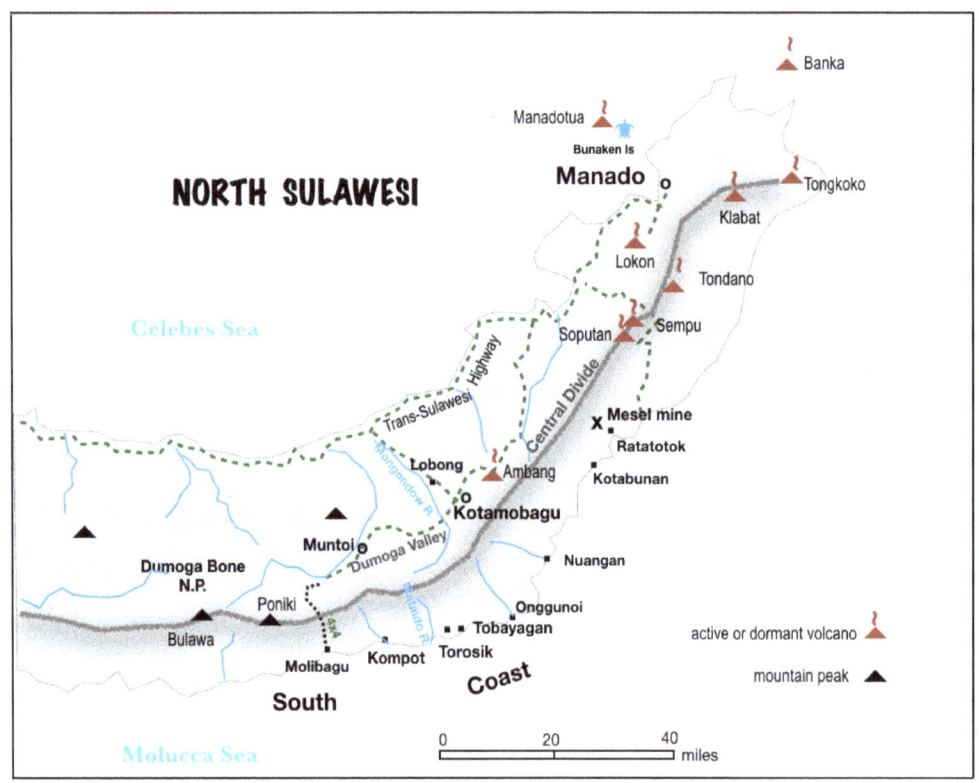

There is all of one paved road along the north coast of the island, officially named the Trans-Sulawesi Highway. That title uses the word "highway" rather loosely. The road is paved but is full of potholes. It is occasionally as much as two lanes wide, and is shared by everything from chickens and ox-drawn carts to buses and big trucks. Much of the road is cut into steep hillsides, undermining the stability of the clay-rich soil and making the route susceptible to landslides. After heavy rains, small landslides often start above the road and cover it with mud and fallen trees. Alternatively, a landslide will begin below the road, cutting upward and sending a section of road sliding down the hillside. In our travels back and forth along the highway, it was routine to encounter missing road sections. The only warning of such a hazard was a small pile of rocks with a couple of green branches stuck in it, set about 50 feet back from the gap in the road. That made for some exciting braking when we encountered one on a blind curve at night. Road maintenance was essentially nonexistent. I remember passing one giant boulder blocking a lane of the highway, which sat there for more than a year. The highway department, such as there may be one, had no intention of moving it – and no means to do so. At least that was one hazard that no longer snuck up on us.

Few roads exist beyond the highway. Most are unimproved dirt (make that mud) trails leading to the villages scattered along the coast. Beyond the villages, we followed abandoned logging roads until they became totally washed out or turned into impassable mud pits. Our chariots were 1960s-1970s Toyota FJ40 Land Cruisers, dubbed "Rambos" by the crews. They were awesome beasts off-road. Even they had their limits in bottomless mud. We spent a considerable amount of time getting them stuck and unstuck.

A very stuck Rambo.

Our journeys up the old logging roads more often than not terminated at broken rudimentary bridges across gullies. The bridges were engineering marvels, constructed from large logs laid across the gully with dirt thrown over the whole affair. They were undoubtedly strong to begin with, but most of the bridges collapsed under their own weight after years of rotting in the jungle. We made far too many hairy crossings on makeshift repairs of those bridges. I preferred to walk across and meet the Land Cruiser on the other side.

Travels Without Charlie – The Natives Ate Him

Crossing a dilapidated makeshift bridge – a tight fit.

The few engineered steel bridges along secondary roads left a lot to be desired. They were perfectly good when new, but decades of neglect left them barely passable. The steel trusses may or may not have rusted through, but it was guaranteed the roadbed was mostly gone – replaced by an array of mismatched boards laid across the rotten wood cross members. Proceed at your own risk! Bamboo bridges across streams were better built and safer.

Truss bridge across the Mongondow River – we elected to take a different route.

Bamboo footbridge at Lobong.

We systematically explored the rivers and streams draining from the highlands, following any signs of mineralization upstream to their sources. Crude regional geological maps and a small amount of geochemical data gleaned from government files guided us. Otherwise, we were on our own. We trained local people, mostly hunters and rattan gatherers, who knew the territory and were keen observers (a prerequisite for survival in the jungle), to prospect. We showed them the rocks and minerals we were looking for, then turned them loose to go find more of the same.

Teams led by young Indonesian geologists followed, surveying their way up the streams, mapping the geology, panning for gold, and collecting stream sediment and rock samples. I alternated between management duties and actively accompanying the teams in the field.

I was joined by Dr. Lew Kleinhans, a fellow CSM graduate (just a couple decades after me), who split duties managing the field crews. Lew and I offset our work schedules of six-weeks-on, two-weeks-off so that one of us was always with the teams while the other was on break. Lew had experience working in Indonesia and was reasonably fluent in the language, which was a big help. He was a constant source of entertainment with his zany antics, which helped to relieve the stress of the challenging and dangerous work.

My role was multifaceted. As the team leader, I made plans, monitored the work, compiled and interpreted data, and communicated progress to upper management. Lew and I also were the cheerleaders for the field crews, providing encouragement and leading by example, despite the grueling work, heat, bugs, and many hazards. We accompanied the crews on the difficult treks into the high country, suffered scrapes and bruises with them, swam across swollen rivers, climbed cliffs, and slept with them in crude jungle camps. My philosophy was to never ask them to do anything I would not do. Over time, we gained the respect of the crews to the point they would risk their lives for us. Foolish fellows!

Lew with his giant hammer – anything for a laugh.

Our first field excursion was on the north side of the central mountains, south of the village of Bakan, where we explored the upper reaches of the Dumoga River drainage. We hired porters from the village to carry our gear and food up a recently reclaimed logging road to the first campsite. I purchased expedition backpacks for the porters to carry 50-pound loads. The porters took two packs, attached them to opposite ends of a bamboo pole, and carried twice the expected load on their shoulders as they marched up the steep muddy trail in bare feet or flip flops. That is what happens when you pay by the kilogram (3000 Rupiah/kilo or 6.3 cents/pound) rather than per load. They are little people but strong and tough as nails.

Porters on the way to Dumoga camp.

We hiked up the trail for three hours to where an advanced team set up a base camp, about 2,300 feet higher than where we started. Our bush camps were primitive but provided the basic necessities for jungle work. Camps typically consisted of a tarp thrown over a raised sleeping platform and a cooking and food preparation area. Except for the tarp, all the materials came from the local jungle. Saplings were cut and pounded into the ground to form the uprights for the platforms. Long horizontal poles were attached to the uprights a few feet off the ground. Branches were laid across the bigger ones, and split bamboo (if available – springy smaller branches if not) was laid across them to make a sleeping platform. The accommodations would not earn many stars on Hotels.com but kept us dry and a safe distance above the nasty ants, beetles, centipedes, and snakes that patrolled the jungle floor at night. We each had a woven mat to lie on and a thin blanket to snuggle in (shades of kindergarten). The crew – Christian, Muslim, and bule – slept harmoniously (sometimes in loud three-part harmony) side-by-side on the platform like canned sardines. Comfortable, it was not. After you have trekked through the jungle all day, you can sleep just about anywhere.

Early morning sleepy heads in camp – my bed was the empty one on the left.

Base camp was in a beautiful place. The logging company did an admirable job of selective logging, taking only certain trees and planting proper replacements. The jungle's impenetrable undergrowth was gone; rays of sunlight actually reached the forest floor in places. Wildlife had already returned in force. We were entertained by Celebes black crested macaques – tail-less monkeys, playing in the trees above us, while colorful lorikeets sang from the treetops and giant hornbills called from their nests hollowed out in tree trunks. It was heartwarming to see a logging job done with proper environmental considerations, a rarity in a corrupt country known for the opposite.

After two days working near base camp, I left my right-hand man, Terkelin, in charge of two survey crews, who would complete work in the area. I foolishly went further into the jungle with Sainun, the crew chief, and a three-man survey team to scout the next campsite and conduct reconnaissance exploration of the area, up the tributaries of the Dumoga River.

Upper Dumoga fly camp – as basic as they come.

Our fly camp was absolutely minimal, just a small sleeping platform with a tarp over it. That was not a big problem – food was. Whether a result of poor planning, lack of communication, or both, we had very little food. Oh, there was plenty of white rice, but not much else. On our first day in the field, I opened the small plastic Tupperware containing my lunch the crew packed for me and found a lovely dried fish's head artistically arranged atop a bed of white rice. It would not have been so bad if the head had not been from a colorful parrotfish, perhaps one I snorkeled with the week before, offshore from our Manado guesthouse. For the next three days, our entire food stock comprised

white rice, dried fish heads, tiny salted & dried minnows (which I somehow managed to eat), a tin of cookies (my savior), and Ovaltine.

On the second day in fly camp, we hiked up a stream, informally named Sungai Aog, not far from camp. The going was tough, with many steep, narrow sections where foaming water tumbled over huge water-worn boulders separated by deep plunge-pools. We made our way through the pools by clinging to the rocks along their sides of the pools or by going around and climbing up and over the rocks surrounding them. Further upstream, we encountered a number of small waterfalls and one large one that involved a treacherous climb. We turned back for the return trip at 4:30 PM, a bit too late. When the sun sets at 1° north of the equator, it doesn't mess around. At precisely 6 PM every day, the sun plunges straight down to the horizon. Twilight lasts all of 10 minutes. The only flashlight we had was back in camp. We hurried to get past the series of waterfalls. We barely made it before complete darkness overtook us. We struggled the rest of the way in the dark, cautiously crawling over and around obstacles in the stream. I missed a handhold or foothold several times and fell into the deep pools below. In the darkness, I heard the splash of the guys behind me doing the same. The survey crew, which was further upstream when we started our retreat, did not reach the waterfalls before dark. They were stuck there. After we arrived back in camp, one of the field assistants took our only flashlight and went back up to lead them down. Miraculously, there were no injuries other than the usual scrapes and bruises from a day of jungle fieldwork.

Sungai Aog: Bad enough in the day - truly treacherous return to camp in the dark.

Back in camp, while being serenaded by the soothing trilling of frogs in the creek next to camp, I made a notation by candlelight in my soggy Rite in The Rain field notebook: *Order headlamps for everyone.*

A year later, we returned to Bakan for follow-up work. The scene that greeted us was not pretty. The place was almost unrecognizable. Large areas had been clear-cut. Banana, clove, and coffee plantations sprang up everywhere in the wake of the clear-cutting. Chain saws could be heard constantly, followed by the crash of forest giants falling, pulling the jungle canopy down with them. This was not an organized big corporation raping the forest. The locals moved in with their "slash-and-burn" farming. Armed with a few chain saws, a portable sawmill to rip logs into rough lumber, and a few oxen to haul the boards down the hill, the local villagers were systematically destroying the jungle one tree at a time. The farmers quickly planted crops to make sure the jungle won't recover. It's all officially illegal, but you can bet all the proper authorities were paid off, and they were looking the other way. It may seem slow on a daily basis, but the cumulative effect of a year of this activity was both shocking and depressing.

Slash....and burn – devastating the jungle one tree at a time.

After completing work at Upper Dumoga, we investigated numerous prospective areas on the north side of the mountain range. Access and infrastructure along the north coast were relatively good. We set up base camps in the villages, renting houses to serve as our lodging and field offices. The families were more than happy to move in with relatives for a while and rent their homes (at what seemed like ridiculously cheap rates to me, but was big money to them). We hired the family members and other villagers to serve as cooks, cleaners, porters, and field assistants. Needless to say, we were welcome in areas where, sadly, there were few employment opportunities. We were a novel attraction in the tiny villages, where life is very simple and essentially the same day-to-day. Having a large group of outsiders, including an odd white man, was quite the curiosity. The villagers, both children and adults, gathered around our place and stared or cheerily shouted the only English they knew, "Hello, mister."

The Minahasan people are infamous for their culinary tastes. If it moves, it belongs on the dinner table, whether it is chicken, bat, cat, rat, or dog. The latter is a local delicacy, *Rinte Wuuk*, or RW (pronounced "air way"), a word you don't want to hear when hors d'oeuvres are served. One of my saddest memories is of pickup trucks heading from coastal villages to Manado with stacked crates full of howling, miserable-looking dogs on their way to the soup pot. The saving grace is that dog on the menu substitutes for the SPCA's spaying and euthanasia programs. Neutering dogs is unheard of here. The island would be overpopulated with starving dogs if not for the predations of their best friend, man.

The villages along the north coast provided my first experience with Islam. Manado is dominantly Christian, a rare exception in Indonesia, which has the largest Muslim population in the world. Outside of Manado, the pleasant ringing of bells from churches was replaced by the obnoxious Islamic call to prayer booming from loudspeakers in the towers of mosques, which are located on just about every block in the villages. Long gone are the talented *muezzins* singing the call to prayer from the mosque's minarets. They have been replaced by scratchy recordings that scream the call to prayer from loudspeakers five times a day (nonstop all day long during the holy month of Ramadan). It would not be so bad if the call was even a little melodic, instead of the irritating screeching, shrieking sound of skinning-cats-alive that blasts at 200 decibels from the mosques.

Most Indonesian Muslims are not particularly religious, except during the month of Ramadan. During Ramadan, Muslims are supposed to refrain from bad behavior such as lying, cheating, and stealing. It is a challenge for most of them to make it through the month. After that, it is back to the same old lying, cheating, and stealing for the next 11 months. Ramadan is the month of fasting. That may sound severe, but fasting is only done during daylight hours. Basically, the people skip lunch. Practicing Muslims rise before dawn to eat a big breakfast before the call to prayer starts at 5:00 AM. As soon as the sun sets, they dive into an elaborate evening feast – not exactly a great weight-loss diet. During the first Ramadan I spent in Sulawesi, many Muslim employees did little or no work. The workers claimed they were exhausted and famished from fasting. That sounded to me like a weak excuse to goof off. The following year, I announced that I would fast with them to share their experience. I fasted for the full month with no ill effects, except losing weight, since I did not get up to eat breakfast at 4:00 AM. I managed to get by on one meal a day, which in jungle camps often consisted of white rice and a couple pieces of tough chicken. Oddly enough, there were no complaints from the workers about being exhausted or famished from fasting during that Ramadan.

Indonesia is a heavily populated country, but rather unevenly so. To alleviate overcrowding in the densely populated islands of Java, Madura, and Bali, the government undertook a program to balance the masses through the policy of Transmigration, whereby impoverished families are uprooted and sent to live on less populated islands. The transplants are given a plot of land (a few acres of jungle or swamp) and a few basic implements (like an axe and shovel) and are expected to develop and farm the land. Added to the difficulty of that task is the conflict inherent in mixing people of totally different cultures. The local people often resent the migrants. Violence has erupted, such as the Sambas conflict of 1999 in Borneo, during which the indigenous Dayaks (not quite yet done with headhunting) rose up and killed thousands of transplanted Madurese, beheading more than 300 people.

We got a rather rude introduction to the transmigration conflict in Sulawesi while working in Dumoga Valley. When we arrived in the village of Mopugat, the local people were busy mining small gold-rich veins in the hills above town. Technically, the mining operations were illegal since the mineral rights were part of Newmont's concession. The miners worked covertly, mostly at night, and were called "ninja miners." We weren't concerned about their small-scale mining, which was mostly done with pick and shovel. But we needed to visit and evaluate the gold showings. Whenever we approached the diggings, the daytime ninjas hid in the jungle or scurried into the darkness of the underground workings like rats in a sewer.

There were two competing groups of ninja miners, the indigenous villagers and the transplanted Javanese. Some conflicts had arisen between the two groups over territory, but nothing too serious.

Around 9 o'clock one night, the lights in the town suddenly went out. Power failures were common. We lit candles and thought no more of the blackout. Power was back in about an hour.

In the morning, we learned the local ninja miners cut the power and attacked the rival Javanese ninjas in their house. Anti-gun campaigners would be happy to know that firearms are banned in Indonesia for everyone except the military, police, and judges (who apparently need them for protection against the people they send up the creek). That didn't help the transplants from Java. The local ninjas stormed the house in the dark. Without firing a shot, they dispatched their rivals in less than a minute – hacked to pieces with machetes.

After that incident, we decided we needed improved security. We hired the Mobile Brigade Corps, affectionately known as the BRIMOB, the paramilitary unit of the Indonesian National Police, to escort us on our field excursions. With such a fancy name and a feared reputation to match, we expected seasoned professionals to be protecting us. What we got was a couple of teenagers in T-shirts, carrying a pistol and an uzi rifle. We finished our work without incident and were glad to move on to less volatile areas.

A sweaty Dr. Lew Kleinhans ready to take on the ninjas.   *BRIMOB standing by.*

Our field days often started with hikes through groves of coconut palms and rice fields in the lowlands, passing into fields of bananas, coffee, chocolate, cloves, and peppers in the foothills. It was sweltering in the fields by mid-morning. The temperature dropped considerably once we entered the jungle, where the canopy provided relief from the burning tropical sun. We hiked along creeks and incredibly muddy jungle trails. The mud was so deep and thick in places that it literally sucked the boots off your feet, leaving you standing on one foot like a flamingo, trying not to topple over into the brown muck while retrieving your lost boot and getting your foot back into it.

In May, we moved south to tackle the real challenge of the exploration concession, the rugged and remote South Coast.

The only habitable part of the South Coast of North Sulawesi is a narrow strip of fertile, flat-lying coastal plain wedged between the ocean and the precipitous central mountain range. Small fishing and subsistence farming villages are scattered along the coastal plain, located along rivers draining south from the cloud-shrouded mountains looming to the north. The rivers flow slowly to the sea across the coastal plain. A short distance upriver, they turn into swift mountain streams, then to rapids and waterfall-filled chasms.

The rivers and streams were our highways. It is incredibly tedious to attempt cross-country travel, hacking your way with machetes through the steep, slippery, densely vegetated jungle. Rock outcrops are few and far between. On the other hand, the drainages had lots of outcrops (sometimes more than we wanted) and were more or less free of jungle undergrowth. The streams also received patches of daylight, which was a welcome respite from the depressing darkness of the rain forest. There was just one drawback – water, often way too much of it.

Our logistical approach was to establish a series of supply and work camps up the major streams, about a half-day trek apart. That allowed us to investigate the main stream and its tributaries near each camp, then to move progressively upstream into the highlands without the necessity of making all-day or longer treks to and from fly camps. The daily routine in camp consisted of surveying our way up each tributary, mapping the geology, and taking rock-chip and stream-sediment samples along the way. Government 1:50,000 scale topographic maps were available. The maps were only used as a guide since they were not very accurate, and we needed much more detail for our work. We used old-fashioned compass-and-tape survey methods to map our way upstream.

Travels Without Charlie – The Natives Ate Him

Panning for gold and sieving stream sediment samples; Compass & tape surveying; Rock-chip sampling.

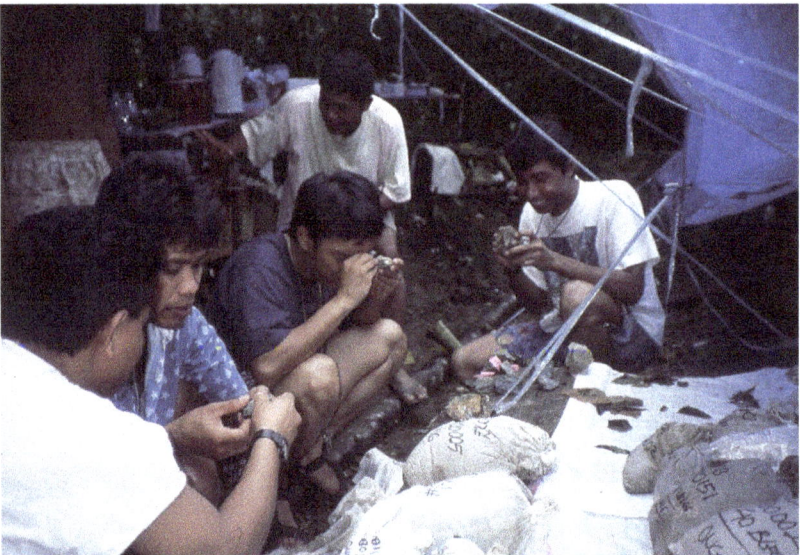

Examining rock samples back in camp.

It goes without saying, when rivers and streams are your highways, you tend to get a wee bit wet. The lower reaches of the rivers were accessible along foot trails. Soon, the trails petered out, and we were in and out of the streams most of the day. The further upstream we went, the steeper and rockier the streams became. In places, we were jammed between vertical rock walls and were forced to climb over and around car-to-house-sized boulders. Placid streams in the lowlands turned into rapid-filled channels, which eventually turned into impassable waterfalls. We made our way upstream by whatever means were available – scaling rocks, shinnying up logs, pulling ourselves up on vines and roots, climbing waterfalls, and resorting to using ropes where necessary (all too often breaking down and using them long after they became necessary).

Working our way up some challenging sections.

More than one way to ascend the waterfalls.

Following the guides. Note the fancy footwear.

Our rubber boots filled to the brim whenever we crossed the streams. The routine upon reaching dry land was to take our boots off and dump out the water before continuing our journey. One day we hiked up a steep slope to reach a high ridge about 2,000 feet above our camp. A couple of hours into the hot, sweaty hike, I stopped to empty the water in the bottom of boots. As I poured the water out, I realized we had not crossed any streams all morning. I was emptying two inches of sweat that drained down my legs into my boots.

Breaking down and using ropes.

If that wasn't bad enough, it rained nearly every afternoon. The only warning was a light breeze blowing through the otherwise still and clammy jungle. Within minutes, big raindrops started falling. Soon, the heavens opened, and it poured. The clear, shallow streams we waded up quickly became raging torrents of muddy water. We turned back as soon as the rain started. To delay could mean being flushed down the muddy stream like turds in a toilet. We struggled back to camp, fighting the rushing waters, falling, swimming at times, and barely escaping being swept along with the floodwaters.

Start of the daily flood – trying to beat the brown torrent.

*Survey crew at the end of the trail – there are limits; upper Milangodaa;
The author is the white guy in the middle.*

Our camp diet was very basic. Basically, it sucked. Sticky white rice was the mainstay – three times a day. A few vegetables, dried fish, and the occasional chicken augmented the rice. Every few days, porters from the local villages made the trek upriver carrying sacks of various supplies and chickens hanging upside down on long poles. Once they arrived in camp, the chickens were tethered on strings and allowed to forage for themselves until their time was up. I could always tell when we were due for another supply run – there would be only one chicken left in camp. Those were the skinniest, toughest chickens you could imagine, but they were a whole lot better than dried fish heads. I relished them.

A typical jungle supply and work camp.

The native workers demanded to be paid every Friday, regardless of where they were. That meant someone from the Manado office had to make the long trek across the island and hike up the rivers to deliver cash to the workers. What the workers planned to do with the money in the middle of the jungle is a mystery. One Friday, when we were camped far up the Onggunoi River, the purser came to camp to deliver the pay. After distributing pay envelopes to the crew, he handed me a large black garbage bag.

"What is this?" I inquired.

"Your money, boss."

I looked inside the bag to find reimbursement for my airfare and other expenses – several hundred million Indonesian rupiahs – about $4,000 worth, the equivalent of several years' salary for the local hires, who worked for about $5/day. Here I was, in the middle of the Sulawesi jungle, sleeping side by side with impoverished machete-wielding natives, and I am supposed to use a veritable fortune in cash for a pillow?

"You have to be kidding me!" I said, "Take this back with you and put it in the safe in Manado. And don't ever do this again. I like my head sitting on my shoulders just like it is now."

I was eager to see an area I thought had excellent gold potential. I decided to go upriver with the first set of porters before a camp was built and before the rest of the geology team mobilized from the last camp. The idea was to start prospecting while the construction team built the camp. I waded upstream accompanied by one field assistant, who spoke no English, and a group of local porters carrying the camp supplies. The trek was going swimmingly until suddenly, the porters stopped, threw down their loads, and refused to go further. This was like a scene out of the original 1936 *Tarzan the Ape Man* movie when Jane's safari reached the Mutia Escarpment. The natives refused to go on, believing the area was sacred and cursed. Going further would mean death. I pondered whether my porters read Tarzan books and had the same fears, were tired, or were simply in the mood for a sit-down strike. At that time, my command of the Indonesian language was rudimentary at best. Lacking verbal communication skills, I tried to relay that we needed to go further by pointing to the map, then pointing upriver. I was not making any progress. The porters were obviously not happy campers; they were getting more agitated. I sat down on a rock to try to sort out a solution. My field assistant opened his pack, took out a warm Bintang beer, and offered it to me. *Oh, what the hell*, I thought. It was only 10 AM, but I was totally frustrated. I accepted his offer. Sitting there sipping my beer halfway to where we needed to be, I finally caught a word or two I understood: *bayar* (pay), *lagi bayar* (more pay). Then it occurred to me what the problem must be. Our expeditor, Alisar, arranged the porters, but somehow there must have been miscommunication on where we were to set up camp. The porters thought we had reached the campsite. They were paid to go only that far. I pointed to the spot on the map where I intended to set up camp and said, "*Camp di sini – lagi bayar, OK?*" saying they would get more money to go the extra distance. Those were magic words. Before I could finish my warm beer, the porters shouldered their loads and were wading up the river merrily singing some native song. Jane and Cheetah would have been proud of me.

Porters merrily resuming the trek up-river.

A few minutes after we resumed our wet march, a tall black bird with a white underbelly and a small triangular yellow head flew across the stream and landed on the branch of a tree. *Awesome*! It was a maleo, one of the oddest birds in the world. Endemic to Sulawesi, the maleo is the worst mother in the bird world. Maleo birds are not fond of wasting time sitting on nests waiting for chicks to hatch. Standing about two feet tall and looking somewhat like a chicken on stilts, the maleo has long claws and partially webbed feet designed for digging. The female maleo digs a hole two-to-three feet deep into which she deposits a single large egg – about six times the size a chicken egg. Then the mother walks away, leaving the chick to hatch, dig its way to the surface, and fly away to fend for itself. The secret to this rather lazy method of raising chicks lies in where the egg is laid. Maleo birds lay their eggs along sandy beaches where the sun heats the sand to the proper incubation temperature. Or they seek hot springs where the sand remains the ideal temperature. The oversize egg allows the chicks to develop fully before hatching and to fly upon emerging from the nest hole. The presence of maleos here made sense – there were several hot springs in the area.

It is a common belief that cockroaches will take over the world if (make that when) humankind is wiped out by nuclear war, an asteroid impact, or a killer virus. Forgot those wimpy bugs. Ants are the real threat. They are not waiting to take over; they already rule the world, especially the jungle. Ants live everywhere except Antarctica and a few barren Arctic islands. There are more than 22,000 species of the little buggers. Indonesia alone has 1,528 identified species. I think I have met them all. They range from itsy bitsy teenie weenie ones to giant suckers more than an inch long. There are black ones, red ones, orange ones, iridescent golden ones. The one thing they have in common, other than six legs, is they all bite.

It never took long for ants to invade our field camps. The scouts somehow found our kitchens as soon as they were set up. A line of ants carrying pilfered food soon extended from the food counter down the leg of the platform, across the floor, and out to some hidden ant metropolis in the jungle. The first morning in one of our camps, I awoke early to the call of nature, rolled off the sleeping platform, slipped on my flip-flops, and started across the floor of the shelter to do my business. Within three steps, about a dozen small ants began biting my feet. I did a jungle jive dance to avoid the offending beasties. When I returned, I saw the source of the problem. I stepped on a two-inch-wide column of ants, leading from the kitchen down to the floor, then under the sleeping platform to the jungle. Thousands of tiny ants were busy hauling off grains of rice and other goodies much larger than themselves. Spaced about three feet apart on the trail were huge black ants, perhaps 50 times larger than the average guys. Those giants moved along as part of the convoy and must represent the 18-wheelers of the ant world. I wondered if the ants kept them as slaves or if they contracted the giants for the heavy lifting.

The following morning, I was prepared for what awaited me. This time I gingerly stepped across the line of ants and went behind camp to see where the ant trail led. Positioning myself on a rock overlooking the convoy, I took aim at the enemy and let the little biting bastards have it full blast. Ah, sweet revenge!

Tropical diseases plagued our operations, with malaria being the main culprit. I arrived at our base camp at Torosik to learn the cook was in the hospital with malaria, the deadly cerebral variety, no less. That evening I noticed small mosquitos nibbling on my arms while I was reviewing maps. *Wonderful!* The *Anopheles* mosquitos, which carry malaria, are tiny nocturnal-biting mosquitos. There was real cause for concern. Fortunately, I was taking Larium pills to ward off malaria. I was one of the very few to take such precautions. Three weeks later, two more workers came down with malaria. At that point, I knew we had a serious problem. I ordered mosquito nets set up in all sleeping areas, *Off* repellent to be distributed to everyone, and Chloroquine to be available for anyone who wanted it.

A week later, another worker was afflicted with malaria. I made the brutal eight-hour trip to Torosik to see what the hell was going on. That evening, as we unloaded our gear from the Land Cruiser, I heard a chorus of frogs serenading us from the ditch along the road. Hmmm…a breeding ground for frogs means a breeding ground for mosquitos as well. A quick check around camp found no sign of mosquito nets or *Off*. I was pissed.

The first order of business was to call a meeting early in the morning with the camp foreman, medic, and safety officer to address the malaria problem. When I asked why there was no *Off* available for use, I was informed the workers didn't like it because it smelled, and it was not strong enough to kill the mosquitos. The concept that repellent is used on the body to prevent being bitten, not to kill the mosquitos, was lost on them. After I presented a dumbed-down explanation of the mosquitos' reproductive cycle, I finally convinced them the likely source of the problem was the stagnant water in the ditch in front of the camp. I directed them to drain and fill the ditch.

Guess what? Problem solved – no more mosquitos and no more malaria after that. Does anyone wonder why 70% of the population suffers from the disease?

An odd foot malady afflicted us in the fly camps. Spending all day, day after day, in wet rubber boots does not do wonders for one's feet. After a week in the fly camps, the skin on the soles of my feet developed small round white spots. I first noticed this one day after taking my bath in the stream next to camp, followed by a liberal application of Neosporin on the numerous scrapes and scratches of the day. The spots grew in size daily, revealing raw flesh below. Slowly, the sores began to merge. Walking became an excruciating activity, which sucked because we spent the entire day hiking. At the end of each stint in fly camps, I limped back to base camp – every step being a painful exercise in determination to reach a proper bed and a cold beer. I never knew if this malady was a variety of foot rot or not. My theory was that walking all day in wet rubber boots, often full of water, produced rubbing between the foot and the boot. In dry boots, friction and heat produce callouses. When your feet are constantly wet, there is friction without heat. Your skin stays soft and eventually wears away. The amazing thing was that after a few days of drying out or, better yet, soaking in salt water (best if done while snorkeling – the cure I pitched to my boss), the soles of the feet magically healed.

Hiking in wet pants every day led to another unique malady – chaffed thighs. Eventually, the skin on the inside of our thighs got raw from the constant abrasion of wet pants against equally wet skin. The symptoms are both evident and amusing. The afflicted assume a "duck-walk" in a futile attempt to keep their pants from rubbing against their thighs. I was not alone in my misery. As I walked along with the team at the end of one day, I noticed many other ducks marching in the group. I started to quack like a duck. Soon the entire flock was quacking and laughing despite the pain.

Not all the time I worked in Sulawesi was spent doing intense jungle work, although that sticks out in my mind. I spent a fair amount of time doing management duties back at the head office in Manado. Located at the northeast end of the island, Manado sits along the shores of a wide bay framed by the conical peaks of three active or "dormant" (meaning maybe active tomorrow) volcanoes. Bunaken Island, a world-famous diving destination with spectacular wall diving along a 3,000-ft drop-off, lies a short boat ride to the north. We slipped in a few snorkeling trips there for R&R during my time with Newmont Minahasa Raya.

The company's guest house was located along the waterfront a few miles south of town. When staying there, my daily routine (at least in the placid water of the dry season) was to rise before dawn and go for a run or do a short workout. I learned the hard way such physical activity left me overheated. Cold showers did little to stop the endless sweating that followed. The solution was to go for a long swim. I donned snorkel gear – mask and snorkel, sans fins – and swam out over scattered patch reefs, viewing lots of fishes, corals, and giant clams beneath me. One morning, on the return leg of my morning aquatic loop, I was swimming over shallow grass beds in the soft, dim light of dawn when I spotted a large, diffuse, dark shape in front of me – heading straight towards me. *Oh, no!* I was aware but in denial that sharks are most active in the early morning. My heart was pounding. *Was this a hungry shark coming for his breakfast?*

As our paths converged, I realized this was not a shark. It was a rare dugong, the South Pacific's equivalent of a manatee. Mr. Dugong swam lazily toward me, and came within about 20 feet of holding a meeting, then turned to my right. I followed behind. After a minute, he came up for a breath of air. I did as well. We both dove. A few seconds later, with a single flip of its tail, it accelerated and disappeared from sight, ending an awesome aquatic experience.

When I was at the main office, I could occasionally call home and have a live conversation with the wife and kids – when the 15-hour time change allowed. The rest of the time, communication with home was via fax. I wrote letters in the jungle camps. Porters carried the letters to the base camps on the return legs of their supply runs. From there, the missives traveled by Land Cruiser over the mountain pass and up the north coast to the Manado office, where the secretary opened the letters, shared them with the office gals, had a good laugh, then faxed them to my wife. Replies, usually detailing the latest disaster to befall the kids, the car, the house, or the bank account, followed the reverse course in a process that often took a week or more to complete. Sadly, the office staff knew what was happening at home long before I did.

View from the guesthouse, Manado Tua volcano on the left.

Flying between Manado and the U.S. required an overnight layover in either Singapore or Denpasar, Bali. Then it was on to Hong Kong or Tokyo for a flight to San Francisco and finally to Reno. I usually opted for the layover in Bali. Bali is very culturally and socially different from the rest of Indonesia. It seems like a different country. The Balinese follow Hinduism instead of Islam. The annoying blasts of caterwauling from the mosque towers are not heard here. Instead, there is a mellow, island-time atmosphere. Smiling faces abound, and small arrangements of flowers and incense pop up all over the streets and sidewalks – little offerings to one or more of the 33 Hindu gods and goddesses.

Bali has always been a favorite vacation destination for Australians, offering a cheap, fun getaway from Down Under – their equivalent of Hawaii. My favorite hangout during my layovers was the Sari Club, a casual sports bar on the main drag in Kuta, where I could drink ice-cold Heineken beer and chat with the young Australians who frequented the bar. You may have heard of it. The Sari Club was one of two bars targeted in the infamous Bali Bombings on Oct 12, 2002. The bombings, part of a jihad by the radical Islamic terrorist group, Jemaah Islamiah, killed 202 people, mostly Australians, plus some Brits and Balinese. To this day, I am haunted by images of those innocent young people being blown to bits – all in the name of freakin' Allah.

I worked for Newmont in Sulawesi for nearly two years. The set work schedule was six weeks on, two weeks off. Long work stretches, 42 days straight, were brutal. Initially, I thought having two weeks off to spend at home with my boys, who were five and seven when I started that crazy job, would be great. What I forgot was that most of the time when I was home, they were in school. I was home only once each summer when we could spend time together traveling, fishing, and camping. For two years, I missed every birthday, anniversary, Thanksgiving, and Fourth of July. That is hard on family life.

There were times when I ended up working for eight or nine weeks straight. There was a reason the official schedule was six weeks on/two weeks off. It was based on years of experience. By week seven, I started to go bonkers. I learned the hard way it was not a good idea to write memos to upper management after six weeks and one day. The solution was to write long scorching letters, complaining about the horrible conditions, mismanagement, et cetera, then delete them from my computer. I got to vent my frustrations without pissing off the boss – for a change.

In July 1996, I was at home packing my bags on the last day of my break before I flew back to Indonesia when I got a call from Pat Cavanaugh, a good friend I worked with at Newmont. He offered me a job working with a new start-up company, White Knight Resources, doing gold exploration in my home state of Nevada. The offer was intriguing, but I told Pat I was committed to Newmont and was going back the next day. The conversation ended with Pat offering to hold the position open in case I changed my mind.

A few weeks later, back in Sulawesi, I was getting weary and wondering not if, but when, I would get greased in a gory head-on collision on the insane roads. We passed enough fatal accidents to know it was only a matter of time. Added to that concern was unrest among the Indonesian employees. A Canadian junior exploration company, Bre-X, claimed to have discovered the world's largest gold deposit on the neighboring island of Borneo (not really, it was a scam – the drill core was salted with gold). A plethora of junior companies descended upon Indonesia, intent on capitalizing on the discovery. None of them had experience in the country or in jungle exploration. They all needed experienced staff. Our employees, whom we spent years training, were offered multiples of their salaries to join the start-ups. We were losing our best workers. The rest of the employees were upset and asking for inflated salaries to stay.

The job was no longer fun. I called Pat Cavanaugh and inquired if the position was still open. The answer was, "Yes." My reply was, "Okay, I'll take it." At the end of the next day, I sheepishly went to my boss, Dave Cole, and explained my decision. We spent the next couple of hours in a debriefing, which became foggier over time as we worked our way through a bottle of Goldschlager I brought along for the occasion.

The following day, I called home. I explained my decision to the wife, who was more or less in favor – mostly, I think because she wouldn't have to take out the trash anymore. My son, Craig, overheard the conversation. I heard him running downstairs where his brother was watching TV, exclaiming, "Eric, Eric, dad is coming home. And he is going to stay!"

That made my day.

I occasionally reflect on those crazy days of jungle exploration. I kind of miss the excitement and adventure of those times.

No, not really.

Chapter 8

THE DROWNING OF REGEX

Bad decisions make for good stories.
— Anonymous

    The team assigned to explore for copper and gold on Newmont's Kotamobagu concession in North Sulawesi, Indonesia, was officially called the Regional Exploration Group, which went by the acronym REGEX, appropriately nicknamed the REJECTS team. In the summer of 1995, our work focused on exploring the South Coast of the island, a truly remote area with very limited access and few inhabitants. Most of the villagers had never seen a white man until we arrived (few westerners had been there since the Dutch left when the Japs invaded in 1942). Sadly, their re-introduction to white folk was through two crazy geologists – Lew Kleinhans, aka "Screwy Lewy," and me.

    We started working the south coast of the island in the spring. By summer, we had a team of 42 employees and a horde of local helpers exploring the island from three different base camps scattered along the coast and a number of fly camps in the highlands.

    One of the crews adopted a puppy (or maybe he adopted them?) from the local villagers, a cute, curly-tailed, brown and white scamp. Appropriately, they named him "Regex" and made him the official mascot of the team. Regex loved the attention, the food (although it was mostly white rice), and the fieldwork. As he got bigger, he went everywhere with the crew.

Regex as a young pup.

    One rough 4x4 trail crosses the central mountain range, providing the only road access from the North Coast to the South Coast. It is passable only during good weather, a rarity. A one-lane dirt road follows the South Coast, connecting fishing and subsistence farming villages: Molibagu, Pinolosian, Mataindo, Torosik, Tobayagan, Dumagin, and Onggunoi. Make that *sometimes* connecting the villages. Most of the settlements are situated along rivers flowing south from the highlands. There were no bridges across the rivers in the mid-1990s. The Provincial government had sort of started to almost begin to think about maybe building some bridges. Crossing the rivers required driving across them at shallow fords, which was possible with regularity only during the dry season. The dry season is short and unpredictable because the South Coast of Sulawesi has not one but two monsoon seasons, which nearly merge into one.

The rivers flood when heavy afternoon rains hit the high country, a daily event during the rainy season(s). Many of the rivers cannot be crossed for weeks during the peak of the monsoon season, isolating the villages. The trick to crossing the rivers was to judge the water depth and avoid any deep channels carved out by the last flood. We made many exciting crossings when the hood of our Toyota Land Cruiser plunged underwater, and the cabin flooded. To my amazement, we always made it across, sometimes accompanied by a fish or two as a bonus.

Lew had been managing the work while I was back home on a much-needed break. I relieved him upon my return in mid-August. His parting words to me were, "It's getting difficult to operate with all the heavy rains." I traveled to our field base in the village of Torosik to check on the crews, arriving just before the big rains hit. It was raining hard on the trip. We barely made it across the Mopungu River in our rusty and trusty 1971 Toyota Land Cruiser. That night it started to rain harder and poured all night. The next morning, a temporary break in the weather was all we needed to be fooled into driving to Tobayagan, the next village along the coast to the east.

We trekked up Posiligan Creek, one of the smaller creeks, into the highlands to sample some veins the prospecting crews found. Of course, as soon as we were halfway there, the heavens opened, and rain came down in buckets. Our little creek turned into a muddy torrent. We wisely turned back. By the time we reached the flat coastal plain, we were wading through knee-deep water across what, just a few hours ago, were barely wet rice paddies. The driver had moved the Land Cruiser to higher ground across the village since the spot where we parked was now under two feet of water. The ride back to Torosik was rather exciting, with trees falling and small landslides happening all around us. When we made it to Torosik, we found the village also flooded. The house we were renting was high and dry – like a small island in a lake. Heavy rain continued unabated for two more days. The water rose steadily, finally stopping just a few inches short of our doorstep.

When the rain ceased and the water subsided, we found that we were stranded between the Mopungu and Tobayagan rivers, which were swollen to the tops of their banks. Our Land Cruisers could go nowhere. No problem, we hired a "speed boat" to take us down the coast, past the swollen Mopungu River, and up the Mataindo River. From there, we planned to trek upstream to see some really cool rocks (what they pay me the big bucks for). As you are probably aware, Indonesia is an island nation, with more than 13,000 of the suckers, not counting the rocks that disappear at high tide. You would expect there to be a lot of seaworthy boats around. Nah, what we got was a dugout outrigger canoe with a 20-HP outboard motor (hence the "speed" part of "speedboat" – speed being a relative term).

The boat trip to Mataindo was uneventful except for passing some splendid scenery – magnificent palm-lined sandy beaches separated by rocky headlands, with not a soul around. It was like a scene from *The Blue Lagoon*, except under gray skies, and sadly minus Brooke Shields in her sexy bikini. Our boat took us up the Mataindo River, then beached at a bend in the river where there was a gap in the impenetrable jungle vegetation. We unloaded and hiked along a mostly washed-out jeep trail following the west bank of the river. After about an hour, we came to what had been a ford across the river. The formerly placid river was now a brown swath about 100 feet wide. The water was moving fast – really fast. The swift water was scary, but it did provide the side benefit of having flushed any saltwater crocodiles, which inhabit the coastal rivers, out to sea for a while.

Bailing water on the way to Mataindo in the outrigger canoe, I mean "speedboat."

Our native crew chief, Sainun, volunteered to see how deep the water was. He waded in and worked his way across. We never learned the depth of the river. Sainun was swept off his feet halfway across and had to swim to the other side, ending up downstream of his intended destination. After a few minutes of debate and soul searching, I made the fateful decision to cross the river. After all, we had come this far; why go back now? I made a quick estimate of how far downstream the current would carry me and chose the appropriate entry point upstream so that I would end up meeting Sainun on the other side. I knew that if the *bule* (white guy) swam across, the rest of the crew would follow. What I didn't factor into the equation was that half the team could not swim! We ended up taking a rope across the river to allow the non-swimmers to cross by pulling themselves along on the rope. It was a grueling task. The current pushed against them and the rope; water surged over their heads. I asked myself why the hell we were doing this. It was too late. The die had been cast. The crew would lose face if they did not follow the old white guy. We all made it safely across, including little Regex, who ended up about 50 yards downstream of us, but came running back with his soggy little tail wagging. I made the executive decision that we would not do this again (except on the return trip home).

Stupidly crossing the east fork of the Mataindo River at flood stage.

Drenched but not yet drowned, we followed the east fork of the Mataindo for a while; then, we picked up a narrow game trail along a tributary, Gentung Creek. Normally this would be a shallow stream, but it was swollen and moving swiftly. At least it was fairly clear and was not the wide obstacle that the main river was. That was good because to reach our destination, we had to cross the stream 12 times when cliffs blocked the way on whatever side of the stream we happened to be on. As we progressed upstream into the steep mountains, the gradient increased; the stream turned into sections of rocky rapids with occasional waterfalls. This made the crossings even more treacherous. We almost lost poor Regex in one of the rapids. Fortunately, he swam like a champ and made it safely to the other side.

After four hours of trudging through the jungle and up the creek, we finally reached our destination. I mapped a few outcrops, and we collected samples. We packed up to repeat the trek in reverse, including all 12 crossings of the tributary and the swim across the Mataindo River. The return trip was easier since we were familiar with the terrain and the stream crossings. At the fifth crossing, one with a particularly nasty set of thigh-deep rapids, shit happened. With half the troupe across and the rest still in the stream, Regex jumped in and started swimming across, just above the crew. The swift current caught him and quickly dragged him past the guys in the stream. Mahmud made a valiant attempt to grab him but missed. The last thing we saw was poor little Regex paddling frantically and looking forlornly upstream as he was swept over a small waterfall and down into an inaccessible, rapid-filled gorge.

We stood there in stunned silence. We lost our little mascot, who brought us much joy and was a great unifying force within the team. Oddly, although the Indonesian people consider dogs a delicacy, everyone loved Regex. I think his spunk, determination, and fearlessness had a lot to do with that. Sainun was particularly saddened since Regex was most attached to him and vice versa.

There was nothing we could do. We slogged our way down the trail, downtrodden but intent on getting to the village before dark. Usually, the guys are cheerful on the way home, joking and singing songs.

Not today; everyone was quiet and pensive.

Wouldn't you know, about 10 minutes later, I heard whoops from the back of the procession. Looking back, I saw a blur of brown and white running down the trail to greet Sainun. The little fellow survived plunging over at least one waterfall and through a series of rocky rapids. After that, he managed to climb out of the gorge and find us. What a tough little pup!

With everyone's spirits lifted, we marched cheerily to the village of Mataindo, where the speedboat was waiting for us. We reached the beach and boarded a little before 6 PM, just as daylight was rapidly fading. The first part of the voyage went smoothly – while we were in the quiet water of the bay, behind the barrier reef. Things changed when we hit heavy surf at the reef edge. The navigator stood at the prow of the canoe and directed the motorman to steer around the breaking waves, barking orders of *kiri* (left) and *kanan* (right). That worked well – for a while. He shouted a loud *terus, terus!* (straight ahead). I looked up to see a 4-foot wave loom out of the half-darkness, about to break right over us – which it did. Fortunately, hollowed-out logs have very few moving parts and float exceedingly well. Our log survived the impact just fine, but it was converted into a semi-submersible bathtub filled to the brim with salt water. The navigator steered us straight into the wave, or we would have capsized. Miraculously, the motor kept sputtering during the ordeal and carried us past the surf. All hands onboard frantically bailed water until we rose to a comfortable height above mean sea level. I once again had to question the sanity of what we were doing, especially since nobody had a life jacket. The spectacular phosphorescence of the sea, which was sending sparklers across the bow and outriggers, enthralled me for the rest of the voyage. I temporarily forgot the danger we had been in.

Heading to sea in the "speedboat," – big breakers loom ahead.

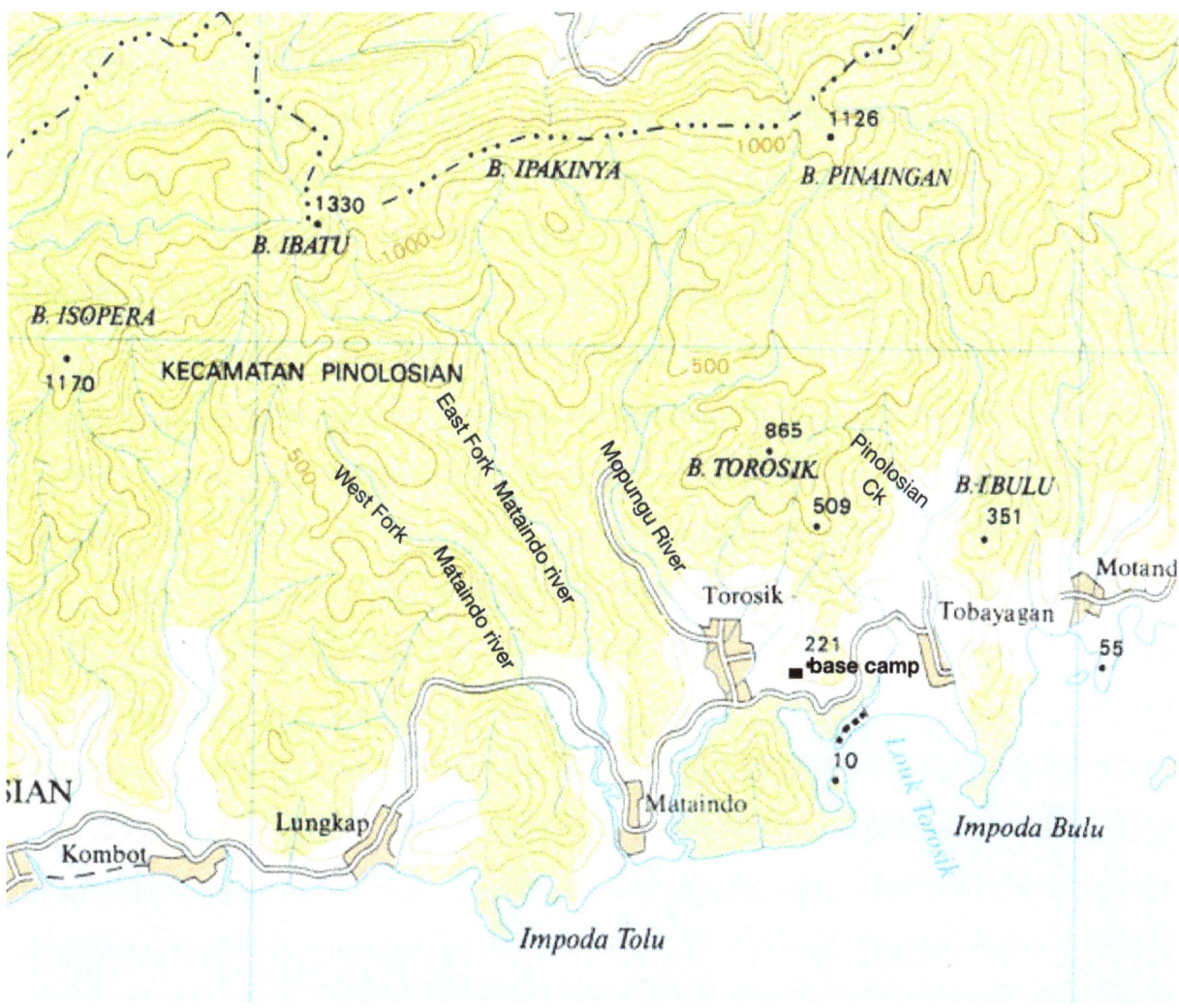

Back in Torosik camp, we spent a day drying out and celebrating Indonesian Independence Day with a goat barbeque and substantial quantities of white rice. The next day, we went back for more fun and adventure. Our canoe picked us up at Torosik Bay early in the morning. We launched in calm waters and made it as far as the mouth of the bay, where the captain wisely made a quick 180 after the second big wave hit us. He apparently learned something from the previous trip. He beached the boat on a sand spit on the far side of the river, next to a few grass huts occupied by a group of fishermen, who were smart enough (definitely smarter than us) to be sitting out the storm at home. One of them volunteered to guide us down the coast via beach and jungle trails. We walked briskly along the beautiful palm-lined beaches we had seen from the canoe two days earlier. The headlands proved more challenging to navigate, given their steep, slippery trails and knee-deep mud in places.

We reached the flood-swollen Mopungu River in about two hours. Once again, a potentially insurmountable obstacle faced us. But our newly appointed guide arranged for a small outrigger canoe to shuttle us across the river.

Miles of unspoiled beach – waiting for a sunny day.

Crossing the Mopungu River in a small dugout canoe.

    We hiked north of the village of Mataindo. Our intended destination was up a tributary of the west fork of the river. Not wanting to repeat the dangerous swim across the lower part of the river, we hacked our way through the jungle with machetes until the river was narrower. Luck was with us. A large tree had fallen across the river, providing a natural bridge for the crossing. Swimming was nearly an option for a couple of us since the log was slippery as snot.

Log crossing of the west fork of the Mataindo River.

After crossing the river, we followed the east side of a small tributary, cutting our way through the jungle undergrowth. We completed our surveying and mapping work quickly, then proceeded to reverse course on our trek back down the tributary and across the flooded Mataindo River. Our little mascot decided to take the day off, which was a good thing because we ended up following the stream on the way back and had to make several crossings through sections of rocky rapids.

We arrived in Mataindo after dark and walked along the road to Torosik. The swollen Mopungu River lay ahead of us and would be impossible to cross. Fortunately, we were able to radio ahead to the Torosik camp manager. He hired a couple of native boys to take us across the river in a makeshift bamboo raft. Huck Finn could relate to this place.

Night crossing of the Mopungu River.

Two days later, I was back at sea in the hired canoe, heading northeast along the coast to our camp at Onggunoi. After surfing our way onto shore, I was met by our field agent, Effendi, and a village boy with an oxcart. My gear went into the oxcart, and we hiked to camp, a little over one hour away. The workers in camp were glad to see me. At first, I was flattered. That is until I learned they had been isolated without supplies for two weeks. They figured if I could get there, so could supplies. The crew was desperately low on food, so I sent word for the speedboat to go back for some critical supplies to keep the camp operating.

Effendi loading the oxen-drawn cart.

Village of Onggunoi.

The weather was actually good for a change. I completed my fieldwork without a glitch. It helped that the team at Onggunoi was chicken and wisely insisted on staying on one side of the Onggunoi River, the one they were already on. I certainly was not going to repeat my mistake and try to convince them otherwise, particularly through leading by bad example. A couple of days later, I joined my favorite oxen taxi for the trek back to the beach to catch the three o'clock dugout for Molibagu. Other than the cart getting hopelessly stuck in mud and having to hire porters to carry my gear and our rock samples the last kilometer, things went relatively smoothly. That is until we loaded the boat and tried to leave. You see, the wiser part of me prevailed (for once). I upgraded our naval fleet to include a larger, safer speed boat. What I got was a 25-foot outrigger canoe with a monstrous (for its size) 40 HP outboard motor.

The bigger, heavier canoe had a deeper draft and was grounded in the sand. The navigator, my assistant, and I had to jump out and push the sucker through the surf. Once the engine finally started (about 5 minutes and 30 pulls later), we were off and flying. The extra 20 HP was noticeable. We arrived safe and sound at Molibagu about an hour and a half later. One of our Land Cruisers met us at the beach. We drove over the rough 4x4 trail crossing the mountains to our field office in the village of Muntoi in Dumoga Valley.

It was only after the trip over the pass that I gained an appreciation for the devastation caused by this series of storms. Trees were down everywhere. Landslides of various proportions partially blocked the mountain road in numerous places. Worse yet, were huge slumps that down-dropped half the road by as much as six feet and were still slowly moving.

One of many small landslides on the pass after the storms– there is a road under there, somewhere.

After overnighting in Muntoi, we drove to the company's headquarters in the relative civilization and blue skies of Manado – yet another harrowing three-hour drive on bad roads. Just to show I am not totally crazy, the first thing I did upon arriving at the office was to meet with the logistical coordinator and tell him I wanted to shut down work on the South Coast and to implement an evacuation plan for the crews stuck there. The next thing I did was order life vests for the team and porters. My trip was successful and ended up being safe, but we pushed the limits and got lucky. The field conditions were simply unsafe. Besides, we weren't accomplishing much. It would be much more effective if we deployed the teams on the relatively dry north coast. What really was scary is the actual monsoon rainy season doesn't start until October. This was August, and we were only experiencing the mild "wet season." Yikes!

After three days of trying to arrange overland evacuation of the field crews, we gave up. It was impossible. Heavy rain continued, and more mudslides and landslides ensued, completely blocking the only road over the mountains. We finally arranged for a WW II-vintage landing craft to pick up the 30 men and the two Land Cruisers stuck on the South Coast.

Time would be of the essence. The rains continued unabated. The office building at the Torosik camp, securely (so we thought) built on a moderately steep hillside, was totally destroyed in a landslide. Fortunately, that happened at noon, while everyone was safely eating lunch in the mess tent.

It was definitely time to get the hell out of there – even Regex agreed.

Tranquil waters of Manado Bay.
– a different world on the lee side of the island, Klabat volcano on the right.

Chapter 9

WILD MAN OF BORNEO

Don't lose your head over a little coal.

I turned to my guide, Bambam, as our motorboat cruised up the muddy Mahakam River, Borneo's equivalent of the mighty Mississippi, and asked if the Dayak natives living in the jungle where we were going still practice headhunting. His reply was, "No. There haven't been any incidents in two or three years." *Oh, joy, they are taking a break*, I thought. *I hope the intermission lasts long enough to finish my work and get the hell out of there*. I thought I was done with Indonesia after my time in Sulawesi with Newmont ended in 1996, but 11 years later, I was back in the Indonesian jungle, this time exploring for coal in Borneo.

We were motoring upriver to visit the Graha coal property in remote East Kalimantan Province. A Singaporean businessman, Mr. Tan, was considering investing $1 million in Kal Energy, the London-based company exploring the property. He wanted to know if the project was legitimate and worthwhile. I had my concerns after reviewing Kal Energy's press releases and technical report. There was definitely some misrepresentation, if not outright fraud, going on. My visit, basically in the role of an industrial spy, was not welcomed by the Kal Energy brass. If the Dayaks didn't make off with my head, odds are the promoters would.

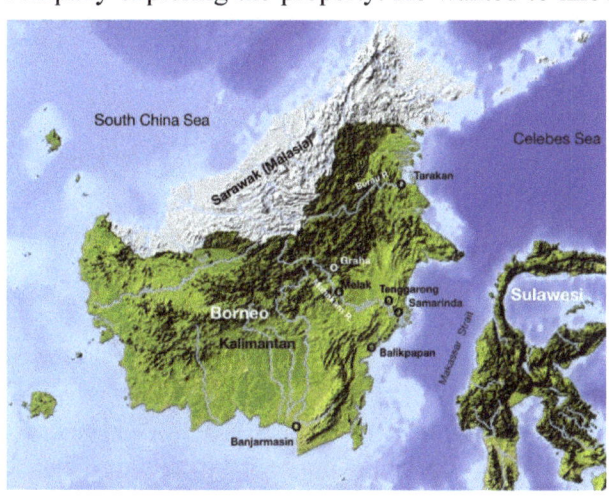

Two days earlier, I met the manager of Kal Energy in Singapore to get his side of the story; then flew to Balikpapan, an oil town on the east coast of Kalimantan, on the island of Borneo. There I met Attila Kovago, the geologist who did the resource calculation for the Graha property. Little did I know, this would be the beginning of a long and curious relationship with the crazy transplanted Hungarian, as our paths were to cross many times over the next five years, in Indonesia, Mongolia, and Australia.

The following morning, I met Bambam, who drove us to Samarinda, at the head of the Mahakam Delta, about 70 miles north of Balikpapan. We followed the Mahakam River from Samarinda upriver to Tenggarong, crossing over the Kutai Kartnegara suspension bridge, built in 2001 to resemble the Golden Gate Bridge (looks can be deceiving – the bridge collapsed in 2011, killing 39 people). We boarded a 16-foot motorboat at Tenggarong and raced up the river, passing canoes, ferryboats, tugboats pulling barges loaded with coal and timber, and skinny long-tail boats zooming along powered by shallow-draft propellers mounted on long poles behind the boats. Small Dayak villages lay scattered along the river, separated by miles of monotonous jungle, broken only by the serpentine meanders of the river. It was the dry season; the river was low. Floating wooden docks compensated for the drop in river level. Shirtless idle Dayak men sat cross-legged on the docks, watching us motor by, while women in floral-patterned dresses washed clothes and dishes. Weather-beaten clapboard-sided houses with thatched roofs lined the top banks of the river.

Dayak village on stilts.　　　　　　　　*Long-tailed boat on the Mahakam.*

    The boring gray of the buildings was occasionally broken by colorful displays of shirts, blouses, and t-shirts hanging out to dry or by the white dome on top of the occasional mosque. High and dry at the moment, the houses sat upon tall stilts, awaiting the annual floods that would turn them into isolated islands. The residents of New Orleans could learn a trick or two from the savvy Borneo natives.

    We stopped to refuel at Melak, the only village of any size along the river and the farthest point upriver that barges can go in the dry season. I transitioned from a slimy Pollywog to a trusty Shellback without fanfare as we crossed the equator just north of the town. After four hours of boating, we arrived at the tiny village of Long Iram, where we pulled up next to a couple of long-tail boats parked at a floating dock with a faded, blue banner advertising *Djarum Super* cigarettes. Getting to dry land required a balancing act along a narrow, precariously sagging, 30-foot-long wood plank connecting the dock to the muddy shore.

    Jonathan O'Dell, a British geologist in charge of the project, met me onshore. We drove off in a mud-splattered Ford SUV for the six-mile drive over abandoned logging roads to the base camp.

    I was struck by the contrast between this part of Kalimantan and Sulawesi, where I worked previously. Gone were the rock-strewn streams cascading from high precipitous mountains to the sea. Instead, small sand-lined tributaries flowed sluggishly across the nearly flat ground – essentially one giant sandbox – slowly making their way to join the Mahakam. The rainforest was also different – the result of multiple logging campaigns, leaving only scraggly third or fourth generation growth. Wildlife was eerily absent in what was left of the once magnificent rainforest, formerly the home of the Borneo elephant, Sumatran rhinoceros, and orangutan.

I spent the next three days reviewing maps and data, observing the drilling procedures, inspecting drill core, looking at outcrops, and reviewing exploration procedures. I was happy to learn that most of the work was being done in a professional manner. One problem became apparent when I hiked to see coal outcrops with Jonathan. From the way he pussyfooted around, it was evident he was not a field geologist. In fact, he was outright afraid of venturing into the jungle. That left the young assistant geologist and the inexperienced Indonesian assistants to fend for themselves – not a good situation.

One of the glaring omissions in Kal Energy's reports was the lack of information on the depth to the coal seams, a critical factor in the economics of the deposit. Back in camp after the first day in the field, I asked to see cross-sections showing the coal seams and the overlying rocks.

"I want to see your cross-sections," I said to Peter, the young Aussie geologist

"We don't have any."

"What?" I replied in shock. "How can you have drilled 40 holes, calculated a resource, and not have cross-sections? Why don't you have them?"

"We don't have the software for that."

"You must be kidding," I barked back, obviously pissed. "You don't need any damn software to make cross-sections. Do you have graph paper?"

"No," was the sheepish answer.

"What the….? Do you have paper… and a ruler?"

"Yes."

"Then go get them," I demanded. "We will make our own cross-section paper."

I showed the young geologist with a degree in geology from some prestigious Australian university what we retro-type geologists learned the first week in Geology 101 class – how to draw a cross-section by hand – without a computer or expensive software.

When I returned from the field the next day, I was pleased to see a set of hand-drawn cross-sections across the property, showing the drill holes and the coal seams intersected. *Well done, young lad*, I thought. Studying the sections, I learned why management had not supplied the computer-dependent younger generation with the software they thought they needed. The depth to the coal was answered – too deep. The stripping ratio would be high. That factor, combined with the low grade of the coal and the fact that barges could not reach the site for half of the year, made the economics of mining the coal unfavorable.

Management was hiding that little detail. They knew the seams were deep. They avoided producing the data that would prove it.

Waterfalls over thick coal outcrop.

Core drilling operation at Graha.

My last task was to fly to Manila to meet the Australian geologist who wrote and signed off on the technical report. The report had tables listing the thickness of coal seams in the drill holes but did not show the depths to the seams. It also lacked cross-sections, which are required by the regulating agencies. It didn't take long to determine the problem. The "expert" coal geologist, whom Kal Energy hired to sign off on the technical report, turned out to be an incompetent drunk. Let me clarify – he was quite competent at drinking; the incompetence was in his professional work.

I wrote my report for the client, noting there was indeed a lot of coal at Graha, but the profitability of starting a mine did not look good. Although there was no actual fraud, there were errors of omission, attempts to hide critical information, and exaggerations made by management to oversell the project. I recommended not proceeding with the investment.

The Graha project ended up going nowhere. For the cost of a couple of airline tickets, a boat ride, and a week of my time, Mr. Tan saved a million bucks.

Chapter 10

BUGGER ME!

Trying not to bend over and take it.

After presenting my negative recommendations on the Graha project to Kal Energy, I thought I would be sent packing. Although they were not happy with my conclusions, they accepted them. They even wanted me to do work for the company on coal and other projects in Indonesia and Mongolia. During the next four years, I bounced around much of Borneo and Sumatra looking for coal acquisition opportunities, initially for Kal Energy, then for the somewhat incestuous companies it morphed into: Bukit Merah Coal and Kumai Energy.

I initially based my work from the oil town of Balikpapan on the south coast of East Kalimantan. I was a one-man show, operating on a shoestring budget. Fortunately, Attila Kovago, the geologist I met during the Graha review, was also based there, working for BHP, the giant Australian mining company. Attila quickly became a good friend and my mentor. I learned a lot about the Indonesian coal industry from Attila. More importantly, I learned about hash.

Attila and I did a lot of hash in Balikpapan. Monday afternoons, we did hash with a group of local businessmen. Both women and men participated on Tuesdays. On Saturdays, entire families joined in the fun. Such activity may sound odd for a guy who never was fond of whacky tobacco. Not really, the hash we were doing was with the Balikpapan Hash House Harriers, the local chapter of the international Hash House Harriers Club, best described as "a drinking club with a bad running habit." Hash has its origins in nearby Malaysia. In 1938, British Army officers decided they needed an activity to provide exercise and help cure the hangovers that arose from the weekend's debauchery. They laid out running courses through the jungle, ending the runs with drinking beer and singing bawdy songs. Hash courses are cleverly designed, so both the fast runners and the slow ones end up at the finish line at about the same time (a prerequisite for drinking beer together). That is accomplished by laying out a series of false leads, wrong turns, and dead ends. The fast runners take the wrong trails and turn back, shouting, "Check back!" to alert the slower runners of the dead end. When they are on the proper path, they shout, "On, on!" That way, most of the runners arrive at the finish line at about the same time, except for the lazy "short-cutting bastards," who run the first few hundred yards before cutting over to the finish. Their deceit is obvious by the lack of sweat on their hash shirts or mud on their running shoes.

After rehydrating with beers, we retired to Sid's Bar, where Attila held court, telling hilarious stories of his upbringing in Hungary and his escape from behind the Iron Curtain. After various failed attempts to escape, Attila booked a flight to Australia but convinced the authorities at the airport that his ticket was to Austria, a neutral country cooperating with the USSR and its satellites. It worked.

My favorite story relates to Attila's first professional job in Australia. He was hired to work with a drilling crew on a coal project. Attila did well. About midway through the program, he was asked to present a progress report to management. Now, you need to realize that when Attila arrived in Australia, he spoke very little English. Being a studious fellow, he learned quickly. The problem was how he learned – from two sources: watching Sesame Street on TV and working with rough Aussie drillers. Attila was only a couple of minutes into his talk when the managers interrupted him, "Stop! You can't talk like that." Attila was taken back. Was his broken English the problem?

Hash obstacle, "On On!"   *Attila serving rehydration beers.*

Attila's speech was riddled with f-bombs and numerous other colorful four-letter words he learned from the drillers. "But that is how the drillers talk," was all he could say in his defense. Fortunately, his speech has improved a lot, although he does have moments of relapse.

When I worked for Newmont in Sulawesi, I was supported by the staff of a major company and a team of helpers, allowing me the luxury of being able to focus on exploration. I was isolated from business dealings with the exception of small transactions, like buying chickens for the field crews. Even that came with a kickback cost. In Indonesia, everyone has his hand out for a cut of any money that changes hands. What we would consider corruption in the U.S. is business as usual to Indonesians. The bribes and small-scale extortion required to get most anything done over there are just considered employee benefits or well-deserved tips. The Foreign Corrupt Practices Act (FCPA) prohibits US firms and employees from offering bribes or under-the-table payments to government officials, putting US companies at a disadvantage to Asian ones, which actually have a budget for bribes. So, instead of bribes, we employed "facilitation payments". Sometimes the bribes, extortion and facilitation antics were amusing.

Dave Cole and I hailed a taxi to go out to dinner during one of our layovers in Bali. On the way to the restaurant, a policeman on a corner waved the driver over. The cabbie got out and talked to the cop for a few minutes. He came back and asked if we could pay our fare early, which we did. As we resumed the ride, we asked the cabdriver:

"What was that all about? Why did the policeman stop you?"

"He was hungry and needed money to buy dinner." The driver made it sound like a perfectly normal situation.

I was talking to Jim Stinson, Newmont's Manager of Lands in the Jakarta office, one day. Jim had just returned from a long trip out of the country. He was upset:

"These damn corrupt people – it cost me $100 to get back into the country."

"Why was that, Jim?"

"My visa expired while I was gone. The Immigration officer gave me a hard time and demanded that I pay him a bribe to let me pass Immigration."

"Jim, did it occur to you that it is a good thing you are in a corrupt country? Anywhere else they would have denied you entry and sent you back on the next plane."

"Well, I guess you have a point."

I arrived at Immigration at the Jakarta airport to catch a flight home after six weeks of traveling around Java, Sumatra, and Borneo. I handed my passport to the agent. He thumbed through it, then asked,

"Where is your arrival card?" – you know, that little card you get when you arrive and wonder what it is for. It was usually in my passport.

"You need your arrival card to depart the country." That would only make sense if it were a "departure card". I overlooked that little detail and checked my wallet and briefcase, but could not find it.

"It must have fallen out of my passport," I explained (no doubt during one of the 100 times I had to show my passport to some official).

"This is a big problem."

"Sorry, but I don't have it. What do I need to do?"

"You have no idea how big a problem this is. You will have to go downtown to Immigration and deal with them and pay a fine. You will miss your flight. This is a really BIG problem."

I was getting the drift and said, "OK for every problem there is a solution. How *much* is the solution to this problem?"

I should have guessed the reply, "How much do you have?"

I was prepared for that and figured two could play his game. I took my wallet, grabbed the cash (about 500,000 rupiah), and tossed it onto the counter in front of the agent.

He freaked, and said, "Don't' do that! Take it back. Do you have any US dollars?"

"No, I exchanged that a long time ago. All I have is Indonesian currency." I was lying; I had US cash and more rupiah in my briefcase, but didn't want to carry it in my wallet.

"OK, then. Now, roll up the money in your hand and pass it to me so nobody sees."

My BIG problem was solved for about $50 US and I was on my way home to a land without corruption – unless you are dealing with politicians.

I learned to deal with the petty corruption. But I was not prepared for the fun and games dealing with the rampant corruption and fraud I found in the coal business – and I was on my own to deal with it. The most corrupt industries in Indonesia are in the natural resources field, specifically forestry and mining. Within mining, coal is the most corrupt sector due to the ability to make fast, easy money. Given that background, it is only fitting I jumped feet first into the coal industry in the Banjarmasin region of South Kalimantan – the most corrupt area in the most corrupt province in the most corrupt industry in one of the most corrupt countries in the world. The fun was about to begin.

The goal was to acquire under-explored coal deposits the company could explore and develop into minable coal reserves. Then we could either start a mine or sell the asset. Unfortunately, we were a bit late to the game. Every coal basin was covered by existing exploration licenses, *Kuasa Pertambangans* (KPs). My work primarily involved making contacts with license owners, negotiating preliminary lease terms, conducting field visits, and performing due diligence on the properties. That approach would not have been much different from doing acquisition work in the U.S., except nearly everyone we dealt with was a liar, a fraudster, or an outright crook. Thus began the most frustrating and least productive period of my generally frustrating and nonproductive career.

Exploration licenses with the best potential are handed out as favors, or bribes, to powerful Generals and politicians. Those KPs were beyond consideration with our low-level connections and finances. Fortunately, there is a lot of good coal in Borneo and Sumatra, so there should have been a good chance of acquiring worthwhile properties from the next rung of KP holders – under-capitalized small to medium-size mining companies and well-connected individuals.

Or so we thought. A typical submittal of a coal KP consisted of a couple pages of the legal KP verbiage along with latitude and longitude coordinates for the license boundaries. That might be accompanied by a few coal analyses, a table of coal seam thicknesses, or a couple crude drill logs showing coal seams. The owners wanted big bucks (hundreds of thousands to millions of dollars) to sell or lease the licenses. Too many of them wanted a cash payment (usually several thousand dollars) upfront before I was allowed to visit the site and conduct due diligence. In many cases the purported owners did not own the property and were not even remotely connected to the owners. They somehow got ahold of the license data and thought they could make a quick buck off the naive *bules*. Those people were obvious scammers, and the properties were rejected.

When we finally were allowed to conduct due diligence, we often found that the analyses or drill data accompanying the license actually came from a different property. Field investigations often revealed that 10-foot-thick coal seams had magically shrunk to one-foot-thick seams – left out in the rain, I guess. Or, analyses from our samples showed much lower-quality coal, which had far less energy content – calorific value (CV) – or higher ash and sulfur content than reported.

The scrubbers on the stacks of power plants and steel mills can only handle up to 1% sulfur in the coal that is burned, or the excess sulfur is blown into the atmosphere, a definite environmental no-no. Coal with more than 1% sulfur content is basically unusable unless it has such high calorific value that it can be blended with low sulfur coal and still be burned with less than 1% sulfur. I sent a young Indonesian geologist, Beben, and his assistant to evaluate and sample a KP west of Berau, which supposedly had thick seams of high CV coal. The data presented by the promoters lacked sulfur analyses. Before sending the team on a lengthy and expensive wild goose chase, I made the comment, "We will be happy to sign a deal, as long as the sulfur is less than 1%." The field geologist reported the seams were indeed thick and looked like good quality coal. Then we got the analyses – 8.3% sulfur! I relayed that deal-killer to the owners, who were shocked and said, "Oh no, we thought it was only 3%." Bugger me! That is still three times the limit. If they had been upfront about that, we could have saved a lot of time, effort, and money.

Anton, my office partner in Jakarta and Kumai Energy's VP of Marketing, and I made a trip to Berau in the far eastern part of Borneo to visit the local Mines Department and review available KPs with the department head. We spent the day touring the office, meeting employees, and looking at maps. There was little discussion about acquiring concessions. That was saved for later. We were invited to come to the department head's home in the evening to discuss specific KPs. Why? Well, because all the good ones somehow ended up being owned by him. When we discussed preliminary terms, we were told, "There will be no receipts, don't even ask."

Having access to good opportunities is largely about contacts. In 2010, the Kumai Energy bosses hooked up with John Lim, alias "The Captain" – an Australian of Malaysian origin. Mr. Lim supposedly had good connections in the Indonesian coal industry. The Captain was a character. An orphan, he was raised by Shaolin Monks and was a martial arts expert. He ended up being an officer in the Singapore Special Forces and later became a ship's Captain, hence his nickname.

I made several trips to various parts of Kalimantan (Borneo) and Sumatra with the Captain. Despite being 63 years old, The Captain was in great shape, possibly in part due to the 500 sit-ups he claimed he did each morning. I am not sure what role that played in his out-of-control libido, but something was certainly messing with it.

The Cap had a wife and children in Perth. He had a second wife in Jakarta. Add to that a girlfriend in Tokyo, one in Singapore, and another one in Jakarta. He was a smooth talker and one helluva juggler. The Captain carried five different cell phones – one for each gal, with different ring tones, of course. On a one-week-long trip across East Kalimantan, we spent countless hours driving crappy roads (more like obstacle courses of potholes loosely connected by patches of asphalt) to reach our destinations. My entertainment consisted of listening to the one music CD the driver brought and played constantly – *Englebert Humperdinck's Greatest Hits* (if I hear "Please release me, let me go" one more time, I may commit hari-kari), or overhearing the Captain yakking on his phones, sweet-talking his ladies to keep them happy in his absence. Not only was he good at juggling the phones, but he was also busy juggling his schedule, "No, honey, I won't be home next weekend. I will see you on the 15th." Yikes! On top of all that, the old horn dog was constantly trying to pick up new girls. Age didn't matter. It was all about the hunt and the conquest. A few months after we finished our field visits, he stopped by the Jakarta office to say hello. He introduced a very attractive 30-something gal, the wife of a prominent Jakarta businessman with whom he was having an affair. The Cap was playing with fire. That was the last time I saw The Captain. I am not sure if he made it to 64.

The author (the white guy, second from left) and The Captain (middle) with Berau Dayaks.

We evaluated several coal licenses in Sumatra, including one near Lampung on the south coast. That one sounded good on paper but ended up being more water than land. Kumai Energy finally picked up a coal concession near Jambi from a legitimate Jakarta businessman, Mr. Harsano. The business terms were reasonable. Our site visit confirmed the reported thickness and quality of the coal outcrops. The next step was to drill, starting with duplicating the results of a few shallow holes drilled previously. That proved easier said than done. Permits needed to be obtained before work could commence. Part of the license fell within a recently developed palm plantation, the scourge of Southeast Asia. At a time when coal mines could not expand by a single acre due to restrictions imposed by some stupid deal made in Tokyo to end climate change before the earth melts, giant palm plantations were springing up all over Indonesia, devastating hundreds of thousands of acres of low-lying forest. Unlike rubber tree plantations, in which rubber trees are part of a diversified forest, palm plantations are monoculture crops, where nothing but palm oil trees can grow, and no animals can survive. Palm oil is a high unit-value crop, so the environment be damned. Any company that wants to use a few acres of palm plantation for another purpose, like drilling a few holes in the ground, can also be damned. Permits came at a high cost and took forever. We never obtained permission to drill on the palm plantation but did drill in another area, where there were supposed to be thick coal seams. Guess what? They were not there. Mr. Harsono was an honest man, but he had been scammed and received falsified data.

South Sumatra coal country - on the soggy side.

It could have been worse. Take, for example, the plight of the small Australian company, Equatorial Coal Ltd. Equatorial made a deal with PT Megacoal Indomine, owned by a wealthy Indonesian businessman, to mine the Alum Duta coal deposit in South Kalimantan. The company started out properly and cautiously, hiring an independent firm to conduct due diligence and verify the quantity and quality of coal on the property. The conclusion was there was a deposit of about four million tons of high-quality coal on the license – good enough to start a mine. Equatorial began preparations to mine the deposit. They funded Megacoal with a $2,000,000 nonrefundable payment to mobilize equipment, build an access road, and construct infrastructure for the mine. In typical Indonesian style, the start of the work was delayed several times over a 12-month period. When requests were made for more financing, the Board of Equatorial got suspicious and hired an independent consultant to investigate, starting with a site visit. The findings were brutal: 1) most of the drill holes outlining the deposit were actually located on an adjacent property, 2) although there was still a good coal deposit on the license, "was" is the proper tense. The entire deposit had been mined during the period when delays in starting the mine were supposedly being incurred. Megacoal got the coal and made off with the $2,000,000. All that Equatorial Coal got was a big hole in the ground and bankruptcy. Maybe they should have had someone on-site, or at least have visited occasionally.

The problems in the coal industry were not limited to fake licenses or thieving mining contractors. Companies that successfully developed mines found themselves out of the cooking pot and into the cannibals' fire. Like sharks following the scent of blood in the water, politicians and bureaucrats flocked to the mines to get their share of the coal bounty. Everyone from the *kepala desa* (village chief), the *Chamat* (district head), the *Bupati* (Regent), to the *Gubenur* (Provincial Governor), as well as the Mines Department, the Forestry Department, the Police, and the Army stopped by regularly seeking "donations." Of course, if their money demands were not met, they would find some trumped-up excuse to shut down the mine – roadblocks being a favored enforcement weapon.

Even shipping coal had its pitfalls. Nearly all coal mines in Borneo are located near the major rivers: the Mahakam, the Berau, the Barito and their major tributaries, because shipping by barge is the only economical way to get the coal to market. That often means several days of barging down the rivers to the ocean ports. Indonesia's version of Pirates of the Caribbean took advantage of the shipping system. Boats pulled up alongside the barges (which were several hundred yards behind the tug boats and out of sight of the captain) at night, and coal pirates used huge dredge siphons to suck the coal off the barges. Coal shipments had a tendency to arrive at port a little on the light side.

Caution — wet road.

Bottomless Sumatran mud.

Berau bridges – we managed to get past — nervously.

Despite the overwhelming frustration of dealing with the rampant corruption, we optimistically continued our efforts to find a suitable coal property to acquire and develop in Borneo and Sumatra, dealing with crooks and scam artists, while braving floods, insanely muddy roads, and washed-out bridges to reach remote jungle locations in search of coal outcrops.

Sadly, we failed in our mission. Kumai Energy didn't get rich developing a coal mine in Indonesia, but didn't go broke trying, either. We saved that for later. The cool, dry Steppes of Mongolia were calling – supposedly offering better opportunities in a somewhat less corrupt business environment…supposedly.

Local guide leading the way; guide, the author, and driver after a long day of trekking through the Borneo jungle.

Chapter 11

THE FIRST CUT IS THE DEEPEST

…and one you will be sure to remember.

Jakarta, sucks! Trust me, I know. I spent the better part of two years based there for Kumai Energy in 2010-2011. *Ibu Kota*, the mother city of Indonesia, is home to more than ten million residents, who are packed like sardines in old rusty cans. In a splendid display of inequality, rickety tin shacks are wedged between 5-star hotels and upscale shopping malls. Beggars line the garbage-strewn streets.

The city is infamous for both its squalor and its traffic congestion – among the worst in the world. There is no mass transit system. To make matters worse, unlike big cities such as London and New York, nobody walks anywhere. There are logical reasons for that. One is that it is constantly hot and humid, making any journey on foot a sweaty one. The other is there are very few sidewalks. Those few are riddled with cracks, holes, and open manholes. Elsewhere, upheaved blocks of concrete, parked motorbikes, and vendor carts make them veritable obstacle courses. As a result, cars, trucks, buses, taxis, hand-drawn carts, and millions of motorcycles jam the streets, constantly squeezing together in an attempt to create an extra lane in every road, which just doesn't work. Rain, which seems to be timed for rush hour daily, further complicates the situation. Jakarta lies just above sea level but is sinking due to drawdown of the water table. When it rains in Jakarta, it pours, and the water (along with the garbage floating in it) goes nowhere.

The worst traffic is on Friday afternoons when half the population tries to get the hell out of the city to flee to the Thousand Islands or the cooler hills of the Western Highlands. If it rains on Friday afternoon, which is almost guaranteed, gridlock ensues. I rarely attempted to leave the city on weekends, partly because I had work to do but mainly because it would take most of the weekend just to get to the city limits and return. I made an exception when my friend, Enih, invited me to visit the village of Cirebon, on the north coast of Java, about 75 miles east of Jakarta, to experience Javanese culture. Of course, it was raining hard as we attempted to leave the city. Although our route followed a good paved highway along the north coast, the trip took six hours – four of them just getting across the flooded city.

I took in the local sights as we toured the village on Saturday morning. Ciberon, like most Javanese villages, is an agrarian town. The locals grow just about everything they need in the fields surrounding the village. Although the people have few possessions, they do have plenty of food.

In the afternoon, the sound of traditional Javanese *gamelan* music: drums, gongs, chimes, and flutes drew us to the main street in town. I was in for a treat, witnessing an event that few westerners get to see, a *khitan* parade. Young girls and boys dressed to the hilt in bright-colored traditional Javanese clothes passed by, riding on floats of lions, tigers, bears, and elephants – each composed of two village men under one large costume. Other kids sat upon enormous birds carried on bamboo poles held by teams of men in royal blue jackets and gold sarongs. Village kids in t-shirts ran enthusiastically after the floats.

    I had been told a little of what the *khitan* was about, but the explanation was delivered in Javanese, the local language. All I understood was it had something to do with a boy growing up. Then the keynote parade entry came by. On a cart sat a pale-faced pantless man, spread-eagled with a large stick (perhaps the original woody) – the end of which was freshly stripped of its bark – sticking out from the middle of his spread eagle.

    Oh yeah, now I got it. The *khitan* is the celebration of a young boy's circumcision, one of the most important occasions in a Javanese Muslim boy's life. Families save money for years or go into debt to finance the celebration, much like Mexican families do for a *quinceanera*, the celebration of a girl's 15th birthday, marking her transition from childhood to womanhood. I would bet the Mexican girls have a lot more fun at their coming-out party than the Javanese boys have. Perhaps the time required to raise the money explains why the family doesn't do this when the victim is a baby. The Javanese wait until the poor fellow is between eight and twelve years old to descend upon his pride and joy with sharp instruments. I guess they want him to remember what he had and lost.

    Ouch, that's gotta hurt.

The First Cut Is the Deepest

This parade entry sure caught the attention of one young maiden.

Chapter 12

SAFETY FIRST

And second, and third, and...

You would think a country founded by hardened convicts would be populated by hardy, independent blokes and sheilas, who would prefer to make their own decisions about how they conduct their lives rather than be mothered by the government. I mistakenly based my expectations on the likes of Crocodile Dundee, wielding his enormous knife to rescue a damsel in distress on the streets of New York City, or Steve Irwin, ignoring both common sense and sanity to wrestle toothy crocodiles and exceedingly venomous snakes, just for fun.

I was shocked and dismayed to learn the truth. Sadly, Australia is deservedly known as the "Nanny State."

With encouragement and a recommendation from my friend, Attila Kovago, I landed a job with a small Australian consulting group, East Coast Exploration, which provided geologists and technicians to the coal industry. I was thrilled to have a job in a civilized country for a change, one where English was spoken (sort of – I did need to learn to speak "Australian"). In early 2012, the industry was booming. There were not enough bodies in Australia to fill jobs in the mining industry, both for labor and skilled technical positions. Australia looked across the big pond to fill positions. At times, mining companies hired entire graduating classes of geologists from universities in the UK and Europe to come work Down Under.

East Coast Exploration hired a dozen recent graduates to work on BHP's coal projects in the Bowen Basin of Queensland. My mission was to mentor the youngsters, attempt to educate them about coal geology, and help develop the skills needed to log and sample drill core.

Before I could start working, I needed remedial training to be made "safe." I had been warned about Australia's over-the-top safety regulations. I had even seen some of it in action in Nevada. Despite that, I was not prepared for what was coming.

Mining geology is no fun compared to exploration work in the hinterlands, where we are pretty much free to do our own thing. Personal responsibility is the rule in the exploration business. Mines, however, have far too many rules and regulations. That is particularly problematic for someone who has difficulty obeying even his own rules.

The Australian approach to safety is crazy. Safety revolves around rules, regulations, and paperwork. I experienced this for the first time with an Australian company working in northern Nevada, way back in 2005. Kennecott Exploration, which was acquired by the Australian mining giant Rio Tinto, joint-ventured White Knight Gold's Lone Mountain project in the Independence Range north of Elko. I took the bosses, John and Gord Leask, to visit the property while Kennecott was drilling. We left the highway and followed a dirt road to the west. About a mile later, we came across a sign on the left side of the road proclaiming, "Danger, heavy equipment ahead". That was followed by a sign warning, "Danger narrow road," and signs every quarter mile warning people to turn back to avoid the imminent danger ahead.

The signs reminded me of the Burma-Shave signs with their clever jingles that lined America's highways in the 1960s. But these signs didn't rhyme and lacked the catchy punchline at the end.

We made our way up the mountain, passing signs spaced more closely together. Around a corner, we came upon a pickup truck parked in the middle of the one-lane dirt trail, blocking it. The deadly, dangerous drill rig was about 200 yards down the road.

John Schaff, a colleague I knew from when I consulted for Kennecott years before, emerged from the cab of the pickup. We exchanged pleasantries, and explained we were there to check progress on the drilling. Gord said he wanted to go look at the drill rig and samples.

"You can't do that," John said.

"What? Why not?"

"You haven't been safety trained. We have strict rules about who can go near drill rigs. You need to attend a week-long safety class to qualify."

The Leask brothers argued this was their project, and they should be able to visit it. Also, they had been working around drill rigs for decades and knew a thing or two about safety.

John finally caved under pressure. He reluctantly said, "OK, I can give you the short course, then you can go to the drill."

John went to the cab of the truck. He opened a file box to retrieve the requisite forms for the safety briefing. Meanwhile, we busied ourselves perusing the stock of safety gear packing the truck: a stretcher, satellite phone, giant first aid kit, defibrillator, and boxes of unidentified equipment.

From the front of the cabin, we heard an "Ah shit!" While rifling through the file box, John suffered a nasty paper cut.

"Damn, now I have to fill out an accident form," John told me, bleeding all over the form for the short safety course. Before John stemmed the flow of blood and finished filling out his accident form, the Leask brothers walked all the way to the drill rig and were on the way back – safely, I might add. So much for drill rig safety training. I think poor John had to go to a remedial file-box training class after that incident.

Back to the Australia story – I flew to Brisbane, where I attended a week-long safety training course run by BMA, the coal division of the mining giant, BHP. I knew I was in trouble when, on the first day in class, the instructor announced, "There is no such thing as common sense." Oh dear, maybe Aussies lack that necessary mental faculty, but we Yankees rely on it regularly. Just to be sure I was not confused about the subject, I googled "common sense" and found the following definition from Mr. Webster: "sound and prudent judgment based on a simple perception of the situation or facts" or "the ability to think and behave in a reasonable way and to make good decisions." My thought was, *Good grief! If they lack basic skills like "sound and prudent judgment" or "the ability to think" down here, how can they possibly be safe around giant pieces of machinery and enormous holes in the ground?* I was in class with all kinds of blokes being trained to work in the mines: mechanics, welders, truck drivers, janitors, you name it. Maybe the instructor was right. I surveyed the room and spied a couple of candidates that might support his claim. Common sense does exist, but not everyone has it, at least not in sufficient quantity. That is good. If we all had lots of common sense, there would be no candidates for the annual Darwin Awards.

I am as safety conscious as anyone and really like to return home at the end of the workday with the same number of digits I had at the start of the day. Except for minor blood-letting and stitches from an incident with a wingless flying machine, I have managed to limit my injuries to scrapes, bruises, and the occasional sprain. That's not a bad record for more than 50 years of fieldwork – and without much formal safety training. Somehow, applying common sense seems to have worked for me.

Sadly, in Australia common sense has been replaced by countless hours of classes, endless rules and regulations, tons of paperwork, excessive restrictions, even outright avoidance of "dangerous" work situations. The net result is needlessly high work inefficiencies. Perhaps they should admit that despite endless rules, regulations, and paperwork, you just can't fix stupid.

After my week of classroom training, I flew to McKay, on the northeast coast of Queensland. There

I attended a driving course, held in a classroom – sans vehicles, instead of on the road. At the end of the class, we drove around some yellow cones in the parking lot – to see if we knew how to steer, I guess. The one important safety factor the instructor forgot to tell this Yankee is that they drive on the wrong side of the road over there. I figured that out on my own (using common sense, I think) as I drove inland to the coal-mining town of Moranbah without incident.

I met my new boss, Brent Delaney, in the evening. We discussed the job and how I could best help the company train the young geologists. Being gung-ho and eager to apply my newfound safety skills, I requested to visit a drill site the next day. Brett said I would have to go as a visitor because I had not yet completed my safety training. He arranged for a field technician to pick me up bright and early in the morning.

I was having a good old time looking at drill core and discussing geology with the junior geologist when a pickup truck arrived. Two official-looking men in BMA uniforms walked over to us. They were safety officers, who were surprised to see me. They asked who I was and why I was there. After a while, they left. I assumed all was well.

Not the case.

By the time I got back to town, the Aussie poop had fallen upon the Outback fan. Brett obtained permission for me to visit the site but had not notified the almighty Site Security Executive. Much to my amazement, this was a very serious infraction. Holy cow, I set a new record – in trouble on the first day on the job and only as a visitor.

I soon learned that, despite my stellar performance in the safety and driving classes and having passed the driving-on-the-wrong-side-of-the-road field test with flying colors, I was far from safe enough to be allowed to work on-site. In fact, it would be a month before I was deemed fit to work on my own – an entire month wasted on safety classes and meetings with no actual accomplishments to show for it.

The real fun began when I finally started working with the drill crews. I am used to spending time with the drillers, watching their work, discussing any problems, giving advice as needed. You can't do that in the Land of Oz. Geologists were not allowed to go near the drill rig. The drill sat behind a 6-foot-high steel fence, a barrier designed to keep marauding geologists a safe distance away. We were required to stay a minimum of 100 feet from the drill rig. If we needed to talk to the driller, we had to do jumping jacks or somersaults to get his attention. Then he could exit the cordoned-off area and come talk to us.

Drills need to circulate water and drilling mud to keep the drill rods lubricated and keep the hole from collapsing. A pit is dug in the ground to hold the mix of water and mud. For safety, the pits are only 4-feet deep and have ramps on both ends, so if someone (other than a midget) fell in, they could climb back out. That is not good enough in Australia. The mud pits were surrounded by 6-foot-high steel fencing adorned with all kinds of warning signs. A life ring, probably retired from some cruise ship, hung on the fence. The safety dudes were taking no chances on accidental mud-pit drownings.

Whatever happened to the good old days, when part of a neophyte geologist's initiation to drilling was being thrown into the mud pit by the drillers? My good friend, Dr. Joe Kowalik, had that happen to him on his first drilling job in Texas. Joe emerged from the pit looking like a mudpuppy but otherwise unharmed – except for his pride. You shouldn't mess with Dr. Joe's pride, as the drill crew was about to learn. Joe was hellbent on revenge. His opportunity came a week later. On the way to the drill site, Joe spied a dead skunk in the middle of the road. He stopped, gingerly picked up the stinking critter, put it in a garbage bag, and threw it in the bed of his truck. When he arrived at the drill site, the drillers were busy drilling. Joe snuck up to their truck, dumped the skunk out of the bag into the cab, and closed the door. The skunk sat there for several hours, cooking in the hot Texas sun until the drillers finished and were set to leave for the day. They opened the truck door, immediately gagged, and ran away in search of fresh air. They never screwed with Joe after that, and his reputation preceded him on future drilling jobs.

One of the major downsides of the job was dealing with the BMA Safety Nazis. All the day-to-day work was performed by contractors, which freed up the BMA employees to spy on the contractors and bust them for any minuscule safety violations, which they did on a daily basis. I am sure they received bonuses or big "at-a-boys" for that. We were required to wear heavy fire-retardant long sleeve shirts, which had to be tucked in, and the sleeves had to be buttoned down. In the 100-degree heat of the summer, just untucking the shirt would lower the effective temperature by 10 degrees. You didn't dare do that since it was a favorite violation of the safety Nazis. One of the geologists was busted because he rolled up his shirt sleeves to keep them dry while washing samples in a bucket. Another geologist was written up for not having his shoes tied tight enough. The culture of constant spying and busting of "violators" created an atmosphere of fear and paranoia that led to bad attitudes and conflict between the BMA employees and the contractors – just what you do not want when working around genuine hazards.

Then there was the endless paperwork. We were supposed to complete a "Take Five" card for every different task performed during the day. The forms required describing the task, identifying any hazards (real or imaginary), suggesting how to avoid them, and afterward explaining how the hazards were successfully avoided. It was supposed to take five minutes to complete the form. Logging core is not normally considered a particularly hazardous task, but we were supposed to address all possibilities, like paper cuts, I guess. Given that we performed many different tasks during a typical workday (logging core, sampling core, carrying core boxes to the truck, and unloading the truck were all different tasks), filling out Take Five cards could involve a lot of time. You can probably guess how many of them I completed. No, the answer is not "zero." I filled out one form for practice in the training class.

Worse than the Take Five cards were the Joint Safety Assessments (JSAs), required for more complicated tasks, like constructing a drill pad or "recovering" a vehicle stuck in the mud. Anywhere else, the latter meant getting a couple guys to push the truck or pulling it with another truck and a tow strap. Nope, paperwork had to come first. A team of safety experts needed to assess the situation, have a conference to decide how to proceed, then write about 10 pages of BS to document the team's brilliant solution to a problem that wasn't a problem at all. Before the work could commence, the paperwork had to be submitted to a supervisor for review and approval. Forget about accomplishing any work that day.

Of course, the paperwork wasn't really about making people safer; it was all about ass-covering for upper management.

Unfortunately, the long arm of the Nanny State extends far from the workplace. It reaches into everyday lives. It is especially prevalent in tourist areas, even in the Outback.

One benefit of the government's work regulations was getting a mandated week-long break after 14 days of work. There was not enough time to fly home and back. I traveled around Australia and saw a lot of the country, but never escaped the safety craziness.

A good friend of mine, who had visited Australia a couple of decades before, told me about a wonderful place to visit, Babinda Boulders, near Cairns in northeastern Queensland. Babinda Creek flows down from Mt. Bartle Frere, at 5,285 feet, the highest peak in Queensland, and plunges through a gorge filled with rounded granite boulders separated by deep pools of cool water – a perfect place to beat the tropical heat while slipping down smooth, natural slides into deep plunge pools.

But no longer. Arriving at the Boulders, a series of signs giving stern warnings not to enter the stream greeted me. The entire area was cordoned off by steel fences, except for one designated swimming area – in a shallow, wide pool above the boulder rapids. There had been several drownings there over the years. The Nanny State's answer to preventing more Aussies from entering the Darwin Awards competition was to seal off the entire area. Of course, I had to climb over the fence and see what the fuss was all about.

Off-limits Babinda Boulders – from the illegal side of the fence.

Western Australia was even worse. When East Coast Exploration transitioned from working for BMA out of Moranbah to starting a new project for Linc Energy near Emerald, my work schedule was thrown off. To get back in synch, I had to take a two-week break instead of one week. Bummer. I flew from Brisbane to Perth on the west coast of Australia. There I rented a gaudily-painted hippy campervan from

Wicked Campers and proceeded to drive up 785 miles of National Route 1, past endless coastal and Outback wonders, on the way to swim with whale sharks at Ningaloo.

The west coast of Australia is spectacular. Unlike the highly developed and crowded Gold Coast on the east side of the continent, settlements are few and far between, and people are scarce (save for the December-January vacation period). Sand is everywhere. Along the coast, the road crosses bright white coral sand blown inland from the beaches. Further inland, red sand derived from millions of years of weathering of iron-rich rock replaces the coral sand. In many places, there is white sand on one side of the road and red sand on the other.

Western Australia offers fantastic hiking along endless deserted beaches, spectacular seaside cliffs, and rugged sandstone gorges. Detracting from the experience are somewhat insulting signs warning of every possible "risk," from slipping and tripping, to falling off cliffs – even "water risk," a rather all-encompassing danger including getting your feet wet.

The signs fail to mention one serious risk – kangaroo risk. Rental vehicles are outfitted with bull bars. Rental agents instruct customers not to swerve for kangaroos, "skippies" in the Australian language, but to hit them. Too many people die trying to avoid the cute fuzzy critters. Not to mention, they are suicidal, waiting along the side of the road until dusk, then bouncing across the road in front of vehicles, as if challenging the drivers to a skippy version of the game of Chicken. Kangaroo carnage is everywhere. I counted an average of three carcasses per mile along one 20-mile section of road.

No "risk" left unannounced.

We worked out of the tiny Outback towns of Moranbah and Emerald. The mining boom was in full force when I arrived in February. The towns could not keep up with housing for the influx of workers. The solution was for fly-in, fly-out schedules. Even workers at fast-food restaurants and motel maids worked rotating schedules, flying back-and-forth from Brisbane. Our routine was to work for two weeks, then drive to the airport, wait for the plane to arrive from Brisbane, meet our replacement as he disembarked, hand him the truck keys, and take his seat on the flight back to Brisbane.

The mining boom was about to be followed by a bust. It didn't help that the Nanny State was a tad on the socialistic side, decidedly green, and short-sighted. The government cast greedy eyes on the big mining companies and their profits, which were indeed high during the boom cycle, primarily driven by China's insatiable demand for iron and coal to make cheap, inferior-quality steel.

Despite having some of the best coal deposits in the world, the Australian mines were losing their competitiveness in the world market due largely to high labor costs and inefficiencies caused by over-the-top safety regulations. Ignoring warnings from the mining industry that boom times do not last, in July, the government passed the 2012 Minerals Resource Rent Tax, aka the Resource Super Profits Tax, imposing a 30% tax on mining profits. Shortly afterward, China drastically cut its imports of raw materials, causing a rapid decline in the price of iron and coal.

The net result of those inefficiencies combined with the tax and lower prices was a rapid bust in the mining cycle. New projects, such as the ones we were working on, were canceled. Less profitable mines closed. The remaining mines struggled to make a profit. In a classic case of killing the goose that was laying golden eggs, the tax boomeranged. Instead of raising the $22.5 billion that was anticipated, a measly $200 million was collected. It took years and billions of dollars for the mining industry and its laid-off workers to recover from the economic damage caused by the stupid, greedy politicians.

Sadly, I was a victim of the downturn. My four-year contract was terminated after a mere eight months. But all was not lost. During my short stint in Australia, I got to see a lot of the Land Down Under and was able to check several items off my bucket list, including snorkeling the Great Barrier reef, swimming with whale sharks, trekking with Komodo dragons in Indonesia, and skiing a New Zealand volcano in August. Not bad.

Clockwise: Kalbarri Cliffs, W. Australia (cliff risk, coastal risk); Skippy and Joey (collision risk); whale shark (snorkeling risk, swimming risk, being-swallowed risk), Ningaloo; hanging out with dragons (being eaten risk), Rinca Island, Indonesia.

SECTION IV

ASIA

Index to Chapters

Chapter 13

NORTH OF THE GREAT WALL

The greatest happiness is to vanquish your enemies; to chase them before you, to rob them of their wealth, to see those dear to them bathed in tears, to clasp to your bosom their wives and daughters.

Chinggis Khaan

In 2007, Kal Energy decided Mongolia was just as good a place to explore for coal and minerals as was Indonesia. Abundant mineral and coal deposits in underexplored areas provided the attraction in both countries. The main difference was the steaming jungle was replaced by frozen steppes. The crooks and scam artists spoke a different language: otherwise, they were identical. I spent the next four years bouncing between Indonesia and Mongolia, looking for coal acquisition opportunities. Why not? They are neighboring countries, aren't they? I was based in Jakarta and traveled to Mongolia when not busy in the jungle. My usual travel schedule had me departing Jakarta for Kuala Lumpur, Malaysia, at the end of the workday. From there, I caught the redeye Malaysia Airlines flight to Beijing. You may be familiar with the flight number, MH 370 – the flight that disappeared over the Indian Ocean and was never found. I am sure I flew many times with the same crazy pilot, fortunately, before he finalized his diabolical plans.

In July 2007, I made my first trip to Mongolia, flying from Manila to Seoul, then on to Ulaanbaatar (UB), the capital of Mongolia and the only city of any size in the country. I was eager to see the fabled land of Genghis Khaan, who, with his Mongol horde, conquered most of the world in the 12th and 13th centuries. As we landed at the airport and taxied down the runway, I looked out the window at a sign on the terminal that read "Chinggis Khaan International Airport." Was I in the wrong place? Did Genghis fall out of favor and get replaced by some relative? Nope. The ruthless leader's proper name is Chinggis Khaan, Genghis being a bastardized western version of his name. First lesson learned,

I picked up my bags and was met by Batuka, a college student serving as the company's translator. We walked across the parking lot to a waiting Toyota Land Cruiser. I threw my bags in the back seat. Noticing that the steering wheel was on the right side, I commented, "I didn't know that Mongolians drove on the left, like in England."

"We don't," replied Batuka. "We drive on the right, like in the U.S."

"But the steering wheel is on the right."

"Yes, there is no standard here. Half the cars have the steering wheel on the left; the others have it on the right."

Wonderful, I thought, *that should make for some exciting times trying to pass big trucks on the open highway.* Second lesson learned.

Mongolia is a vast landlocked land of sweeping steppes, sand-blasted deserts, forested hills, and high snow-capped mountains wedged between its communist neighbors, China and Russia, who have vied for control of the country for hundreds of years. Russia controlled the country for 70 years before Mongolia gained true independence as a democratic Parliamentarian Republic in 1992. In 2007, the young democracy was still learning how to operate effectively. Any policy could change overnight on a whim of the legislators, as it did in 2006 when Parliament slapped a 68% royalty on the production of gold and copper – by far the highest tax in the world. That stupid law stopped both exploration and mining dead in their tracks. Part of our reason for being there in 2007 was the hope that we would be well-positioned when the law was repealed. That was inevitable. Just the timing was unknown.

Back in the glory days of the USSR, the Russians conducted aggressive programs exploring their satellite countries for mineral and energy resources to feed the mother country. Mongolia was no exception. When the Soviet Union fell apart in 1990, the Russians left Mongolia overnight, abandoning everything from offices and schools to jets sitting on runways. Fortunately, also left behind was a treasure trove of archives documenting all the exploration work the Soviets conducted, stored in the offices of the Mongolian Department of Minerals in Ulaanbaatar.

The problem for us was that the information was all in Russian. Batuka, our young translator, was fluent in Mongolian and English, but not Russian. The older generations spoke Russian, but the younger people shunned it. We hired an elderly translator, who was relatively fluent in Mongolian, Russian, and English, to translate some documents. That almost worked. The geological terminology, essentially a fourth language, totally confused him. The solution was to find one of the few old geologists left in the country who had worked for the Russians and could read Russian geological literature and translate it to English.

While reviewing old reports, we were introduced to Altan, the head of the Minerals Department in Ulaanbaatar. Altan put us in touch with a company that held a license on a coking coal prospect called Bayanjargalan. Based on the available data, the property looked like a good one. Kal Energy hired my friend, Attila Kovago, to visit the property and assess the coal potential. Attila was impressed. We decided to pursue a deal.

The Bayanjargalan prospect derives its name from the nearest town, Bayanjargalan, which lies about 100 miles southeast of Ulaanbaatar. I accompanied Mark Stinson, the project manager, and Martin, one of Kal Energy's Directors, on a visit to the property. These two make quite a Mutt and Jeff combination. Martin is a prim-and-proper London financial-type known for his pinstripe suits, who wouldn't break 120 pounds dripping wet. Mark, a Kiwi and a former star rugby prop, is a heavy equipment operator, standing 6'3" and weighing in at 280. His shoulders span at least two Martin widths, and his hands are like hams. Mutt and Jeff, and I left the Bayangol Hotel in Ulaanbaatar and followed the property owners eastward across town through stop-and-go traffic. Ulaanbaatar has some of the worst traffic in the world, in part because the drivers have neither training nor discipline. Lane markers are considered a form of artwork. Aggressive drivers take over any lane they can by force and intimidation, snarling traffic.

Once we hit the highway, our guides sped away, leaving us to fend for ourselves. Fortunately, Mark had been to the site a couple of weeks before. Unfortunately, he doesn't have a particularly good memory. After two-and-a-half hours of driving over crappy dirt trails, we finally came to the town of Bayanjargalan. Mark's recollection ended there. I knew the project was about 10 miles northeast of the town. I also learned from viewing satellite images that the coal occurs in Permian sediments overlying older granites. The sediments are folded into a series of giant W's readily visible from space, like billboards advertising Wendy's hamburgers to hungry aliens wanting to know the answer to the famous question, "Where's the beef?"

Downtown Bayanjargalan consists of several old run-down concrete buildings and a few rather weather-beaten wooden ones nestled in the bottom of a gully. The town is wedged between knobby hills of granite and looks like something out of a John Wayne movie. The Duke would be at home here. In the center of town, just a stone's throw from the gas pumps, stands an over-sized statue of a horse. Mongolian cowboys revere their steeds just as much as John Wayne or Roy Rogers did, although they prefer honoring them with concrete statues rather than stuffing them and putting them on display like trophy elk.

After half an hour of wandering around in the general direction of our destination, we spied a lone ger on a ridge top. Altan and the license owners were waiting for us there. We greeted the crew, then inspected some trenches cut into the coal seam. Along the trend of the trenches lay a series of small holes with accompanying small piles of coal. These were the original discovery holes. Our partners do not get credit for finding the coal. That honor goes to marmots, you know – those furry groundhog-like critters. They dug burrows through the dirt into the coal and threw it out onto their spoil piles. Attila was so impressed with the ability of the marmots to find coal he decided they were better than geologists. After his first visit to the property, he handed Mark and me a list that read, "Rodney M., Nigel M., Roderick M., etc." When we asked what the list was for, Attila informed us it was the company's roster of "diggers" – marmots. He was going to hire the whole Marmot family to explore and dig up the coal.

After inspecting the trenches, we were ushered into the ger for a traditional Mongolian lunch. Cheese, bread, *aaruul* (curds), and slabs of beef with equal amounts of glistening white fat were laid out on a table. In a blue bowl was a pile of small bones with dark meat intermingled with blobs of disgusting semi-transparent fat. Out of the corner of the pot stuck a scraggly tail. Oh no, it was a marmot's tail. One of the diggers ended up in the pot! Mongolians consider marmot a delicacy. I can verify it is an acquired taste. I picked out a meaty thighbone and did my best to avoid the slimy fat pervading the pot, much to the dismay of our hosts. The best part, so they say, is the fat. They claim it is good for you since it contains no cholesterol. Right.

Mongolians have a clever traditional way of hunting marmots. The little buggers are exceedingly wary and dive into their burrows at the slightest sign of danger. Their downfall is that they are also very curious. The traditional hunt begins with the hunter donning a white suit with floppy bunny ears. With a rifle slung over his shoulder, he scurries around in a crouch in an attempt to get closer to the marmot's burrow. Once he is in rifle range, he waves a white yak tail to get the marmot's attention. Curiosity kills more than cats. The marmot stands up to check out this odd critter, and "Bang!" he gets it right between the eyes. Marmot hunting is illegal now. Being an environmentally conscious type, I assumed the marmots had been over-hunted and were now being protected. Not the case. Hunting is banned because the dirty little devils carry the Black Plague, that insidious disease that wiped out half of Europe back in the Dark Ages. Marmot hunters claim they can tell plague-bearing marmots by their behavior and don't shoot the sick ones. Sure, those would be the ones scratching their fleas.

I was only halfway through my leg of marmot when the top came off the bottle of Chinggis vodka. I usually dread the vodka toasts, especially so early in the day. Today I welcomed them to purge the taste of plague from my palate and soften the loss of one of our dedicated diggers. The toasts ensued, this time in a ceremonial silver cup. Silver has a special significance when drinking Mongolian vodka. We were told that Chinggis Khaan used to let his drink run down to the silver ring on his finger to see if the vodka had been poisoned. Apparently, the designer poison of the 12th century tarnished silver – and fast. I wasn't too concerned about our vodka being poisoned. I was more worried about alcohol poisoning in general.

Travels Without Charlie – The Natives Ate Him

Clockwise l to r: Attila, head of the Diggers clan; Rodney surveying his territory from his digging; Oh no! I recognize that tail. Is that you, Rodney?

We polished off about three-quarters of the bottle before we finished dining and stumbled out of the ger. Following lunch, the plan was to look at some more trenches but not until after target practice. Isn't that what you always do after drinking heavily and consuming plague-ridden rodents? Jagalsaikhan, an ex-military type who always dresses in full camo, opened the back of his Land Cruiser and pulled out an arsenal the Taliban would die for. Mark took the AK-47 while Martin was issued a thirty-aught-six. Jagalsaikhan was playing with an antique Russian sniper rifle. Altan proudly handed me a shiny black pistol. Handguns are illegal in Mongolia, so this was an especially prized possession. My ignorance of firearms must have shown because Altan soon grabbed the butt of the pistol and removed the clip. We all felt better after that.

After the hillside across from camp was filled with lead, nearly all of which missed the intended target, we piled into two Land Cruisers and drove over the hill to inspect some more trenches. After a few minutes, I caught something out of the corner of my eye. A fox was ambling across the slope below us. I made the mistake of saying, "Oh look, a fox!" The geological investigation ended abruptly as the inebriated hunters ran asses over elbows and jumped into the two Land Cruisers. With loaded rifles hanging out the windows, the chase was on.

With Jagalsaikhan at the wheel, the lead Land Cruiser careened down the hillside in pursuit of the fox. I rode in the following vehicle. Rifles hung out of the passenger windows, and shots were fired wildly in the ever-changing general direction of the fox as the Land Cruisers bounced over the rolling grassy terrain. The fox zigged and zagged, with the vehicles following its every move, nearly colliding more than once. Sanity finally prevailed. We backed off in fear of being caught in a crossfire. We followed behind at a safe distance until the fox outwitted the hunters and disappeared into a narrow gully.

I thought the craziness was over, but as we crested the hill on the other side of the gully, the hunters spied a small herd of saiga antelope on the ridge to the west. The hunt was back on, with a different quarry. We chased the poor antelope for half an hour, occasionally stopping for the hunters to take a few shots at the herd. Puffs of dust rose where the bullets hit. The antelope had little to worry about.

The sun was setting in the west as the antelope hunt came to an unsuccessful end. We had been driving aimlessly across the hills and gullies for nearly an hour and had no idea where we were. I once again used geo-navigating to find our way back. We were in the Permian sandstones. That meant we needed to go further west to meet the granite contact, then turn south. We headed south by west and soon saw Bayanjargalan in the distance.

The site visit with the bosses, which ended up being more of a hunting expedition, didn't answer many of my technical questions. I noted some drill holes near the trenches. When I inquired about them, the property owners told us the holes were old, and they did not have the data for them. That seemed odd. In true Mongolian form, our partners-to-be were playing games. We received only partial data before signing an agreement. We were told the rest of the data would come after signing and paying. That did not sound like a good idea to me. The company signed a lease agreement anyway.

A clause in the agreement allowed a short due diligence period in which we could drill some core holes to verify the coal seams before making a rather large payment to finalize the deal. Attila and I scrambled to contract a drilling company, schedule geophysical logging of the holes, arrange for assaying samples, and obtain permits for shipping the samples to a lab outside the country (easier said than done).

Drilling got underway in mid-October and went smoothly, except when the diesel froze solid in the fuel lines when a blast of cold air from Siberia rolled through on the second night. Sadly, but not unexpectedly, the drilling encountered only thin coal seams with no commercial value. When confronted with our displeasure, our partners magically found the drilling data they claimed they did not have – little or no coal in the holes. Without question, our skanky Mongolian partners were hoping we would not be able to complete the drilling during the due diligence period and we would have to make the next payment. The company dropped the property like a hot potato. Fortunately, Kal Energy made only a small upfront payment. The lying Mongols did not get the $1 million check they expected.

Altan and Attila at the drill rig on a frigid October night; Attila measuring coal – all one foot of it.

With the Bayanjargalan prospect dead and buried, my attention turned to finding a replacement – one that actually had coal. Over the next four years, I reviewed information on countless coal prospects, some of it translated from Russian, some of it in undecipherable Cyrillic, but with maps and cross-sections I could understand. I made numerous trips to the countryside to visit properties, most of which turned into wild goose chases. Fortunately, the chase was always interesting.

In early May 2010, I left the Indonesian jungle behind and flew to Ulaanbaatar to embark on a 10-day journey across the Gobi Desert to evaluate a dozen mining licenses held by Belgravia Mining, a small Mongolian company. Suvvda, a young Mongolian lady working for Belgravia, who spoke some English, was my guide.

With Sanzai at the wheel, we traveled east of UB on a decent paved road, which turned to a dirt track about 30 miles east of town. The following nine days of travel were mostly over barely discernable dirt tracks across the steppes and desert, or were cross-country travel using dead reckoning to reach our destinations. We visited a couple of coal properties in the steppes southeast of UB, then turned south and entered the Gobi Desert at Khuut. From there, we began cutting a large arc across the southern Gobi, culminating at the town of Altai, some 750 miles to the west.

Our routes across Mongolia.

Between UB and Altai, we passed through only one town of significance, Dalanzagrad, which had an airport, schools, stores, and accommodations. My room at the Dalanzagrad Hotel was on the 5th floor of a four-story building – not because the builders couldn't count, because "four" in Chinese means "death," and a lot of Chinese tourists stay in Dalanzagrad on their way to visit the nearby Flaming Cliffs. We stayed in tiny soums (villages) the rest of the time, sharing a single room in vacant government buildings, school dormitories, or in gers – as guests of the local herders.

Our route largely followed broad east-west trending valleys. The barren flats of the desert were occasionally broken by badlands of eroded red sandstone, making for a welcome change in scenery. We crossed several minor mountain ranges on rough 4x4 trails. Sadly, conversation was limited, at least from my perspective. Suvvda and the driver carried on lively discussions in Mongolian but rarely bothered to translate anything to English, making for hours of boring travel.

Signs of civilization were almost nonexistent, save for the occasional ger set back against rocky hills in an attempt to break the fierce winter winds. There were no road signs, billboards (thankfully), truck stops, or rest areas for the entire trip. We passed very few vehicles. Roaming herds of camels and horses appeared as we approached soums. Herders on horseback or riding dirt bikes were our only human contacts. We always stopped to chat with them and to get directions to the nearest soum.

Somewhere south of Dalanzagrad, we encountered a special Gobi treat. In front of us we could see a yellow-brown cloud heading our way. An intense blast of wind rocked the vehicle. We were soon enveloped in a whirly dervish of dust and sand, reducing visibility to zero. We were forced to stop and wait for the sandstorm to pass, thankful we were safe inside a modern vehicle. Being outside in such a choking, stinging tempest would not be fun.

The winter of 2009-2010 was a particularly harsh one in Mongolia. Weeks of -50° F temperatures with relentless winds took a heavy toll on livestock. Piles of dead sheep and goats greeted us along the side of the road outside of each soum. It was a sad reminder of the harsh environment in which the rugged descendants of Chinggis Khaan and their life-supporting herds manage to survive – or don't.

Clockwise from upper left: typical section of highway across the Gobi Desert; 4x4 road across mountains (with carcass of camel that didn't make it through the winter); Gobi girl at the family ger; Suvvda and the author in the Gobi Desert (Biger Mountains in the background).

    Dusk was approaching as we finished our visit to the Zavsar prospect in the far southern reaches of the Gobi. We had at least another three-hour drive to reach the nearest soum where we might find accommodations for the night. Off to the west, we spied a few camels clustering around a well. We drove toward the camels. A herd of goats appeared in front of a couple gers backed up against a rocky hill. We drove to the gers and met the occupants, a middle-aged lady, her husband, and their daughter. They invited us for conversation over bowls of the house specialty, fermented camel's milk, which the lady poured from a big wooden butter churner. It looked and tasted like lumpy sourdough starter with a zesty fizz and a bit of a zippy kick. In typical Mongolian hospitality, after a dinner of *bouz* and several more bowls of kick-ass fermented camel's milk, we were invited to spend the night there. The six of us slept lined up like dominoes on the floor of the ger.

    In the morning, we watched the herder's daughter collect fleece from the goats. Mongolia is famed for its high-quality cashmere wool. The best wool comes from Zaala Jinst white goats, which live in the cold arid climate of the Gobi. Instead of shearing the goats, the herders collect the fleece by hand combing, which causes less stress on the animals. The goats seemed to enjoy being relieved of their heavy winter coat by the combing, much as dogs do.

Clockwise from upper left: Gobi camel herd; ger camp – our home for the night; serving up fermented camel's milk; cashmere goat waiting to be combed.

We worked our way across the Gobi, visiting Belgravia's properties along the way, ending in the town of Altai. There we enjoyed two amenities we had missed for the past seven days – internet access and a motel room with a hot shower.

The trip back was on slightly better roads, with some graded or even short paved sections between Altai and Bayanhongor. Our destination for the last night of the trip was Arvaikheer, the next soum to the east of Bayanhongor. We had a late start and got confused about which trail to take from Bayanhongor, causing delays. As a result, we reached the foot of the Hangayn mountains late in the afternoon. Arvaikheer lay on the other side of a high pass along a faint 4x4 trail. It started snowing halfway up the ascent. The storm intensified as we climbed to the pass. Snow accumulated rapidly, obliterating the trail. Darkness fell shortly before we made the summit in a raging blizzard with near-zero visibility. The road disappeared under a foot of fresh snow. With no idea of where the road was, two of us walked ahead of the Land Cruiser with flashlights, kicking the snow to find the trail. The vehicle followed once we established the path. It was slow going. We finally worked our way down the east side of the pass to where the snow decreased, and we could see the trail.

Feeling for the jeep trail over the pass.

We arrived in Arvikheer at midnight, exhausted but relieved that the nerve-wracking trip was over. Darkness enveloped the town; the storm caused a power outage. Sanzai knocked on the door of the only motel in town and woke the proprietor. We scored the last available room and celebrated with a flashlight-lit late dinner of salted fish, Altai Gobi beer, and Chingiss vodka.

The next day we drove back to UB on snow-packed roads blasted by high winds, creating low visibility from blowing and drifting snow. Massive trucks hauling goods south to China burst out of whiteouts like giant gray ghosts, scaring the hell out of the driver and his passengers.

Three of the properties I evaluated actually had good coal potential. I recommended acquiring them. Kumai Energy's management pursued the acquisitions only to learn Belgravia did not actually own the properties. Imagine that.

In between trips to the field, I hung out in an apartment in downtown UB that the company rented. The building was brand new, but you would not know it. It was already falling apart. My apartment was on the 12th floor, overlooking Peace Street, the main drag in town, with a good view of the timber-clad mountains to the north. I always took the stairs to my apartment, partly for exercise but mostly to avoid being trapped in the totally unreliable elevator. I heard people screaming and pounding on the elevator doors in hopes of escaping them too many times to make the mistake of riding that contraption. The water in the sink was the color of root beer. How brand-new pipes could be that rusted is beyond me (but maybe they were not new, just newly installed). The apartment had no heat. I was glad to head south to Indonesia in November before the place turned to ice.

The following summer, Kumai Energy was offered the opportunity to buy a large brown coal deposit in eastern Mongolia. The property was owned by a company headed by a geologist and geophysicist, both of whom previously worked on the project for the Russians. I made arrangements to visit the property. The plan was to leave UB at 10 AM to make the nine-hour drive to Choibalsan before late at night. In classic Mongolian form, the owners were not ready to leave until 3:00 PM. We drove east of town and stopped on a low pass in the hills overlooking UB. A large *ovoo*, a ceremonial pile of rocks with a wooden pole in the center wrapped in blue silk ribbons and flags, was on the left side of the road. The Mongols got out of the vehicle, grabbed a couple of loose rocks, and circled the *ovoo* three times in a clockwise direction, adding the stones to the pile. Tradition has it doing so will protect you on your journey.

We could use some protection.

Before we got back on the highway, I felt a tap on my left shoulder. The geologist handed me a glass half full of vodka. *Oh no, here we go again.* I was asked whether I wanted water or juice for a chaser. I chose the juice since it was in an opaque bottle – something that would come in very handy later.

The vodka flowed freely throughout the trip. I applied sleight-of-hand to survive the ordeal. I downed the juice, leaving me with an empty secret weapon. Joining each toast is obligatory. I took a drink, held it in my mouth for a while, then spit most of the vodka into the empty juice bottle while pretending to take a drink of the chaser. When we made pit stops, I slyly emptied the juice bottle. Despite that effort, I was feeling no pain by the halfway mark of the trip.

The road turned from pavement to graded gravel, then to dirt. The further we drove, the less visible were tracks in the hard soil. Eventually, we lost all signs of the trail and wandered around aimlessly in the darkness, not even sure which direction we were heading. We stopped to look for some bearings. I was confident I could orient us by the stars and guide us on our way eastward. I looked up in search of the Little Dipper and the North Star. The high altitude, dry, clean air, and lack of unnatural light sources make the Mongolian steppes the perfect place to stargaze. However, I was unable to identify any constellations, possibly due to the density of the stars. It didn't help that I was seeing two of each of them, and looking up made my head spin.

We drove further, occasionally stopping to look and listen. Eventually, we heard a dog barking off to our right. We drove in that direction and came across a lone ger. The driver went in to get directions. We were back on the trail shortly.

We arrived at the hotel in Choibalsan a little after midnight. My only thought was to hit the sack as soon as possible. I had barely opened my pack to retrieve my toothbrush when there was a knock on the door. The geologist appeared with a water glass half full of vodka. He invited me to join the guys for a nightcap. I told him I had some work to do before going to the property in the morning and would drink the nightcap while I did the work. The door had barely closed behind him when the vodka went straight down the drain.

The geophysicist's parents lived in Choibalsan. A breakfast feast was awaiting us at their ger. Bread, *aaruul*, *boortsog* (traditional cookies), and bowls of *suute tsai* (milk tea) were laid out on a table in the middle of their large ger, the walls of which was lined with ornate rugs. In the center of the table was a bottle of Bolor vodka. *Yup, barely eight hours since the last toast, and here we go again.* Down the gullet went a few toasts.

Breakfast of champions – bowls of Bolor vodka; consecrating the discovery with a vodka toast.

After breakfast, we drove across town to visit a government official to get his blessing on developing a mine. His office was small and sparsely furnished. He sat at an old wooden desk with drawers on each side. We were invited to sit in chairs in front of the official while pleasantries were exchanged. Before long, he reached into one of the drawers and pulled out – you guessed it – a bottle of vodka. *Oh no, here we go yet again.*

The next stop was a field visit to the coal property. Upon arrival at the discovery site, where the geologist found pieces of coal, he immediately produced a bottle of Chinggis vodka. A toast was made to the discovery. Then, a ceremonial sacrifice of vodka was tossed over the geologist's shoulder to consecrate the ground. I also secretly blessed the parched grass with my share of the booze.

Shortly after the ceremony, a teenage girl rode up to us, carrying a long stick used to herd sheep and goats, epitomizing life on the Mongolian steppes. Across the steppes, we were constantly reminded of the importance of horses to the nomad life – as steeds, sources of milk for yogurt and cheese, and for sport.

Clockwise from upper left: Mongol cowboys; teenage girl herder; milking mares; herd on the move.

By late summer 2011, the company's coffers were running low. Kumai Energy desperately needed a good coal property the company could promote to raise money. To the rescue came Lucky Strike Mining Company, which offered a supposedly huge coal deposit, Narin Kharghait, for sale. The property reportedly had 120-feet-thick coal seams and contained at least 1.1 billion tons of coal. That is a lot of coal.

Narin Kharghait lies in the far southwestern corner of Mongolia, just north of the Chinese border. Four of us: my translator, a geologist from Lucky Strike, the driver, and I loaded into a Land Cruiser for the four-day, 1,000-mile drive to the site, most of it over barely discernable trails. We stayed in motels in soums the first two nights. Northeast of the small settlement of Bayanhangor, we came to the Baidrag River, flowing from the Khangai Mountains to its terminus at saline Buuntsagaan Lake, a haven for migratory birds including pelicans and rare swan geese. I was convinced we needed a boat to cross the wide river. The driver didn't hesitate to plunge the Land Cruiser into the river and motor across the wide expanse of water, which fortunately was shallower than it looked. We had a lunch of *tsuivan* (fried mutton with noodles) at a tiny restaurant on the far side of the river before resuming our trek across the desert.

Fording the Baidrag River,

5 Star restaurant on the other side.

On the third night, we camped in the middle of a wide valley after wandering around aimlessly for more than an hour – lost in the dark with no landmarks to guide us. We set up our tents by starlight, followed by a quick dinner of rehydrated noodles and a nightcap of vodka to conclude the day's long journey. The morning presented us with a spectacular yellow sunrise and confirmation of which way was indeed east. We arrived at Narin Kharghait in the afternoon and set up our humble camp, comprising three small tents nestled in a narrow valley. We were a hundred miles from the nearest trace of civilization. It would be hard to find a place more remote than this.

Our field camp at Narin Kharghait.

Digging a trench to sample weathered coal.

The property evaluation went well. The Mongolian geologist and I found several coal seams, which we measured and sampled. However, the 120-foot-thick coal seams promoted by the owners turned out to be between three and 15 feet thick.

Back in UB, in between fielding calls from pissed-off vendors, who had not been paid, I wrote a report detailing the findings of the field investigation. Although there was a lot of coal at Narin Kharghait, there was nowhere near what the vendors claimed. Furthermore, the rugged topography and steep dips of the folded and faulted coal seams significantly reduced the amount of coal that could be mined profitably. In the report, I included a section estimating the amount of mineable coal – between one and four million tons, not 1.1 billion tons.

I sent the report off to Kumai's managers. A few days later, I received a call. The bosses liked the report. They wanted to lease the property and use the report to promote it. However, they requested I remove one section from it – the estimate of mineable coal. They wanted me to falsify the report so they could lie to investors and raise the money they desperately needed. My reply was swift and to the point: "No worries, I can remove that section. Then my name goes off the report – and yours goes on it."

After dealing with skanky, lying Mongolian and Indonesian crooks for the past four years, it appeared I was now dealing with skanky lying Australian scammers. That was the last straw. I packed my bags and caught a flight to Jakarta. Back at the office, I conferred with my partner, Anton, who was experiencing similar problems and was planning to quit. Seeing the bold writing on the wall, I booked a flight back to the U.S., where I composed a letter of resignation.

I was done with both the Mongolian crooks and the dang Aussies, who had learned too many tricks from Chinggis Khaan's descendants

Chapter 14

ONE DAY OF MONGOLIA

....and life goes on.

Arguably the most famous and most copied Mongolian painting is a turn-of-the-century mural by B. Sharav, titled, "One Day of Mongolia", which depicts a typical day in rural Mongolia. The painting features nomads moving and setting up gers; people celebrating and drinking too much; herders herding sheep, goats, and horses; hunters hunting antelope, archers arching; wrestlers wrestling; riders racing horses; and camels transporting goods. My favorite knock-off features a somewhat perverted Mongolian version of "Where's Waldo?" If you look closely, you will see, amongst the hordes of people and herds of animals: horses horsing around, rams ramming ewes, humpty-backed camels a-humping, and even a half-naked Mongolian couple frolicking in the grass of an oasis. What the heck? It is just One Day in Mongolia. And life must go on.

This story features the events of one day in modern Mongolia, specifically Saturday, October 6, 2007. The day's adventure shares a number of activities with the famous painting, minus the good weather and the procreation fun – dang it!

The day starts before dawn when the alarm goes off at an unwelcome 5:30 AM – not that 5:30 is normally such an ungodly hour to rise, but given the conditions, it is most unwelcome. It had been a rough night. The bed at the Tansag Hotel was hard as a rock; there were no sheets or blankets, just a thin bedspread. The night had been unseasonably cold – and rainy, *very* rainy. The wind forced the rain through the gaps in the poorly sealed window of the old Soviet-built cement-block building. Frigid water dripped down onto the bed, adding insult to already cold injury. I drag myself out of bed, find the light switch and fumble around to put on multiple layers of clothes in anticipation of a cold, wet day in the field. We plan to travel about 90 miles northeast of Baruun-Urt to look at some gold prospects, return to Baruun-Urt, then drive 370 miles back to Ulaanbaatar, a 10-hour drive under good conditions. It is an ambitious plan at best, more like an insane plan under current conditions.

The lack of sleep makes for a rough start, but it is causing only minor pain compared to the effects of the previous evening's debauchery. Four of us made the trip from Ulaanbaatar in north-central Mongolia to the provincial capital of Baruun-Urt in the eastern part of the country. With us are Tuga, the young Mongolian driver behind the wheel of his Toyota Land Cruiser; Batuka, a student at the Mongolia International University who serves as our interpreter and a constant source of entertainment; Aawa, our guide; and Yours truly.

After last night's dinner and very many greetings of and by Aawa, who knows everyone in town, we visited the local disco to take in some nightlife. The bar sits above one of the two restaurants in town. It consists of a row of booths on each side, a small bar at the far end, and a dance floor in the middle. Disco music was blaring, and a few people were dancing as we slipped into the corner booth down at the end by the bar. Before I could say, "I'll have an Altan Gobi," – not bad Mongolian brew – Aawa caught the bartender's eye. A bottle of vodka and four glasses were plunked down on the table. *Oh dear, here we go.* The Russians, who occupied Mongolia for nearly 70 years, trained the Mongols well in the art of vodka drinking. Rule #1: you don't order vodka by the drink – you buy a full bottle. Rule #2: you don't drink vodka by the shot – it is served in tall glasses. Rule #3: make lots of toasts to ensure that everyone gets equally blotto. The toasts began: a toast to the success of our trip, a toast to Mongolia, one for good business relations, and so on. I have been through this before. After a morning of feeling like death warmed over following my first evening out with our Mongolian partners, I developed a survival plan. Part one is to grab the inside corner position of the booth. As in any battle scenario, the lay of the land needs to be to your advantage. Part two is to make full use of your weaponry. The Mongols like to drink their vodka with a chaser, usually a tall glass of water. The first toast is a mandatory down-the-gullet chug. After that, you must drink a little at each toast, then sip at will, as long as you don't fall behind your comrades' all-too-fast consumption rate. I took a sip with the toast. Then, at the first diversion, I poured the rest of the glass into the water glass or onto the floor (hence the strategic position in the corner of the booth). I hid my drinking indiscretion by keeping my hand wrapped around the empty vodka glass. At the next toast, I lifted the glass high and downed the "rest' of the glass. Dirty pool, I know, but this was truly a survival situation.

Aawa, who measures 5' 10'' – in all directions – is famous in these parts. He is the champion wrestler of the province, and wrestling is one of the Mongolian people's passions. Traditional Mongolian wrestling is rather bizarre. Two big brutes dressed in blue speedos, red halter-tops, and cowboy boots grab each other by the shorts and try to throw each other onto the mat to win the match. Lacking a throw, a wrestler can win by giving his opponent such an enormous wedgie that he calls "uncle" (which sounds more like *argggg* in Mongolian) in a newly acquired high-pitched voice. I think Aawa's specialty is the Atomic Wedgie, which has won him many a match while giving his opponent a chance for a second career singing as a soprano on Mongolian Idol. Wherever we go, Aawa is greeted as a celebrity. This gives the rest of our party a degree of respect, like a reverse form of guilt by association (of which I am much more familiar).

Traditional Mongolian wrestlers trying to give each other wedgies.

I am almost dressed and not feeling much like braving the cold outdoors when, *blink*, out go the lights. Peering out the rain-drenched window, I see the entire town is black. Divine Intervention! This is a no-brainer. I am back in the sack in an instant, snuggling in my inadequate bed cover and trying to avoid the wet spot along the window side of the bed. We didn't gas up last night. We are not going anywhere until the power comes back on. The wind and rain batter the window, and the loose tin on the roof flaps loudly, but I manage to doze off after a few minutes.

About a quarter past six, there is a knock on the door. Batuka tells me we need to get going. It is now light. I finish dressing and walk down the hall to Batuka and Tuga's quarters. Aawa is there with them, looking out the window at the rain splashing on huge puddles across the street.

"We have a long day planned and need to get going now," Batuka says.

"Just one problem," I interject.

"What's that?"

"We didn't get gas last night, and the pumps run on electricity."

"Shit!"

Yep, our plans for an early start are stuffed. Aawa is already curled up on his bed, feeling the ill effects of last night even more than I, the sneaky Yank. We decide to get a few more winks, then Aawa will look for a generator to get one of the gas pumps running.

At 9:00, Batuka comes by to tell me the gas station has a backup generator running, but it has only a little gas. We have to fill up quickly before the generator shuts down. In the short drive to the station, I wonder why they don't pump a little gas to fuel the generator. But hey, this is Mongolia. Who am I to tell them how to run their country? Nobody asked me anyway. We fill up in a shallow lake with a couple of gas pumps floating in the center and finally head out of town. Aawa, the guide, is at the wheel, and Tuga, the driver, is crashed in the back seat, saving his driving energy for the long trip home and trying to recover from his hangover.

A soggy, powerless Baruun-Urt early in the morning.

The road leading north is an unimproved dirt track. Every low spot is now a mud hole. Brown water and mud splash over the windshield as we charge ahead. A little further on, we realize we spent the night at the south edge of a massive cold front. We are now driving straight into it. The rain is turning to sleet and snow. Mud holes turn to frozen ponds, and snow is on the ground. The wind is fierce. A herd of horses is visible off to the left, butts into the wind with tails down to conserve body heat.

As we progress northward, we leave the plains behind and enter a land of gently rolling hills. It is snowing harder. The ground turns to a solid blanket of white. Soon, we are in several inches of snow. The wind is blowing hard, and snowdrifts are starting to build. I question whether we should continue but keep my thoughts to myself. After being cooped up in the smog and congestion of Ulaanbaatar for ten days attending meetings and making business arrangements, I was anxious to get out into the countryside of Mongolia. We had been having gorgeous fall weather – bright blue skies and warm temperatures. Only the low sun angle foretold of impending winter. I often joked the weather would be great until I went to the field; then it would turn to shit. Had I only known how right I would be!

The snow comes harder, but we push on. The road is now an indistinct linear depression in the fresh snow. We come across a sedan just off the road. Aawa pulls alongside and honks the horn. Three Mongol heads pop up. These guys were heading to Baruun-Urt from some small village up north when they ran off the road and got stuck. We pile out of the Land Cruiser and try to push the car out, but to no avail. Tuga gets out a cable and pulls the car loose. A quick, "*Bayarlaa,*" and the travelers are on the way again. No Mongol worth his salt would abandon another in such brutal conditions.

The guys are chatting along in rather animated conversations in Mongol. Occasionally Batuka translates a portion so that I am not entirely left out. Despite my concerted efforts, I have learned very little of the language. It is a bitch. For starters, they use a bastardized version of the Cyrillic or Russian alphabet, a bizarre mixture of letters that, I am convinced, was developed by and for dyslexics. Twenty-six letters were not enough for these people. They had to add an extra 11. There are several standard letters, a bunch of backward ones, some Greek letters, and a few characters that appear to have stepped out of the bar in Star Wars. And to make things worse, half of the "normal" letters mean something else (H is N, R is G, etc.). Add to that Mongolian pronunciation, and you have a near-impossible challenge. The language is full of hissy-swishy, shhh, tshhh, chhhh, and shhhchhh slurs. It often sounds like something spoken by drunks with marbles in their mouths. I was almost fluent after the second bottle of vodka last night.

As we continue our journey, the wind and snow intensify. Aawa knows the area well, but there are no landmarks to guide him in this land of white. Even in good conditions, forays like this often turn into wild goose chases since we never have detailed topo maps (they are government secrets) or even directions to where we are going. I do have approximate GPS coordinates of our destination. I attempt to provide some general guidance. Soon we leave the road and travel cross-country – no big deal since the road was not much, to begin with, and we can't find it under the snow, anyway.

Yesterday, we traveled east from Ulaanbaatar along a road that was barely faint tracks in the grass and which constantly bifurcated and later reunited (everyone in Mongolia has a better idea of where the trail should go). The green valleys and tree-covered mountains of north-central Mongolia slowly gave way to broad valleys with a few low hills, then eventually to nearly flat plains. This is the Steppe country of Mongolia. It is reminiscent of the plains of eastern Montana or Wyoming, but with sparse vegetation and totally lacking fences. Only short grasses and a few low bushes grow here. Despite the paucity of what looks like vegetation, this is excellent grazing land due to the high nutritional value of the forage.

Aawa drives like a madman across the rolling terrain. Snowdrifts explode, and their debris avalanches down the windshield, blinding us as we crash through them. After about an hour, the faint form of a lone ger appears on a ridge to the west; we head toward it. As we approach the ger, the family's guard dog, a big husky-shepherd mix, is barking threateningly and running around. A woman in traditional Mongolian dress pokes her head out the door and invites Aawa in. I get out to take some pictures, and I call the dog over. He is a frisky young fellow and just wants some attention. We play tag for a few minutes. Aawa emerges from the ger, confident that he finally knows where we are. We plunge northeast down the ridge.

Gers in the steppes along the "highway" to Baruun Urt – pre-storm.

Now we are bouncing along in increasingly deep snow. The wind is howling; visibility is near zero. The storm is not letting up. I am getting a little nervous. I know better than to be out in these conditions, especially since we are not really prepared for it. Batuka tells me he brought his sleeping bag along. I quip, "Great, except there are four of us, and it will get a bit tight in there." Aawa would more than fill it by himself. He laughs, surely not knowing what it would really be like to be stuck out here in a raging blizzard. Toyota Land Cruisers are not in the same fuel efficiency category as the Prius, especially when in 4WD all day. Our travel plans stretch the fuel capacity of the vehicle. If we get stuck here overnight, we will undoubtedly run out of gas, just occasionally idling the engine to stay warm. My mind wanders. I recall the last time the gales of November came early. The Edmund Fitzgerald, a load of iron ore, and 29 good men went straight to the bottom of Lake Superior.

Half an hour later, the ghostly outline of a couple of gers appears on a windswept ridge to the left. We make a beeline for it. Once again, Aawa goes in to get directions. We resume our insane quest. I am wondering what the heck I am going to do if we do reach our destination. There is too much snow on the ground to see any rocks. But, what the hell, we have come this far; we may as well complete the mission. At least I can say I was there.

After a while, Aawa spots something familiar to the west. We take a sharp left, launch across a gully and head up a hill. Big snowdrifts block our way. Aawa tries to steer around them. In these whiteout conditions, it is impossible to tell where the drifts are. The inevitable happens – we get stuck in a snowdrift that comes up level with the hood. After a few minutes of spinning the wheels, Toyota engineering comes to the rescue; Aawa pushes the Differential Lock button, and we slowly back out of the drift. We go further up the hill until the Land Cruiser comes to a halt. Aawa points to something off to the right. A thin dark line is visible through the blowing snow, about 50 yards away. It is the dump of a trench that was dug earlier this year. There is rumored to be gold in them thar rocks.

I grab my rock hammer, button up tight, and head for the trench. The snow is knee-deep. That is until I break through the wind-pack and sink up to my waist. I wallow for a few steps, then pull myself up onto the snow and walk on my knees the rest of the way to the trench, making sure I don't break through the supporting crust. Aawa follows, but he is having a hard time. Weighing in at 250-some pounds, Aawa cannot stay on top of the snow. He struggles in waist-deep drifts. I reach the trench, now filled to the top with snow. On top of the dump, I break a few frozen rocks. The wind is whipping snow in my face; my eyes are tearing, and I really can't tell what the heck I am seeing. I collect a few rocks and head back to the Land Cruiser. Mission accomplished – I think.

We visit a second trench 200 yards below the first one, then, we decide to retreat to the last gers we visited. Aawa is disoriented; our tracks are already obliterated. We are going in the wrong direction, so I take a compass bearing. We swing around and head in the right general direction. I now have verses of an old Jim Reeves country-and-western ballad, *The Blizzard*, in my head. As the song goes, a cowboy is riding through a snowstorm to reach the safety and warmth of his girlfriend Mary Anne's place, which is seven miles away. The going is tough, but he and his lame horse, Dan, push on because it is only five more miles to Mary Anne's, then only three more miles to Mary Anne's. Eventually, exhausted and hypothermic, he tells Dan that it is OK to stop and have a rest since they are only 100 yards from Mary Anne's.

They found the cowboy and Dan the next morning – just a hundred yards from Mary Anne's.
I try to convince myself that we will not become the subjects of a Mongolian version of the song:
    Yes, they found them there on the Steppes, their hands frozen to the Land Cruiser
    They were just a hundred yards from Aawa's place.

Now, out of desperation, we use the GPS track function to find our way back. We are in a complete whiteout and would be totally lost if left to our own devices. Modern technology pays off. After 20 minutes of tense blind travel, a ger finally emerges out of the confusing white maze where land and sky are as one.

Carts used to transport gers and a not-so-happy horse.

We exit our 21st-century conveyance and step back hundreds of years in time. Gers, round tents made of felt, date back prior to Marco Polo's time. They have remained essentially unchanged over the centuries, a testament to how perfectly adapted they are to the nomadic lifestyle and the harsh climate of the Mongolian Steppes, where temperatures can range from 100° F in the summer to -50° F in the winter. We enter the ger, making sure not to step on the threshold, as this is believed to contain the spirit of the house. It would be a great offense to tread on that. Once inside, we are greeted by the residents of the ger, an elderly couple and their two-year-old granddaughter. In the center of the ger is a small earthen stove sunken into the floor, which is used both for warmth and cooking. A metal box filled with dried dung, nature's greenest fuel (in more ways than one), sits in front of the stove. Between the two is a bowl of murky water containing a large bone with bits of old dried meat hanging from it.

The ger is sparsely furnished. On one side is a bed. Ornately decorated wooden boxes and chests, used as seats and to store goods during migrations, line the other wall and the back of the ger. The only changes in gers since the time of Chinggis Khaan are wooden doors, replacing the original felt ones, and linoleum on the floor instead of furs. The whole affair can be dismantled and packed up in a couple of hours. It can be hauled off on the backs of camels or in carts resembling gypsy wagons drawn by horses, yaks, or camels.

While the men talk, the lady of the ger sets about preparing a late lunch for her distinguished guests. It is a tradition in Mongolian nomadic society that all visitors are welcome to both food and shelter. In fact, travelers are welcome to make themselves at home even when the residents are not there. Food is always left out for such visitors. Perhaps this explains why my Mongol companions were not overly concerned about the possibility of being stranded in the middle of nowhere. The only other people I am aware of who have offered such hospitality to strangers were the Highland Scots. But that changed in 1692, when the Campbells, staying as guests of the Chief of the McDonalds of Glencoe, slaughtered 37 members of the McDonald clan while they slept. Apparently, the Mongolians have not had such a blemished history. Nor have they been big on clan warfare ever since Chinggis Khaan united them as the infamous Mongol Horde and went about conquering most of the world.

A large metal bowl resembling a wok sits on the stove. In it is rather dirty water. My first thought is that this soup will be the death of me. Thankfully, our hostess uses the water to finish cleaning the pot and discards the dirty water – dodged a bullet on that one. Into the pot goes a combination of unpasteurized milk and water. The lady throws in a large square piece of cloth and begins stirring the mixture.

Batuka, who is sitting next to me on a small wooden box, asks, "What was that she threw in?"

"Looks like a sock to me," I answered. "Maybe she is multi-tasking and doing her laundry at the same time as making soup."

Batuka asks our hostess what was in the sock. She answers, "*suute tsai.*" She is making traditional Mongolian milk tea. When the tea is ready, we are offered a bowl full of bread, homemade butter (*tsotsgiin tos*), and an unidentified thick off-white crust. I ask what the mystery crust is. I am informed it is *aaruul* or curds.

I have wondered what curds were ever since kindergarten when we read that Little Miss Muffet sat on her tuffet eating them. So, I give the curds a try. Not bad, a sour milk flavor, but quite edible for dried curdled milk. The homemade butter is also not bad – on the bread, that is. A large chunk of it also goes into the milk tea, floating in my bowl like a slimy rotating yellow island. Nomads need a lot of calories in the winter. This is one way they get them. Our hostess is now mixing flour and water to make dough. She kneads it on a board sitting on the floor of the ger. She keeps the fire burning by periodically adding pieces of dung, using tongs to pick up the dung. But more than once, I see her grab a piece of cow shit by hand, then go right back to kneading the dough. This should be especially tasty bread.

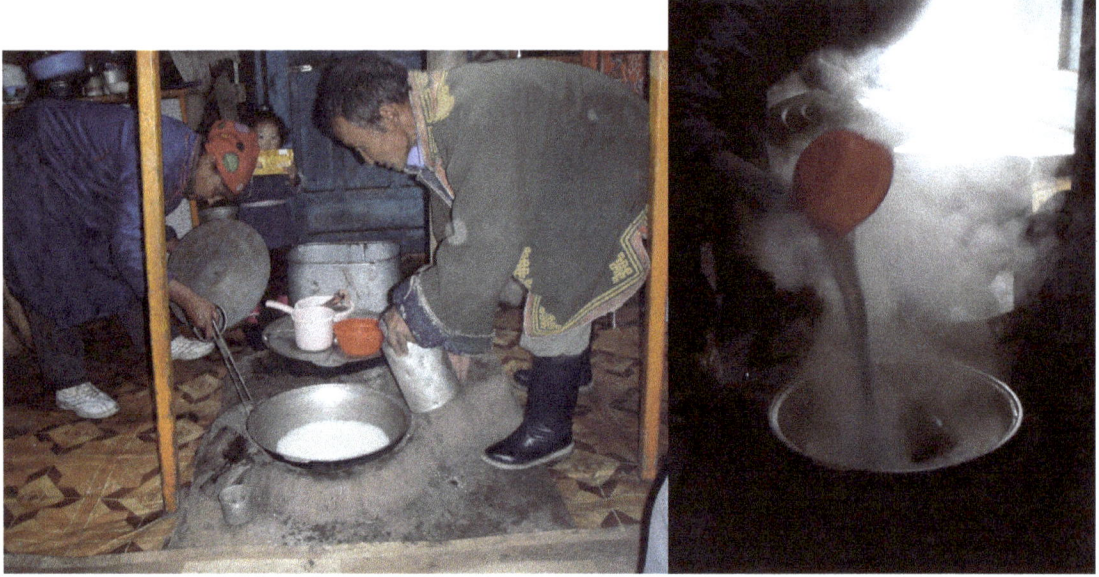

Stoking the fire with dried dung;   making milk tea.

The shy little girl quickly becomes the center of attention. We learn that her parents are off herding their livestock in some remote corner of the steppe, so she spends nearly all the time with her grandparents.

Aawa goes out to the Land Cruiser and comes back with a package of cookies to give the girl. She is a happy camper now. We ask if there are other children around for her to play with. Grandma says, "No." They tried to enroll her in a pre-school in town, but she fought and scratched the other children. Sadly, it looks like she will grow up without many childhood social skills. She will, however, know the nomadic ways.

It is surprisingly warm in the ger. The walls are wood-lattice panels arranged in a circle. Poles are lashed to the top of the wall panels and leaned toward the center in teepee style, converging on the central vent hole at the top to form the upper portion of the ger. Layers of thick felt are wrapped around the skeleton. The felt is excellent insulation, and the round shape is incredibly stable in high winds. In fact, we forget that we came here to escape a blizzard. The only evidence of the storm is light snow sifting in through the vent hole and drifting down onto the hearth. The little girl looks up and yells something. Grandma laughs and says she is telling the snow, "Stop that."

The lady of the ger is now cutting the dough into thin strips. I finally realize she is making noodles, not bread. We are having homemade noodle soup. Into the pot go the noodles, a handful of vegetables, and (oh no) the big bone with meat scraps that had been soaking in yucky water. We are served the soup in short order. A small bowl filled with a brown substance is passed around. It is crushed dried grass, cow food. The local Mongols are proud of their grass, claiming this variety grows only in Suhbaatar Province and produces the tastiest beef in the country. What is good for the cows must be good for their predators – in goes a pinch of grass. The soup is delicious; I even go back for seconds. After lunch, we say our farewell. Our gracious hostess gives each of us a handful of dried curds as we exit the ger. Truer hospitality I have rarely experienced. Why is it that those who have the least give the most?

Kneading the dough. *Happy camper.*

Aawa & granddad in the ger.

Saying farewell.

Outside, the storm rages on. Aawa is determined that we visit the other prospects on our way back south. I would prefer to just deadhead it back, but I concede since the other areas are not too far off of our return path – if we can find them. We travel cross-country. The windshield is fogged over, making it nearly impossible to see from my side. Aawa has a tiny hole that he is looking through. We crash through drifts and occasionally cross unseen gullies way too fast, launching the Toyota in the air. I worry we could hit a boulder and take out the differential, but large rocks seem to be scarce around these parts. Gullies are more common. We hit one hard, and the front end of the Toyota plows into the far side, leaving the left side of the bumper bent and dangling. Tuga is not happy but isn't really upset either. It's just another day in Mongolia, and shit happens. Besides, he knows who will end up paying for the damage. After an hour of this, we somehow find the next prospect. A few rocks dug out of a now buried trench poke through the snow. Once again, I get out and break a few token rocks. The rest of the party stays in the warmth of the Land Cruiser. Getting any useful information is out of the question. I go through the motions just for effect and bring back a few prize pieces of rock.

Off we go again, heading blindly into the blizzard, following a compass bearing cross-country to the next prospect. As we head south, the terrain flattens a little and the snow decreases. We can finally see a little. Through the snowy haze, we spy a group of large animals – bigger than cows or horses. As we get closer, I realize they are camels. It seems odd to see critters customarily associated with burning desert sands out in the snow, but they are well adapted to these conditions. Topping a rounded hill, we spot the mast of a drill rig. This is our next destination, a neighboring molybdenum prospect, being drilled by a Chinese company. Not far from the rig is a ger where the drillers are staying. Wisely, they are hanging out in the ger and not drilling today. After a quick visit with the drillers, Aawa drives over to the drill. I inspect their frozen core. A few pirated samples later, we are on our way home.

But wouldn't you know, by now, we would be low on gas. We can't make it back to Baruun Urt. We take a straight shot south to intersect the road leading to the outpost of Sukhbaatar, which lies north and east of Baruun Urt, where we can get gas. It is about 25 miles to the village, most of which will be cross-country travel. Aawa puts the pedal to the metal. We fly across hill and dale. Crossing a low ridge, we surprise a small herd of antelope in a swale. They take off; we race to intercept them to take pictures, then let them go on their way. The small herds out this way are just remnants of the great herds of thousands of saiga antelope that once roamed the steppes before hunting and overgrazing took their toll.

We have traveled cross-country for a long time and need to intersect the "road" to Sukhbaatar. We are on a course to cross it, but the faint tracks may be hopelessly buried in snow. If we overshoot it, we could end up too far south and might run out of gas before reaching our pit stop. I am on pins and needles at this point, having no desire to become a human Popsicle out on the Mongolian steppes tonight. The usual erratic banging from hitting clumps of weeds and gopher holes is interrupted by two sharp close-spaced bumps. I signal the driver to slow down and turn back. Sure enough, those bumps are the nearly invisible two-track to Sukhbaatar. We take a turn to the east and are on our way back to civilization.

Dusk is falling as we finally approach Sukhbaatar. The village is not much more than a ger encampment, with wooden fences surrounding the compound and encircling individual gers. The fences are there to break the fierce winds of the steppes. We pull up to a ger, and Aawa goes in to ask for directions to the gas station. A few gers down the trail and three more to the left, we arrive at the home of the gas station attendant. He comes out with a crank in his hand. He pumps by hand and dispenses gas the old-fashioned way. Winners Corner this is not but it is full service.

We complete the final 20 miles of crappy two-track trail to Baruun-Urt without incident. Arriving at the Tansag Hotel at 8 PM, we realize our next challenge. We don't have reservations, there are several cars in front of the hotel, and there are only six rooms. We may end up sleeping in the car after all. Fortunately, they have one room left. I take that room; Batuka and Tuga make a frantic search of town and score the last available room. Aawa goes to stay with friends. Yet another bullet is dodged.

Sunday is another day, a very different one. Clear blue skies greet us. We decide to make a quick trip to the last two prospects, near Sukhbaatar, in the low-snow zone. What a difference to see what is in front of us! I am optimistic we can actually accomplish something today. That unbridled enthusiasm comes from *inside* the Land Cruiser. After a fair amount of wandering and reconnoitering, we find the first prospect. I jump out and am almost blown over by the wind. Crap! The wind chill is well below zero. I struggle against the biting wind and manage to break a few frozen rocks loose from their winter coffins. This is definitely not fun. I finish my evaluation in record time. Then, we head south to the next area, passing small herds of camels and horses that somehow seem unphased by the bitter conditions. And they should. This is just an introduction to the brutal 50-below weather that will follow. An even quicker visit to the remaining prospect finds us on the road back to Baruun-Urt.

Aawa trying to figure out where we are – and where we are going.

Frosty humpty-backed camels.

We have a quick brunch at the hotel restaurant and tell Aawa's ever-present fans tales of our adventure. Across from us, a group of locals is having breakfast. Although it is only 10:30 AM, a bottle of vodka is being delivered to their table. Looks like just another Sunday in Mongolia.

The first hundred miles of the return trip are on a two-track dirt road. It is now a maze of deep mud holes and half-melted snowdrifts. Slow going. After at least five repeats, my Mongolian companions finally tire of the two cassette tapes in Tuga's collection, both eclectic mixtures of top 40, disco, house, Mongolian pop, and an occasional Russian song. Now they are singing Mongolian folk songs. Not bad singing, actually, and they all seem to know the words. They want me to sing. No way! I can't carry a tune in a bucket with a lid on it. I finally agree to lip-synch a song. Back in goes the tape, and I am "singing" to the Global Deejays' disco remix of "Sounds of San Francisco," one of the songs we danced to at the disco late Friday night. All goes well for a few verses until the electronic mix kicks in, and I totally lost. We all laugh. Entertainment comes cheap out on the steppes.

I look forward to reaching the paved road to make better time and arrive in Ulaanbaatar at a reasonable hour. We finally hit the pavement. I forgot one thing. Mongolia has few snowplows, or at least not enough to plow this far out in the countryside. It is now after dark. The road is a maze of frozen ruts. There is one meandering path ("lane" does not quite apply) through the maze. Travel outside of that is bone-jarring bouncing over frozen mashed potatoes. A game of Mongolian Chicken ensues. As vehicles approach from opposing directions, each one stays in the main rut until the last second, then one eventually yields and hits the frozen ruts. We win more than our share of Chicken fights. We finally make it back to the Khan Palace Hotel in Ulaanbaatar just before midnight. I drag myself off to my room, thankful for my warm bed and a safe conclusion to this Mongolian adventure.

Postscript

A week later, over a dinner of traditional Mongolian fare, including such delicacies as ox tongue, lamb's tail, horse meat, and *nermel* (vodka made from fermented yogurt), we learn that nine people in eastern Mongolia perished in that storm. The unexpected blizzard caught many herders out in the open on foot with no protection from the fierce winds. Some became disoriented in the blizzard and did not make it back to the safety of their gers. Had it not been for good old Toyota reliability, satellite technology, and perhaps some guidance from above, the toll could have easily been 13. In America, a storm with such consequences would make headlines: "Nine Dead as Blizzard Sweeps Plains." But here in the land of the rugged descendants of Chinggis Khaan, it was just One Day of Mongolia. And life goes on.

Chapter 15

WHERE REINDEER FLY
Now, Dasher! Now, Dancer! Now, Prancer and Vixen,
On, Chingiss! On, Tului! On, Kublai! and Jochi!

If you think Santa Claus has a lock on sleighs pulled by reindeer or on the flying reindeer market, you are mistaken. In the far northern reaches of Mongolia, along the Russian border, live the Tsaatan or Dukha people. The Tsaatan, like their Scandinavian counterparts in Norway, are nomadic reindeer herders. They move from camp to camp to graze their reindeer in the high taiga during the summer months. In the winter, the Tsaatan people retreat to lower camps or cabins near the settlement of Tsagaan Nuur, where the reindeer forage for lichen and moss under the snow.

Rather than returning to the U.S. for my scheduled break in July 2011, I elected to stay in Mongolia and play tourist. My older son, Eric, who had just graduated from college, flew to Ulaanbaatar to join me on a Mongolian expedition.

Eric arrived on June 28, suffering from jet lag and culture shock. I showed him around town the next day, visiting local attractions. We toured the Chojin Lama Buddhist Temple, one of the few to survive the Great Purge of 1937 when the communists destroyed about 6,000 temples and killed more than 20,000 Buddhist lamas. The temple changed my view of Buddhism as a peaceful religion. Murals on the upper walls portrayed the gruesome side of Buddhism, featuring severed heads, limbs, and castrated bodies hanging on a rope; and a caravan of naked men on hands and knees carrying heavy loads while being whipped by two black masters. Yikes! After lunch, we took in an outdoor concert at Sukhbaatar Square. The highlight was when a group of 1,000 fiddlers played the *morin khuur*, the traditional Mongolian fiddle.

By the time cocktail hour was approaching, we were seriously parched. We walked east along Peace Street, past the 12-foot-tall bronze statue of Lenin – the badass Russian guy – in front of the Ulaanbaatar Hotel. Our destination was the Tse Bar, a few blocks down Peace Street, something I just had to show Eric. The front of the bar is unassuming, just a big blue "Tse" sign over the door against a gray background. Inside is a different story. Swastikas are plastered all over the place, including on the tables. Posters of Nazi soldiers doing *sieg heil*s to *der Fuhrer* line the walls. This is not a joke. It is serious shit. The Tse Bar is the meeting place and watering hole of the Mongolian Nazi Party. If being wedged between the Communist countries of Russia and China were not enough, do the Mongols also have to embrace the Nazis?

The next day, we launched our Mongolian adventure with a short trip to the Gobi Desert. We flew from Ulaanbaatar to Dalanzagrad, then traveled by Russian 4x4 van to a ger camp west of town. We toured the Flaming Cliffs, where Roy Chapman Andrews and his team discovered the first dinosaur eggs during the Central Asiatic Expedition of 1922. We also visited Yolo Am (Vulture Valley) along the north flank of the Altai mountains in search of elusive ibex and snow leopards but didn't see any. Eric got to experience Mongolian nomadic culture – sleeping in a traditional ger, riding a horribly uncomfortable camel, and drinking *airag* – fermented mare's milk. I was proud of the young lad for not hurling his cookies, unlike his dad did on his first attempt.

As bad luck would have it, the CEO and CFO of Kumai Energy, the Australian company I was working for, decided the middle of my planned work break would be the perfect time to come up from Jakarta for a meeting to review the exploration program. I needed something for Eric to do while I was stuck in the meeting. What could be better for a young Yankee in Mongolia than to spend the Fourth of July at a shooting range – a fully stocked military one, at that! Eric went off with my translator to spend the day firing an AK-47 and a rocket launcher at the shooting range east of town, owned by a retired Army General. It was one of the highlights of his trip. He certainly had more fun than I did in the meetings.

Rambo of Mongolia.

Eric firing the rocket launcher.

The Gobi trip was just a prelude to the primary adventure, a visit to the Tsaatan people. We took the 1½-hour MIAT Airlines flight from Ulaanbaatar to Moron. There, we met our guide, Gerelchuluun, who told us to call him Jerry, giving our twisted tongues a needed break. After obtaining supplies in Moron's chaotic open-air market, we began a two-day journey north in an old Russian 4x4 van, a roomy but overly stiff and terribly uncomfortable conveyance. Shocks and springs must be optional in Russia. The van took us to the outpost of Ulaan Uul, seven hours of bumpy road away, where we spent the night at the van driver's house. Ulaan Uul reminded me of frontier towns in Alaska – dirt streets lined by rustic log cabins, but with the added Mongolian touch of hitching posts in front of the buildings.

After a quick breakfast of milk tea and biscuits, we climbed back into the van for another six grueling hours of driving over rough 4x4 trails. We banged up and down steep hills; forded rocky, crystal-clear mountain streams; crossed rivers on rickety wooden bridges that looked like roller-coaster tracks; and got lost when the "road" occasionally disappeared altogether. The pain of travel was partly alleviated by the incredible beauty of the Darhat Valley, a broad expanse of rolling grassy parks ringed by evergreen forests, reminiscent of Montana but with nary a fence to be seen. The rugged peaks of the Khoridol Saridag Mountains rose sharply along the east side of the valley, framing a near-perfect picture. Widely scattered gers and cabins were the only signs of humanity we passed.

We reached the tiny village of Renchinlhumbe in mid-afternoon. Finding the trail to our destination, a ranch about another two hours northwest of town, proved to be a challenge. After asking directions from several herders we met along the way, we finally arrived at the ranch shortly after 5 PM. We exited the old van feeling like we had been stuffed into a tumble dryer for the better part of a day.

Darhat Valley and Khoridol Saridag mountains.

Drunken bridge near Renchinlhumbe.

We spent the night in a cozy ger at the ranch, resting for the challenge of the following day – a seven-hour horseback ride up and over a high mountain pass to where the Tsaatan were camped, or maybe were no longer camped. The reindeer herders move their camps every few weeks to ensure the best grazing for their herds, making the camps moving targets.

In the morning, while the wrangler was busy rounding up horses to carry the riders and supplies to deliver to the Tsaatan people, Eric and I walked around the ranch, marveling at the wooly prehistoric-looking yaks that roamed the meadows – little changed from their Pleistocene ancestors. We finally saddled up around 10:00 – Eric, Jerry, I, the wrangler, and three Mongolian missionaries, "Devil dodgers," as Eric called them, who were hell-bent on converting the shamanistic Tsaatan to some new-fangled religion.

Mongolian horses are short, stocky beasts, somewhere between ponies and horses in size. They are incredibly strong and have phenomenal endurance. They are also as stubborn as mules. The Mongolian people revere their horses and depend on them for three purposes: racing (for the best few), riding (for most of them), and dinner (for the ones that don't fit racing or riding). We were about to learn where our steeds belonged (hint – not the first two categories).

We rode off through meadows of high grass and subalpine flowers, something the horses could not resist. Both Eric's horse and mine were far more interested in filling their bellies than in moving along. We were constantly pulling their heavy heads up with the reins and prodding them to keep moving. I nicknamed mine "Purina." I kept threatening I would turn him into dog food if he didn't cooperate. Eric's horse was worse. We named him "Anchor" because he was one. Eventually, the wrangler and Jerry took turns towing Anchor, or we would have never reached our destination.

Despite their propensity for being feedbags, I had to hand it to these horses for being tough and incredibly sure-footed. The tundra is underlain by permafrost. The top layer melts a little each summer, creating a layer of black slippery muck. Much of the first part of our route was slogging through the soggy melted permafrost. The horses often plunged to their knees in the muck and frequently stumbled but recovered and plowed ahead without falling. Well, with one exception. My steed, Purina, stumbled, tried to recover, but went down in the bog. I slid off his back and landed on my feet, shin-deep in mud. Purina struggled back onto his feet. I somehow managed to pull off a Roy Rogers vaulting mount from the back (not quite the feat it may sound for a horse that is only ten hands high with about four of them stuck in the muck), and we continued on our way.

Horseflies bit us, and small pestering flies buzzed around our heads and tried to fly up our nostrils. Oddly, we only encountered them when we were in full sunshine. I theorized the flies were solar-powered. I prayed for more clouds to mess with recharging their batteries.

After several hours of struggling across the lowland tundra, we started to climb higher and drier as we entered taiga forest with its stunted fir and spruce trees, many of them tilted at various angles in the unstable permafrost – "drunken forests." At one point, and quite unexpectedly, I might add, ole Purina decided he had enough of this riding stuff and just laid down in the trail, fortunately in a dry section. I landed on my feet and stood next to what I thought was a dead or dying horse. I announced, "Horse down!" for lack of better words. Despite his looks, Purina was indeed alive, just resting. After a couple of minutes, he got back up. Not wanting to be left behind and not wanting to cause the plug to go down a second time, I cautiously mounted up. We rode up the trail as if nothing happened.

Socializing with passing travelers.   *Feedbag feeding during lunch break.*

We wound our way up above timberline to a mountain pass and stopped there to rest, taking in a spectacular 360-degree mountain vista. Snowfields glistened on the Sayan Mountains to the north, across the border in Siberia, not more than five miles away. We were getting close to our destination, but our guide was not sure how close. He had not been to the Tsaatan camp in a couple of weeks. So, the camp may not be where he last saw it. In that case, we would have to go hunting for it, making for a longer ride, something my posterior portions were not looking forward to.

The first part of the trail below the pass was very steep. We dismounted and walked the horses for a while before remounting. At the bottom of the hill, we followed the stream to the east, eventually crossed it, then proceeded up a smaller tributary. Before long, we spotted a wisp of smoke rising above the valley. We breathed a sigh of relief; the camp was where it should be. Our ride was about over.

The backside of the pass - Tsaatan camp is around the corner on the right & up the valley.

Our little procession arrived in the Tsaatan camp at about 6 PM. Eric and I set up our tent near the north end of the encampment, which contained about a dozen teepees scattered across the valley. Stunted fir trees dotted the green mountain slopes surrounding the camp, located just below timberline. A few reindeer wandered about, their antlers still in velvet and sporting a patchwork of fur as they shed their thick winter coats. Others sat tethered to stakes. As we prepared to cook dinner over our camp stove, we heard shouts. We looked up the valley to see a group of Tsaatan men on horseback driving a herd of reindeer down to camp. Each morning the herd is taken to the high meadows to graze. In the evening, they are brought back to camp, where they are safe from the hungry wolves that roam the tundra. Upon arrival in camp, the Tsataan ladies separate the reindeer according to family ownership. Then they take them to their respective teepees, where they milk the does.

Our dinner preparations were progressing slowly when Jerry came by and invited us to join him for dinner in the teepee next door. We gladly accepted since it looked like it would take our tiny stove a long time to cook anything. The Tsaatan teepees are nearly identical to those of the Plains Indians. A series of long poles are set in the ground arranged in a circle, then leaned inward and joined at the top to form an inverted cone. An outer waterproof covering, made of hides sewn together or of canvas, is wrapped around the frame of poles to provide an outer waterproof cover. A few poles are leaned against the outer side to help hold the covering down. Inside the teepee, Jerry introduced us to our hosts, an elderly couple – wrinkled as prunes and sporting a few token teeth – sitting cross-legged on the dirt and grass floor of the ger with a bowl of *tsaat tse* in their hands. The man of the house wore a deel, a traditional full-length Mongolian herder's overcoat, and calf-length riding boots with holes worn through the soles. His wife wore what looked like a dirty and tattered housecoat and was barefoot.

We were offered *tsaat tse* (milk tea). A big meaty bone was passed around for us to cut off a piece as a chewy appetizer. Jerry explained the bone came from an elk a hunting party recently shot. Each of the five families living in the valley received their portion of the prize. Drying slices of elk meat hung from strings tied across the teepee frame poles. We brought our stock of sausage, cheese, and a few vegetables to add to a communal dinner cooked over the cast iron stove that sat in the middle of the teepee floor.

The Tsaatan people follow shamanism, nature worship. The religious leaders are shamans, who dress much like the medicine men of the American Plains Indians and do many similar rituals. Our hostess was a shamanka, a female shaman. Jerry translated as our hosts told stories of the history of the reindeer people, life in the taiga, and their struggles with the Mongolian government, which has been trying to force them off the land to preserve it as wilderness, where hunting is banned. We learned the Tsataan trace their roots back thousands of years, when their ancestors, members of the Tuv clan, flew to Siberia with their reindeer. Large upright slabs of granite, 4,000-year-old "deer stones" with carved images of flying reindeer, attest to the airborne migration legend. Could this be where the legend of Santa Claus and his team of flying reindeer originated?

The Tsaatan use their reindeer for the staple of their diet – dairy products: milk, cheese, and yogurt. They rarely slaughter reindeer for meat but do make use of their hides when they die. They also use reindeer hair to make thread for sewing. Antlers are carved for sale to tourists or are sold to the Chinese for traditional medicine. The Tsaatan men ride the reindeer and use them as pack animals. In the winter, the bigger males pull sleighs – once again shades of ole St. Nick.

The following morning, I exited our tent and walked up the valley about 100 yards to start the day off properly – taking a leak. Before I could finish my business, a reindeer came running up from behind me and started licking the ground at my feet. *Weird critters,* I thought to myself. On the second morning, I again walked out of camp to relieve myself. A reindeer followed me. This one was even weirder. It stood in front of me and drank straight from the faucet. That freaked me out. I later learned the reindeer are starved for salt and will get it any way they can. They have learned that people are reliable sources of salt, although by somewhat gross methods.

Feeding salt to the babies.

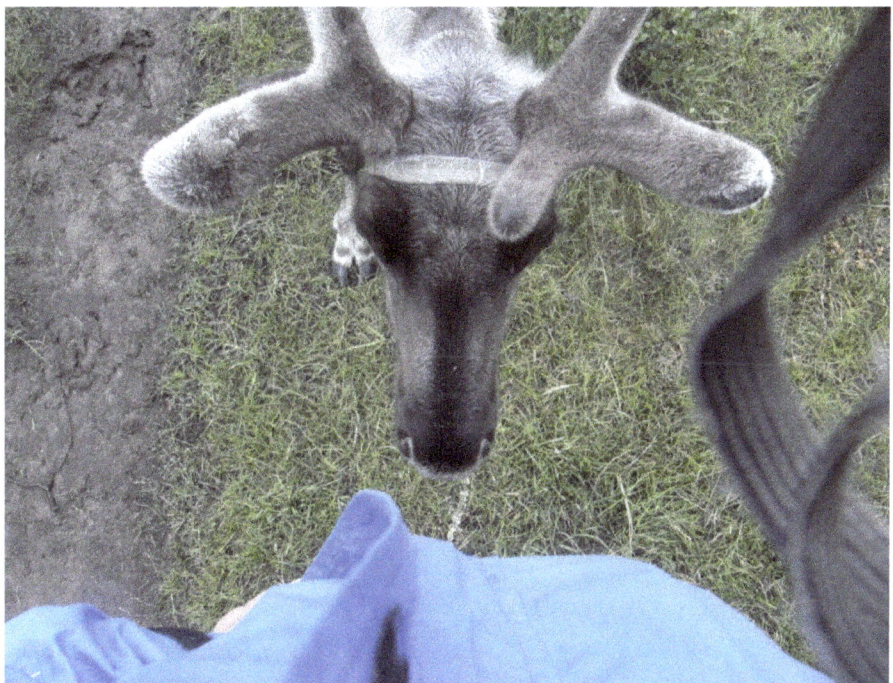
Salt straight from the source.

    A drizzling rain started. Instead of hiking up to the high country to see the grazing reindeer herd, as we had planned, we spent the day visiting with our hosts and a few other Tsataan families. One of the highlights of the soggy day was spending time with our hosts' dogs. The Tsaatan's dogs are not pets, but they are treated better than most western pets. They are essential to the survival of the reindeer people. Trained as hunting companions, the dogs track and corner game for the hunters to shoot. Our hosts had two dogs, a sleepy old fellow, who was retired, and an energetic younger one.

Eric dubbed the young dog Black Wolf because he looked like one – and was black. These are not the heavyset Malamute mixes of the Mongolian Steppes. Instead, they are sleek, long-legged, fast dogs with great endurance, not much removed from wolves in appearance or function. Eric fell in love with Black Wolf, who spent more time with us than with his masters. He would have loved to take Black Wolf home.

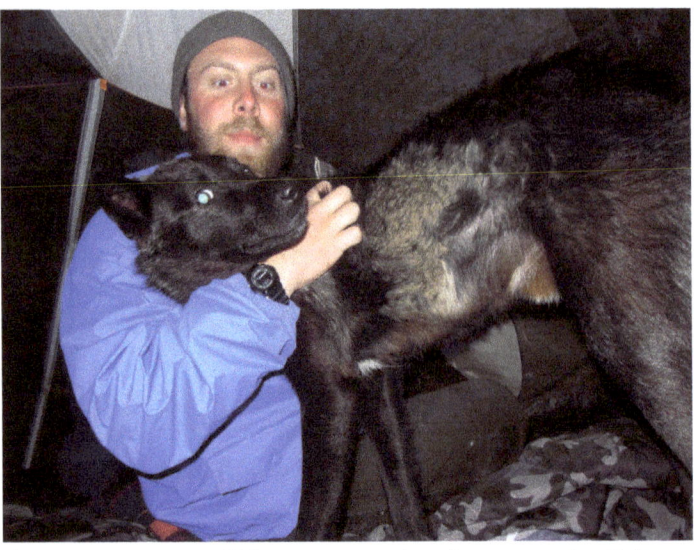

Eric disappeared around noon and was missing for a couple of hours before I started searching for him. I found him in a teepee on the far end of the encampment, talking to a couple of young gals, Rebecca and Emily, fellow Yanks. Rebecca was a biologist who was trying to organize a study on wolverines in the area. Her friend, Emily, was a photographer who amazingly knew some of Eric's friends in Santa Cruz, California. Small world, indeed! The gals heard that missionaries had arrived in camp. They purposely avoided us, thinking we were the missionaries. Eric and I got a good chuckle out of that.

When the rain stopped in the late afternoon, the camp residents quickly set up the local arts and crafts fair, featuring a variety of antler carvings laid out on a couple of blankets. We bought a few carvings to help the local economy. Over time, they have become cherished souvenirs.

Three of the four generations of Tsaatan.

We were back in the company of our Tsaatan hosts in the evening when the missionaries came into the teepee. The head Devil Dodger worked on the old lady for a while, trying to explain why modern religion was so much better than the one which guided her Tuv ancestors for thousands of years. She was not messing around – going for the gold by trying to convert the shamanka, the clan's religious leader and chief medicine lady.

The missionary was not making much progress. She turned to us and started a conversation about God and creation. I told her I preferred the shamanka's stories to those of the Bible. She finally got to spring the question she thought would trip me up.

"But someone had to create the universe. Who was that?"

I had been waiting for this moment. I unbuttoned my flannel shirt to reveal my T-shirt, which had "Pastafarian" in big yellow letters on a red banner across it, a drawing of the Flying Spaghetti Monster, and "Church of the Flying Spaghetti Monster" across the top.

"Why, the Flying Spaghetti Monster!" I proclaimed and commenced to proselytize her with the tenets of Pastafarianism, such as Friday is the holy day and is dedicated to drinking beer; the difference between Heaven and Hell is that Heaven is populated with hot babes and cold beer, while Hell has hot, stale beer and ugly women. Pastafarianism is more or less a takeoff on Rastafarianism but is much more fun. It is at least as much a real religion as geology is a real science. Go ahead, google it. The head Devil Dodger wisely gave up on converting us hopeless souls and went back to working on the pagans.

Our time with the reindeer people came to an end the next morning. Before we left, I handed Jerry a high-quality hunting knife, which I brought from the U.S., to present to our host in appreciation of his hospitality. The old man was thrilled and had tears in his eyes. Jerry told us he said the knife was the best present he ever received.

Eric, the author, and our hosts in front of their teepee (Devil Dodger in background).

We said our goodbyes and walked to the south side of the camp, where the wrangler was saddling up our horses for the long ride back to the ranch. I was pleased to see I was assigned a different horse. Perhaps the wrangler noticed my uncomplimentary comments about Purina, like, "This isn't a horse, it's a lawnmower." Eric still had old Anchor, who once again lived up to his namesake. Anchor made it halfway up the pass before he had to be towed by the wrangler. Shortly after reaching the pass, the wrangler abandoned Eric and Anchor. We were falling far behind the rest of the procession. I put a rope on Anchor and towed him. My new horse was great; he wanted to get back home and pulled like a tow truck so, we named him accordingly. The term "barn horse" applies even to horses that have never seen a barn. After passing through the boggy part of the trail and reaching firm ground for the final couple of miles of the ride, our steeds, even stubborn Anchor, suddenly gained renewed vigor and began galloping home. It was all I could do to stay in the uncomfortable worn-out Russian saddle.

The van driver met us at the ranch. Then, we began the long drive back to Ulaan Uul. Rebecca and Emily, who rode out with us, joined us as our guests for the journey since we had chartered the van – or so we thought.

    The driver took us back via a different route, along the west side of Darhat Valley, passing the village of Tsaagan Nuur. There is no bridge to cross the wide Sished River before Tsaagan Nuur. Instead, a ferry made of a raft propelled by two strong men pulling on a cable serves for the crossing. At the far side of the river, we were accosted by a group of obviously intoxicated men, the toll keepers. They were the Mongolian equivalent of the ugly troll under the bridge the Billy Goat Gruffs had to cross. A big argument ensued over the toll, apparently because we had two extra passengers. As the gals tried to negotiate their fare, the van driver suddenly got pissed off and threw Rebecca and Emily out of the van, forcing them to walk to town. Eric and I were upset, but there was no use in arguing with the driver using our nonexistent Mongolian. Jerry was not about to go there.

    The hassle over the girls put a cloud over the rest of the trip. To make matters worse, the driver stopped and picked up several people on the way back until the van was full. We didn't begrudge giving people a free ride in the middle of nowhere, but not after making room by kicking out our guests.

    We arrived in Tsaagan Nuur in time to take in the wrestling competition at the local Naadam festival. The weather was perfect for the outdoor games, sunny and warm with barely a breeze. Wrestling matches went on for most of the afternoon. The matches, featuring burly men in brightly-colored heavy-duty speedos and half a shirt on their backs, were agonizingly long events filled with more pomp and ceremony than actual wrestling. Two contestants walk around with their arms in the air, pretending to fly like eagles in what looks more like Tai Chi practice than a wrestling match. The birdmen size each other up and occasionally make a feint at actually contacting their opponent, but back off at the last second, or make contact but quickly break away. Meanwhile, the head judge, dressed in a full-length ornate deel and wearing a round hat with a high topknot, follows them around, carrying a second hat in his hand. Once the wrestlers finally engage, it usually takes a short time before one contestant throws his opponent to the ground. More dancing like an eagle follows. The winner gets the hat.

After overnighting at Ulaan Uul, we bounced our way up an even rougher 4x4 trail over a high pass in the Khoridol Saridag Mountains to reach Hatgal at the south end of Lake Hovsgol. There we took in a larger Naadam celebration, featuring the three manly events: wrestling, archery, and horseback riding.

Archery, which consists of men dressed in their finest and most colorful deels pulling back on big bows and shooting arrows across an open field, is not much of a spectator sport. It would be a lot more exciting if they were shooting Chinggis Khan-style at dummies of soldiers, better yet, at an entire fake army.

The best event of Naadam is the horse race. This is not your typical quick race around a track; it is a cross-country endurance race, usually 10-20 miles long, across rough terrain. Races are divided into age categories – from two to seven years (the horses' ages, not the riders', but there is little difference). Although the race is billed as one of the three "manly games," the jockeys are just boys, as young as five years old. It is a true test of the skill and endurance of both man (make that boy) and beast. To add further challenge and excitement, the kids ride bareback.

Late in the afternoon, Eric and I didn't want anything more to do with the butthead of a van driver. We stocked up on cheese, salami, a loaf of bread, and plenty of beer from the local store. We asked to be dropped off at Lake Hovsgol, where we could camp in peace. Hovsgol is a magnificent lake, the Mongolian equivalent of Lake Tahoe. It is enormous, measuring 85 miles long by 30 miles wide, and is up to 1,000 feet deep. Hovsgol is known for its exceptional clarity and water purity. We set up our tent in the trees a short distance back from the beach on the west shore of the lake. We had the place to ourselves until a group of Mongols, who obviously spent too much time at the Naadam beer tent, pitched camp about 50 yards from us. The Mongols grew louder and louder while we ate our dinner and quietly drank our Tiger beers. I was expecting to put up with the cacophony until the wee hours when Eric announced, "They will be quiet in half an hour." The young lad, fresh from college drinking days, was incredibly tuned to the finer points of intoxication. His prediction came true nearly to the minute. We spent the rest of the evening sitting around our camp in peace, telling stories and sharing some special father-son moments over beers.

Early in the morning, we awoke to the sound of something chomping outside our tent. Peering out the door, we saw the source of the sound, a crazy yak finishing the last of our salami and cheese that we foolishly left out overnight. Rats, that was supposed to be our lunch. Fortunately, there was a tourist camp about a mile up the road. We walked to the camp and ate lunch while we cursed that damn carnivorous yak. We burned off our lunch calories by renting canoes and paddling along the shore of the placid lake under intense blue skies.

Back in camp, we took down our tent and sat on the beach, drinking the last of our beers, watching small waves lap against the shore, and taking in the majestic scenery while we waited for the van to pick us up. Then it was back to Hovsgol to spend the night at Jerry's cabin, where we watched boring Mongolian wrestling on his TV until the wee hours of the morning.

The carnivorous yak.

After a quick breakfast, we boarded the dreaded Russian van with the moron behind the wheel to drive back to Moron to catch our flight to Ulaanbaatar. As we said our farewell to Jerry, for a tip I gave him a hunting knife similar to the one I gave our Tsaatan host. The asshole van driver didn't get shit.

Chapter 16

SOUTH OF THE GREAT WALL

Bu zuo si jiu bu hui si, No zuo no die
(If you don't do stupid things, they won't come back to bite you in the ass. But if you do, they most certainly will).   Chinese proverb

    Faint ghosts of modern glass-enshrined high-rise buildings swimming in a thick brown and yellow cloud of ozone, nitrous oxides, and particulate matter greeted me as Air China Flight 986 slowly descended through the low clouds. Welcome to Beijing and the Peoples' Republic of China.

    June 2008 – Beijing was preparing for the Summer Olympics. The smothering smog shrouding the city was soon to be banned along with half the cars, all the heavy industry within 50 miles of the city, the porn peddlers (who accosted me with pirated movies every time I walked out of the Kuntai Royal hotel, "*Girwy movie, Mista? Weally sexy ladies fo you?*"), the prostitutes (shipped off to Abu Dhabi and other foreign ports), and the stray dogs and cats (which ended up in cooking pots). The last of the smog was cleverly but only temporarily removed by cloud seeding. China was determined to make Beijing the showcase of the country for 17 glorious days in August. Then it was back to normal.

South of the Great Wall

Mr. Li, the driver for Goviex, the small exploration company for which I would be working, met me at the brand-spanking-new Beijing airport. After I checked into the hotel, Mr. Li whisked me off to have dinner with the company's President, the Chief Operating Officer, and the Chief Geologist – Jing Tu Xiao at, of all places, a Mexican restaurant. I somehow maintained a semblance of coherency during dinner, despite the deleterious effects of 18-hours of travel, including a very uncomfortable 12-hour flight and 15 hours of time change. Just as jet lag was about to reduce my brain to mush, the party moved on to the after-dinner entertainment. Mr. Li deposited us at the steps of a magnificent four-story building with a façade featuring ionic columns, surprisingly out of place for what I expected in China. The next thing I knew, I was standing naked in an elegant tiled sauna pool with three guys I met just two hours earlier, surrounded by little Chinese guys holding towels and offering services I probably didn't want (I was advised not to go for the skin scrub, which apparently left you raw and bleeding).

Later, a cute little Chinese gal, who might have weighed 80 pounds dripping wet, led me upstairs to the massage parlor. How she was able to exert 500 psi pressure on my weary muscles and bones remains a mystery to this day. Chinese deep-tissue massage is a punishing combination of pain and pleasure, which must have been invented by sadomasochists, but is immensely popular. Beijing is full of multi-story massage parlors, where families go and sometimes spend the entire night having massages and relaxing. I left feeling bruised more than relaxed from my massage, but it did help with the jet lag.

Mexican food, cute girls, massages – China is looking good so far.

Over the next six months, I would see more of rural China than most westerners see in a lifetime (and more than I would like to see again) – from steaming subtropical Guizhou in southern China to the frozen northernmost tip of Manchuria, and a lot of country in between. The initial plan was to fly down to Guiyang, the capital of Guizhou province in southern China, in two days to meet with officials of China National Gold Company (CNGC), a state-owned enterprise. Following the meetings, we planned to visit several gold properties in the Golden Triangle, just north of the Vietnam border, hoping to arrange joint venture exploration of the better prospects.

But heavy rains in the past month caused severe flooding (an annual event, it turns out), washing out roads, damaging buildings, and causing multiple deaths (252 lives lost in southern and eastern China, more than 57 in Guizhou alone). After nearly a week's delay waiting for the floodwaters to recede, Jing and I boarded a China Southern Airways flight to Guiyang, where we met a group of government geologists at the CNGC office. After a couple of hours listening to discussions in Mandarin with little translation for the token white guy, we broke for lunch. The restaurant down the road offered an intriguing selection of Chinese fare, but that was not the primary objective of lunch – drinking ample amounts of Zhujiang beer was, as one-by-one the hosts took turns toasting their guests, over and over.

Back at the office, I sat at a cubicle and continued my futile quest to interpret a series of geological maps covered in undecipherable kung fu characters. After about an hour of frustration, I got up to talk to Jing, only to find him asleep in his chair and the CNGC personnel out cold with their heads on their desks, sleeping off the effects of the liquid lunch. So, this is how China, the great competitor to the western world, operates.

Sleeping Beauty and the Seven Dwarves awoke just in time to call it a day and prepare for the next event, an important dinner with the executives of CNGC. The venue was an upscale restaurant perched above the Nanming River on the edge of town. As with all official functions in China, a spacious private room was reserved for the dinner. A giant glass lazy Susan with various unidentified dishes sat in the center of a round table. About 20 GNGC officials and staff sat around the table. As guests of honor, Jing and I sat at the head of the table (if a round table has such a thing) on either side of the chief dude, Zhou Dung Ping ("Just call me, 'Don', " although Dung was more appropriate), a big fellow with a matching beer gut. Dung poured a clear liquid into our wine glasses. I asked what it was and was told "maotai, white wine," the local delicacy. I applied my newfound Chinese drinking etiquette learned during the lunch beer fest; you don't drink on your own, only when toasting or being toasted. Lots of toasting ensued. I about gagged on the first toast. This was not fine chardonnay. Instead of the anticipated buttery oak flavors with perhaps a hint of vanilla and toffee, I got heavy-bodied brake fluid with notes of turpentine and acetone. And it was strong, much stronger than wine should be. As toast after toast went down my numbed gullet, I failed to notice that the other toasters were drinking from tiny ceramic glasses, more like thimbles. Only Dung and I were drinking from big wine glasses. After a while, things got a bit blurry. I soldiered on, not wanting to tarnish the reputation of the good ole USA drinking team.

I do not remember finishing dinner or leaving the restaurant, let alone what followed. I awoke the following morning, still fully dressed, minus shoes, to an awful stench. The floor of the Regal Hotel was not looking so regal. There lay the half-digested remains of the dinner I didn't know I ate, swimming in an odiferous sea of rejected maotai. I felt like I had been severely beaten around the head and ears by a gang of ninjas. Showering did little to return cognitive function. I stumbled downstairs to meet Jing and Zhao Cheng Hai, a young Chinese geologist, who would serve as my assistant and translator on the three-week excursion across southern China we were about to embark upon.

I greeted Jing with, "I don't remember anything after about the ninth or tenth toast last night. I hope I didn't do anything stupid."

"No, you were clear," was the response. I never really understood what he meant by that. The maotai was clear, but I suspect I was more than a little cloudy.

"What was that wine we drank last night? It kicked my ass."

Maotai, I was pleased to learn too late, is not really wine. The Chinese call it wine for lack of a better word. Maotai is the Chinese equivalent of vodka or gin, fermented and distilled sorghum with a dash of cyanide, arsenic, and strychnine thrown in for an extra kick. It is 106 proof poison. Good old Dung and I were downing glasses of it like it was water. No wonder I felt like shit.

As we were loading our suitcases and field gear into the SUV, one of the CNGC geologists carried a case of maotai to the vehicle. *Oh no, not that crap,* I thought. Zhao informed me that Dung, a true aficionado of the stuff, and I toasted each other until he finally yelled "uncle." Since I drank so much of it at the dinner and bested the champ, the guys decided I must really like it. They brought along a "sufficient supply" for the trip. *Wonderful!*

We traveled south of Guiyang to the town of Luodian, one of the places hardest hit by the torrential rains and flooding. The effects were immediately evident. The main roads were down to one lane or less in places where they had washed out or collapsed from underneath. Bridges across the rivers were gone. Giant chunks of concrete with bent rebar sticking out of them lay in the rivers – the only remnants of the bridges. Houses that once lined the rivers and side streams were wiped out, torn off their foundations. Others were ripped in half, the remainders hanging precariously on gouged-out cliffs with furniture and belongings dangling from the new grossly exaggerated open floor plans. It was a sad sight.

The June flood in Guizhou damaged more than 140,000 homes, of which about 45,000 were completely destroyed. Fifty-seven people died in the floods, and 1.3 million people were displaced. Those staggering statistics are partly the result of above-normal rain during the annual monsoon season of "normal" heavy rains and ensuing floods. The other factor is that people live everywhere in China – in floodplains, on the banks of seasonally flooded rivers, on overly steep unstable hillsides – everywhere. It is hard for normal weather, let alone natural disasters, to not have enormous human consequences.

Flood devastation – Guizhou, June 2008.

We spent the next week navigating the remains of the Guizhou road system as we attempted to visit mineral licenses in the Golden Triangle. The dirt backroads we followed to our work areas were in even worse shape than the main roads. Landslides, mudslides, and washouts obliterated the roads in places, forcing us to walk to our destinations or abort the trip altogether. Boulders and fallen trees blocked the roads in many places. We spent a lot of time clearing debris from the road. Outside of a small mountain village, we came across an elderly couple, busily working to move boulders in a mudslide blocking the road, with only a pickaxe to assist them. We offered to help. The feisty lady, thin as a rail and as wrinkled as an elephant's hide, refused, arguing maintaining that section of road was her duty. We helped move one heavy boulder so we could drive through. She insisted that cleaning up the rest of the mudslide was her job; we should butt out of her business.

Don't mess with little old ladies wielding pickaxes.

Road damage – Guizhou.

Landslides took a heavy toll on the villagers during the torrential rains. Several hillside villages were partially destroyed when the steep slopes they were built upon collapsed. Those calamities followed on the heels of the horrific Great Wenchuan earthquake of May 12, during which over 80,000 people perished; more than 7,000 poorly built or poorly located school buildings in Sichuan province collapsed; between 5,335 and 19,065 students lost their lives – the lower number being "official" government statistics versus local body counts. When the residents of one remote village learned an American geologist was in the area, one from the Foremost School of Mineral Engineering no less, they asked the government geologists if I could come to find a safe place to relocate their school. I gladly accepted. After 38 years, I would finally get to put my GE401 Engineering Geology course to use (if I remembered any of it).

We drove on rough, slippery, muddy roads for more than an hour to reach the village, perched high on a steep hillside. As we walked through the town to visit the school, we passed a house with something cooking on a spit over an open fire. I paid little attention to it at the time.

The schoolhouse was a one-room brick hut with a thatched roof. In the school was a scattering of old desks and not enough chairs for the 20-some kids, ranging from toddlers to 10-year-olds, who milled around, some sitting, some standing. Most of the students were paying attention to the teacher, who was writing on an old blackboard. The others seemed more interested in the strange white man standing at the doorway.

After visiting the school, we walked to the north end of the village, which crossed a gully descending steeply to the broad valley below. One by one, small landslides were eating their way up the gulch, right to the edge of the town. I didn't need to explain the danger to the villagers; that is why they brought me there. I discussed moving the settlement to a safer place. I was told they only cared about the school. Yes, the Commies love their children, too.

Our entourage: the CNGC crew, the village chief, a few curious locals, including a granddad with a two-year-old on his back, and I spent a couple of hours roaming the surrounding hills searching for a reasonably safe location for the school. About half a mile south of the village was a somewhat flatter (better described as less steep) spot just above the road where solid bedrock was exposed. I declared this the best and safest location for the school. I instructed the village chief to remove the crops planted above the site and plant trees instead. Deforestation is one of the major contributors to slope instability. The rural people have devastated the forests to grow crops, often on impossibly steep terraced slopes. I also suggested building water bars above the site to divert sheet wash from heavy rains away from the schoolhouse.

The villagers were elated. They would begin construction of the new school immediately. Well, almost immediately. A celebration was in order.

A feast was laid out in the village center, on a concrete patio next to a brick building, which served as the community center. On the way there, we passed the house where the barbeque spit had been. There was nothing cooking there now. Bowls of homemade maotai were served from a plastic gas can (how appropriate). I recognized most of the entrees on the table: chicken (identified by the clawed foot sticking out of the bowl), vegetable soup, beans, rice. One dish was chopped mystery meat. I asked Zhou if it was what I thought it might be. He nodded in the affirmative. Now I knew what had been cooking on the spit. There would be one less dog barking in the village that night. I opted for the soup and veggies and left the chicken feet and barbequed hound for the other guests.

Oh, no!..what happened to Charlie?

It didn't take long for word of the school relocation to spread. A week later, I was asked to help relocate another village, this one up the Beipan Jiang River from Wangmo. A landslide swept away part of the village. After a suitable new location was agreed upon, the mayor announced another celebration. Instead of holding the event in the remnants of the village or in the blue relief tents that largely replaced it, the local officials held a dinner in town. The usual toasts ensued. I was older and wiser and stuck with the thimble-sized shots of maotai rather than the super-sized ones. Following dinner, the entourage retired to a local karaoke bar for after-dinner drinks, song, and dance.

Temporary village after the flood and landslide.

I laughed as we approached the bar. The black double doors featured vertical Chinese characters on the sides and a smiling Santa Claus face on each door with "Merry Christmas" written across the old gentleman's white beard. Santa would be out of place in southern China, even it was not June. I wondered if the local patrons had any idea what Santa was all about. Odds are they liked it just because it was a sign of western civilization and capitalism, something they were happily embracing. Inside, the bar had a jukebox and a decent-size wood dance floor. We sat in plastic chairs around small tables along the side of the dance floor and listened while the local officials belted out their favorite Chinese pop songs. When they were done, coins went into the jukebox, and the dancing began. We took turns dancing with the only other people in the place, a pretty young lady wearing a green knee-length polka dot dress, whom I was told was the proprietor's wife, and her sister – casually dressed in calf-length pants and a white blouse. Since there were only two gals and six guys, I spent a lot of time quietly sitting with the Chinese officials. Lively conversation was not an option. We watched the others dance but generally were bored.

To break the boredom, I guess, or maybe out of respect for a distinguished foreigner, or for some odd local custom, the man next to me stood up, reached out his hand, and *Oh, no*, motioned to the dance floor. *What should I do? Damn, it was a slow song. Oh crap, what could I do?* My maotai-numbed mind flashed to a hilarious Rodney Carrington song, "Dancing with a Man." Rodney got pulled onto the dance floor by a babe, only to realize too late that "she" had calluses on her hands and a rather deep voice.

Unlike Rodney, I didn't *think* I was dancing with a man; I *knew* I was dancin' with a man! My dance partner was an important local official who would lose face if I refused. Besides, what happens in Wangmo, stays in Wangmo – doesn't it? Well, not if you take pictures and put them in a book, it doesn't.

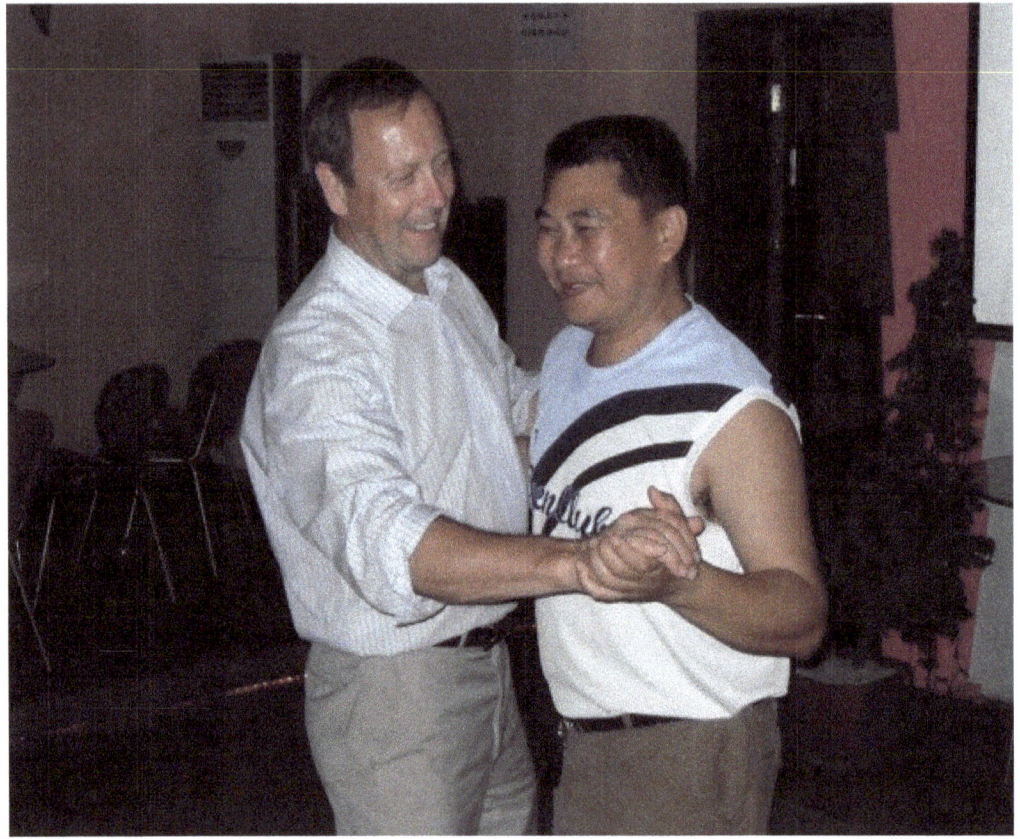
Dancing with a man.

We completed our work in the Wangmo area and moved further west to visit prospects in the mountains between Xingren and Kunming. This part of China is karst country, underlain by deeply weathered limestone, which produces caves, sinkholes, and spectacular limestone towers. Rivers disappear into vertical walls of limestone, and others issue forth from the mouths of caves like magic. The residents are the Miao people, an ethnic minority, related to the Hmong people of Vietnam and Laos. The area is mountainous and isolated, with no industry except farming rice and vegetables, and raising chickens and pigs, which wander the villages everywhere (swine flu, anyone?). This is truly ancient China, where there are few vehicles and even fewer roads. We walked long distances between villages to reach many of our destinations.

I survived the tour of southern China (partly because most of the case of maotai also survived) and returned to Beijing to review the results of our fieldwork. After I gave presentations to the managers of Goviex and the brass of China National Gold Company, my work was done for a while. I flew back to the U.S. while Goviex and CNGC negotiated terms for potential joint ventures.

Returning home after a trip to the village market, Zhongpan, Guizhou.

I returned to China in October for two months of nonstop travel investigating gold prospects in Nei Mongolia along the Mongolia border, in Gansu Province of central China, and Heilongjiang along the Siberian border.

The first mission was evaluating several mineral licenses in Nei Mongolia (Inner Mongolia). My companion was Yaote, a staff geologist for Goviex. We flew from Beijing to Xian, then on to Jiuyuguan in the northwest corner of Gansu Province. In the morning, Mr. Wu, who owned the exploration licenses, drove us north of town in his prized Jeep Wagoneer, a bit of an oddity in China. After crossing the barely recognizable ruins of the Great Wall of China, merely a hump in the road after 1,000 years of neglect and pilferage of bricks for more important building projects, we passed farms of wheat, oats, and corn along the course of the nearly dry Ruo Shui River, then turned northwest into Nei Mongolia and the southern Gobi Desert, a desolate land of bare rock and sand.

We spent the next four days in the field, mapping and sampling Mr. Wu's prospects, based out of the tiny village of Saihan Toroi at the edge of the desert. Our accommodations at the local motel were meager, to say the least. My bed was hard as a rock, made of a wood platform with a one-inch-thick "mattress," which I think was stuffed with gravel. There was no shower, but that was OK. There were no towels and no hot water. Big black plastic tanks on the motel roof are supposed to provide solar hot water. It was October, the sun was low in the sky, and the tanks were full of cold water. A communal squat toilet down the hall served for the toilet facilities.

Despite being overall low quality, the motel had one of the few decent restaurants I had the pleasure to dine at in China. Breakfast was simple – spicy noodles. Dinners were multi-plate extravaganzas. One evening, while we were dining, an obviously intoxicated Chinaman dressed in a buttoned-down high-collar black Mao suit came over to our table and started talking to Mr. Wu. He seemed very agitated, gesticulating wildly, and shouting. I took him for the chairman of the local chapter of the Communist Party. I assumed he was pissed off about something, maybe the presence of a damn capitalist in his town. He left and came back a few minutes later with his 10-year-old son. I knew I was in trouble then – he wanted to show his son what a despicable Yankee spy looked like. Maybe he wanted to spit on me (instead of the floor for a change) to show his contempt for Yanks. Much to my amazement, what he wanted was a picture of his son and him with me, the guest of honor in town. I obliged. He walked away happy as a commie clam at high tide. I knew intonation and inflection are as important as words in Mandarin, but I had no clue such aggressive tones could mean something totally different.

Mr. Wu (left) and Yoate at one of his Gobi Desert woofers.

Unfortunately, Mr. Wu did not woo us with his woofers. We flew back to Xian to start the next leg of the expedition, evaluating CNGC's properties in Gansu Province.

We caught a Hainan Airlines flight from Chengdu to Jiuhuang airport near the town of Jiuzhaijou. Stormy weather delayed the flight. We took off into heavy cloud cover. As we approached our destination, spectacular snow-capped glacier-sculpted peaks poked out of the clouds. This was definitely a different part of China than hot, humid subtropical Guizhou or the relatively flat desert of Inner Mongolia.

The runway came up quickly as the airbus dropped out of the clouds. We seemed to be coming in too fast for a typical final approach. The plane wandered all over the runway. It felt almost like it was still flying before coming to a hard-braking stop near the end of the runway, beyond which lay a 1,000-foot drop-off into Jiuzhaijou Valley. We essentially *were* still flying. Sitting on the edge of the Tibetan Plateau at 11,312 feet elevation, the Jiuhuang airport is one of the highest in the world. The air is so thin that planes cannot make normal approaches and must come in "hot" to maintain control, then hit the brakes hard to stop before the end of the runway. Taking off is equally exciting. The plane roars down the short runway and drops off the end just as it becomes fully airborne.

A CGNC driver picked us up. We drove down a series of tight switchbacks to the valley below. Long strings of bright yellow, blue, white, green, and red cloth rectangles – Tibetan prayer flags – lined the road, flapping in the breeze, while furry yaks wandered the hillsides – and the roadway. A few near collisions ensued. We were in northern Sichuan Province. The influence of Tibetan Buddhism was everywhere, including the houses, which sported upward curving tiled roofs characteristic of Buddhist architecture.

Jiuzhaijou is one of the most beautiful and unspoiled areas in China (one of the very few unspoiled ones, actually), set aside as the Jiuzhaijou Valley Scenic and Historic Interest Area in 1984. Natural attractions include pristine streams, waterfalls, and alpine lakes with 14,000-foot snow-draped peaks in the background, wonders that most Chinese people never have the opportunity to see.

Alpine splendor of Juizhaijou

Unfortunately, the natural attractions of Jiuzhaijou were not on our to-do list. We overnighted at Nanping in Jiuzhaijou Valley. The next day, we continued our journey, climbing over a 12,000-foot pass crossing the Min Shan Mountains to arrive at Wenxian in southern Gansu Province. Blue tents greeted us as we drove down the valley. The epicenter of the Great Wenchuan earthquake lay 145 miles to the southwest, but aftershocks from the magnitude-8.0 earthquake spread out for 175 miles northeast along the Longmenshan fault all the way to the Gansu border. Several plus-5.0 quakes, and a 6.0 aftershock with its epicenter only 35 miles southeast of Wenxian, rocked the area and caused significant damage. Five months after the big temblor, people were still living in tents provided by the government, either because their houses were severely damaged, or they were afraid to return to them. We felt a few strong aftershocks during our time in southern Gansu.

We spent the next ten days traveling around Wen County, visiting gold mines and prospects of China National Gold Company. This part of Gansu is very mountainous. Tiny villages occupy narrow valleys or are plastered against steep mountainsides that are terraced as far as the eye can see. Only nearly vertical cliffs seem to have escaped the farmer's hoe.

We hiked through remote villages and climbed high into the hills to visit small mines being explored by CNGC. It was like being in mining camps in the States 150 years ago. The equipment: explosives, drill steel, compressors, and generators, was hauled along steep, narrow foot trails by mules or on the backs of the miners. Two men hauled broken ore and waste out of the mines – one pulling and the other pushing a big wheelbarrow.

High country trail.

For part of our time in Wen County, our lodging was a schoolhouse in Houjiazui, a small village southeast of the town of Wenxian. "Recently former" school would be more appropriate. The earthquakes left the building cracked and unstable. The school was temporarily moved to an undamaged government building located across a large courtyard. Although the old schoolhouse was condemned for use as a school, it was deemed perfectly adequate for housing geologists.

A husband-and-wife team taught 31 children ranging in age from four to about 12 in the one-room school. The teachers lived with their 6-year-old son in a small house between the new school and the abandoned one where we stayed. We got to know them well. They were thrilled to have a genuine English-speaking person in town, apparently, the first one ever to visit. I was invited to give an English lesson to the class. That was a cool experience. Unlike the ramshackle school in Guizhou, where the kids dressed in dirty rags and the school lacked essentials such as pencils, paper, and chairs; this school was clean and organized. The children were well dressed (especially if you like Chairman-Mao outfits) and attentive. I was amazed they already knew the English alphabet and numbers. In Guizhou, I was struck by the sadness and absent look in the children's eyes just about everywhere we went. Here, the children were happy and smiling. That evening, I was invited to share dinner with the teachers in their tiny house. The highlight of the evening was learning how to eat soup the polite Chinese way – slurping loudly, instead of using a spoon the quiet western way.

31 kids, two teachers, one Yankee, and a pet dog (not for the soup pot).

After dining with the teachers and their cherished only son, I pondered the effects of China's one-child policy. The forced limitation on breeding was accomplishing the government's primary goal, capping the explosive growth of the Chinese population. It also had the side effect of slowly destroying the agrarian lifestyle of the rural Chinese. Farmers around the world have traditionally produced large families to work the family farm. In a part of the world where there is no mechanized equipment, and all labor is performed by people, horses, or water buffalo, that cannot be accomplished with only one child per family. Farmers are struggling. To make matters worse, when the children grow up, they tend to migrate to the cities, leaving the mountain towns with even fewer laborers. There are also enormous social costs. The government basically cut the branches off the family tree. Imagine a world where you have no brothers or sisters, no cousins, aunts, or uncles. A family reunion would consist of you and your parents. That is what China will look like after a generation or two of the one-child policy.

The World Health Organization estimates that 700 people per day are killed on China's roads. That hardly makes a dent in the country's burgeoning population of about 1.4 quadrillion people but is a staggering statistic given the number of vehicles on the road and the total miles driven. Some deaths can be attributed to poorly designed or poorly maintained roads, of which there is no shortage. But the real culprit is operator error. I once thought the problem was inexperience and poor training. However, after too many hours riding with trained "professional drivers" I know something else is amiss. When Chinese people get behind the wheel of a car, something snaps; there is a disconnection between the brain and the rest of the body, particularly the hands and feet that are supposed to prevent the vehicle from becoming a deadly missile. The professional driver CNGC provided when we worked in Guizhou served as a driver in the Chinese Army. He came away with all of one good habit, frequently stopping to rest and prevent fatigue. However, I suspect the main reason was to have a cigarette break since I forbade smoking in the vehicle. He routinely raced along muddy, ridiculously narrow, frighteningly twisted roads. Riding shotgun, I too often had the unnerving view of the front right wheel of the SUV coming within inches of the drop-off at the outside edge of the road as the madman sped around blind corners. If his mission was to scare the crap out of me, he succeeded admirably. Despite those suicidal efforts, the prize for driving stupidity went to our driver in Gansu. When Zhou and I were going back to Jiuzhaijou along the busy two-lane highway, following the tightly meandering headwaters of the Yellow River, the driver wanted to pass a big truck. He put the peddle to the floor as we entered the inside curve but couldn't get past the truck. Instead of backing off at the end of the curve, he continued to try passing around the blind outside curve. Once again, the SUV didn't have the power to pass. The crazy bastard continued through the next inside curve and around a *second* blind outside curve before passing the truck. I was convinced the death toll for that day would be 703.

China is a vast country, the fourth largest in the world, with lots of mountainous terrain. New cities seem to pop up overnight like mushrooms. Flying is the only way to get around the country. There are about as many Chinese airlines as there are cities to serve. I think we flew on most of them: Air China, China Southern, China Eastern, Hainan Air, Kunming Airlines, Lucky Air, Chengdu Air, and Grand China Express, to name a few. There were also a few I wish we had not flown. By far, the worst flight I have ever been on was from Xian to Tianshui on Kunpeng Airlines, when we went to visit the Dragon Mountain property near Lixian. The plane was a Bombardier CRJ 200, a 44-passenger jet, which is not a bad plane – if it is maintained, that is. This one looked good. It had a fresh coat of red-and-white paint. I think that was all that held it together. The flight was delayed for three hours, with no explanation. I assume it was to let the paint finish drying. Jing, Yaote, and I boarded the plane with one other passenger. Perhaps the Chinese knew something we didn't. The plane roared down the runway, shaking like a malaria patient with a breaking fever. I thought it would fall apart before it became airborne, but the paint did its job.

Tianshui lies in a narrow valley between high mountains. Clouds filled the valley. The plane descended; the landing gear was lowered; then the plane climbed steeply, aborting the landing. We flew on to Lanzhou, the next stop on the plane's itinerary. When we touched down, the plane again shook violently, and loud screeching came from the landing gear. It sounded and felt like the plane's wheels were roller skates with worn-out bearings, and they were about to fall off.

When the plane came to a screeching stop at the gate, our fellow passenger, whose destination was Lanzhou (the lucky bastard), exited the plane. Jing and Yaote begged the flight attendant to let us also leave the plane. We were more than willing, desperate actually, to get off that piece of dilapidated junk, forfeit our tickets, and take a slow train to Tianshui. But in China, rules are rules. There was no way we were allowed to change our flight plans. One-and-a-half hours later, we endured one more frightening takeoff followed by an equally terrifying landing at Tianshui. When we finished our work evaluating the Dragon Mountain property, we took the long, safe overnight train ride back to Xian.

Despite the fact that Mao Zedong tortured or executed millions of his own people and starved a mere 45 million peasants to death, jolly old Chairman Mao is still a revered figure in Mainland China today. One of the things he is credited with is emancipating the women of China, making them the equals of men.

As Mao put it, "Women hold up half the sky." I have seen them holding up their half: swinging sledgehammers to make little rocks out of big ones, digging boulders out of mudslides, hoeing fields, carrying heavy loads that should be reserved for mules. Yes, the great defender of women did them a wonderful service. Be careful women's libbers; you could achieve true equality someday, as have the human mules we met in Guizhou.

Hoe, hoe, hoe, it's off to work we go.

A lady's life, achieving equality (with mules); Guizhou.

Of course, Mao's men have never done their half of holding up the sky. They are too busy playing checkers and generally goofing off while the women work.

Mao's men hard at work – Wushan, September 2008

In mid-November, Jing, Yoate, and I flew to Heihei to evaluate a gold project, Zheng Guang, that an Australian company offered for sale or joint venture. Heihei lies in northeastern Heilongjiang Province (formerly Manchuria), nearly at the northernmost tip of China. The city sits on the southwest bank of the mighty Amur River (Heilongjiang or "Black Dragon River" to the Chinese), across from the Siberian town of Blagovescensk. Heihei sports an average annual temperature of just below freezing. The temperature was a smidgen below average when we arrived, hovering around 0° F. As usual, my backward schedule had us working in sweltering southernmost China in June and July and freezing northernmost China in November.

We drove 75 miles west of Heihei to the project area, passing through rolling hills of boreal forest. Along the Nen Jiang River, I noticed odd pine trees. They were turning bright yellow; many had dropped their needles. I assumed some disease was killing the trees, much like spruce budworm and pine beetle have devastated pine forests in the western U.S. The trees turned out to be perfectly healthy Siberian larch – a deciduous coniferous tree, in other words, a non-evergreen evergreen. Makes sense, doesn't it? Each fall, the needles turn yellow and fall off like the leaves of a maple tree but grow back in the spring.

Jing, Yaote, and I toured the property with our guide, then spent the night at a small dingy motel in Heibaoshan, a dirty, cold, dreary coal mining town with a population of about 10,000, where everything and everybody is covered in coal dust. The precise population varies with the all-too-often deaths in the unsafe, poorly regulated underground coal mines. China records an average of 5000 deaths annually from fires, explosions, cave-ins, and floods in its coal mines. Heilongjiang is no exception. Almost a year to the day after our visit, 108 miners were killed in an explosion at the nearby Xin Xiang coal mine. Appalling as official statistics are, they are far short of the actual number of deaths. There are many small mines, mostly illegal operations run by shady operators, who don't report deaths in their mines. At times they don't even bother to recover the bodies. They simply replace the workers and keep mining, following Chairman Mao's sympathetic philosophy: "China is such a populous nation, it is not as if we cannot do without a few people."

Dusty, smoggy Heibaoshan. *5-star, coal-dusted restaurant in Heibaoshan.*

We spent the next day inspecting drill core stacked outside in the company's storage yard. The temperature was -10° F, which made for interesting work. When we sprayed water on the core to better observe the details, the water immediately froze, coating the drill core in a thin layer of clear ice. I am a dedicated rock-licker, but there was no way I was going to lick that core. I never forgot the time my older brother talked me into licking the pole on our swing set after a Canadian cold snap plunged the temperature to below zero. Nor have I forgiven him. He laughed, but I thought I would spend the rest of the winter stuck there, unable to call for help beyond a muffled, "Hehhp, my tonnnn iss 'tuck." We alternated between inspecting the core and trips to thaw out and sip hot tea in the nearby office tent, heated by a coal-burning stove.

We returned to Heihei, where we planned to spend the night, then take a morning flight back to Beijing. In the morning, we awoke to a world of white and eerie quiet. The town and the airport were shut down by a major snowstorm. All flights were canceled. The storm lasted for two days, trapping us but providing the opportunity to explore the city. Heihei is a bustling tourist destination – for the Russians, who flock there for the cheap made-in-China goods, food, and drink (most important to the Russians).

Our hotel had a big indoor pool and a sauna. In the evening, I went to the sauna and found a group of young Russian men sweating away but carefully replacing the lost fluids with liters of beer. They noticed I didn't look exceptionally Chinese and assumed I was a fellow Russian tourist. They started talking to me in Russian. Of course, I didn't understand a single word. I sheepishly told them I was not Russian but American, not knowing what to expect. The guys were soldiers on leave. To my amazement, they were thrilled to meet a genuine Yankee. We had a great time chatting in their broken English, drinking beer, telling jokes, and sweating together. Maybe politicians should follow our lead.

On our last morning, we ate breakfast in the hotel dining room. An elaborate buffet of Chinese, Russian, and western food was laid out. On our left sat a group of thin, cute Chinese flight attendants daintily enjoying a breakfast of tea and rice porridge. Off to the right was a group of rather stout, make that rotund, middle-age Russian ladies. Plates of every kind of food – pastries, omelettes, ham, bacon, sausages, toast, pancakes – were on the table. A bottle of vodka and several liters of beer completed the feast. What a contrast!

Everywhere we went in China, the people were extremely friendly and accommodating. Big cities, such as Beijing, displayed the effects of a booming economy with increasing freedom for businessmen to be successful. Well-dressed young Chinese entrepreneurs driving expensive sports cars were a notable sign of the newfound wealth. In contrast, little change had come to the people in rural China, except the loss of many young people to the cities. The peasants had not seen many westerners and were keen on learning about the world outside China. Likewise, I was interested in learning about China and its transition from an exceedingly isolated Communist country to a slightly less isolated Communist country, but with a new degree of pseudo-capitalism thrown in. As I departed China at the end of November, returning to a country that was about to be led by a far-left-leaning socialist Senator from Illinois, I couldn't help but think China was becoming more like the United States, and the U.S. was becoming more like China daily.

China was winning.

Meeting of the Gansu chapter of the Chinese Communist Politburo.

Chapter 17

KUNG FU FOOD

When hungry, order more Chinese food.
- Fortune cookie saying

I would be remiss in writing about China without mentioning the fantastic cuisine. For the uninitiated, it is important to understand food in China is very different from the Chinese food we know in the United States. The main difference is our version of Chinese cuisine is edible, even worth eating now and then – although you will be hungry an hour later. Not the case in China. Oh sure, 5-star restaurants in big cities such as Beijing and Shanghai serve good food under reasonably sanitary conditions. I ate some excellent meals in Beijing. One of my Asian gastronomic highlights was eating Peking Duck in the famous Quanjude Roast Duck Restaurant. Our duck, #115,652,750 as my official Commemoration Card says (they have been counting on their abacus since the third year of Tongzhi, Qing Dynasty – 1864 for the rest of us), was expertly sliced and served with sweet sauce right at our table by a waiter wearing a sparkling-clean head-to-toe white chef's outfit. But get beyond the city gates, and you better pack a jar of peanut butter or plan to go hungry. Sometimes the food is not so bad if you fish around in the big communal bowl in the center of the table long enough to find some meat amongst the intestines, lungs, brains, stomach, liver, and the head and feet of whatever critter is being served.

The Chinese boast that they invented the culinary arts. There actually is some support for that claim. Over the past 5,000 years, the emperors of just about every dynasty left records of the extravagant feasts they held, replete with hundreds of dishes, including exotic items such as tiger, wolf, elephant, bear paw, and sparrow (a multitude of which is needed for an entrée). In true "if it moves, it must be eaten" style, the Zhou Dynasty (1046-256 BC) reportedly had greater than 100 kinds of animals on its menu. Try making a list of 100 animals you would like to eat. I can't get past about a dozen. But, of course, that is a list of what I would *like* to eat. After years of working in exotic locales such as Indonesia, Mongolia, and China, the list of critters I have eaten but didn't really want to eat is longer than the list of ones I wanted to eat. Then there are the animals I don't mind eating, but not those extraneous parts: tasty testicles, awful offal, or the thinly sliced bull's dick served for lunch at a Chinese restaurant in Ulaanbaatar (hey, it was on the menu; I had to try it) – way too tough and chewy for my liking.

The Chinese prefer fresh food. It is common to see the evening's specialties swimming or crawling around in tanks conspicuously displayed in the restaurants. I peered into the window of a fine dining establishment on my evening walk in Beijing and thought I was visiting the local aquarium. A variety of exotic fish and eels, several kinds of turtles, and a 6-foot-long crocodile with its jaws secured by big rubber bands swam nervously in a huge glass tank. I couldn't tell you how many people it takes to consume a crocodile, but I think asking the chef to slice off a piece of its tail for a snack would seriously piss off the beast. You better bring some friends if you are hankering for cooked croc.

It is customary to pick your entrée from the tank or have the waitress bring one to your table for your approval. At one of the better restaurants in Hohhot, we sat next to three large glass tanks containing eels and some sort of catfish. The water was disgustingly murky. One of the fish was listing badly to the right, and another one was floating belly up in the top tank. Now I know why it is a good idea to personally choose your dinner entrée.

While our dinner of yak stew was simmering in a shiny pot over a butane stove on our table in Jiuzhaijou, we watched as a cute Tibetan girl brought a large, very alive fish to a table. As she approached the diners, the fish had a change of heart and decided it didn't want to be dinner. It did a backflip and jumped out of her hands, landing unceremoniously on the dirty floor, where it flopped around like... well, a fish out of water. The diners approved of its antics. Into the pot, it went.

Chinese delicacies: honey roasted sparrows; red worms, anyone?; turtle soup to be; baked giant mealworms.

We ate at several upscale restaurants while we were snowbound in Heihei. As we entered one restaurant, we passed tanks full of frogs, turtles, eels, sea urchins, and some kind of giant red worms. What really caught my attention was the daily special – a plate with a sea cucumber in the center flanked by a piece of broccoli and a large pink flower (for accent, I think, but maybe it was supposed to be eaten), with two big chicken feet complete with claws arching around the arrangement to complete the artistic presentation. We went with the spicy noodle bowl.

Speaking of imperial banquets serving sparrows, I have no idea how it got ordered, but in Beijing, a plate of half a dozen tiny sparrows, glazed in sweet sauce and sitting on a bed of lettuce with their mouths open as if singing their last melody, arrived at our table. Of course, I had to try one. They were crunchy, especially the beaks. The all-time exotic award goes to a restaurant in Wangmo, where the appetizer was baked giant mealworms. Even crunchier than sparrows they were, but they could have used some of that sweet sauce.

It is not just the strange entrees that make food in China so unappetizing. The Chinese are experts at preparing a potpourri of unpalatable dishes prepared under appallingly filthy conditions and served with an unappealing presentation. There you go, everything you need to know in one succinct sentence.

Stores in the mountain villages hung their meat selection, along with the critters' insides, outside so everyone could inspect them. Flies buzzed around and hung out on the meat. I knew what those perverted flies were doing – screwing and laying eggs. If you have ever come across a few-days-old roadkill, you know what comes next. In Gansu, the entrance to the best restaurant in Wenxian was also the butcher's shop. Buckets full of legs, ribs, and guts greeted us on the entry steps. A man in dirty clothes squatted next to a bloody stump on which he chopped up the evening's special cuts of meat with a hatchet. I damn near became an instant vegan. We ate at open-air restaurants with dirt, grease, and guts plastered on the walls. The glass lazy Susans in far too many restaurants were so dirty you couldn't see through them. The floors of dining establishments looked like the insides of garbage cans – covered in dropped food, napkins, and spit.

I think the Chinese are afflicted with colds 365 days a year. They never seem to lack nasty honkers to spit on the sidewalks or floors. In a fancy restaurant in Hohhot, we watched as a Chinaman worked up a big loogie, leaned over, and hocked the disgusting thing onto the floor. The glistening goober of sticky snot lay there like a gelatinous crouching tiger – waiting for a victim. When the dinner party got up to leave, one of the diners stepped on the damn thing. It stuck to his shoe and stretched out as he stepped forward, then snapped back to his heel. The action, step-stretch-snap, step-stretch-snap, repeated as he walked out – like the Slinky toy I had as a kid. I almost tossed my cookies.

I have a fondness for salads. Since the Chinese have gardens everywhere and grow a variety of leafy vegetables, you would expect to see the occasional salad on the menu (not that I could actually read the darn things). Not the case. For more than a month, every vegetable I was offered was boiled. I was really hankering for something green and crisp when we arrived in Hohhot. As we entered the dining room of the restaurant around the corner from our hotel, I was thrilled to see a table full of a wide selection of greens, as well as a variety of sliced meats. I thought I was in salad bar heaven. A big pot was placed on our table and set upon a butane stove, the kind you would use for fondue. A plate of sliced meats, greens, and veggies was brought to our table – and dumped into the boiling pot. Oh no, we had to use our chopsticks to fish out the limp remains of the once-crisp greens, drained of their vitamins and nutrients. I was bummed – once again deprived of crisp fresh greens. *What is wrong with these people?* I thought.

Well, now I know what is wrong with them. And I am glad I was denied my fresh veggies. You see, the Chinese still use human excrement as fertilizer. It is a multi-tasking thing. They need to fertilize their crops. Since most rural villages lack either sewer or septic systems, it makes sense to kill two birds (and the occasional diner) with one stone and put all that crap to good use.

And you wonder why the annual flu starts in China?

Unappetizing appetizers – pickled snakes & lizards.

SECTION V

CLOSER TO HOME
(well, mostly)

Index to chapters

Chapter 18

ROCKY

"Go on, fight the champ. Yeah, I'll fight him. Get my face kicked in."
Rocky — Sylvester Stallone

Being a charter member of the Peace-Love generation, I have never been big on pugilistic pursuits. I would like to say I was a lover, not a fighter, but I was not very good at that either. Despite my best efforts at peaceful coexistence, there were times when I had to get physical to stand up for my beliefs, defend my honor, save damsels in distress – or just try to survive an impending ass-kicking.

During the summer of 1978, Wyoming, the Cowboy State, provided a welcome, somewhat civilized break from the rigors of working the Alaska wilderness. Exxon Minerals set up a field office and living quarters in Laramie, taking over the fifth-floor penthouse apartment on top of the old folks' home. We got a great deal on rent since the old geezers didn't want the penthouse. The elevator only went to the 4th floor, making delivery of Meals on Wheels to the upper floor difficult. Plus, the old folks got nosebleeds starting around 7,200 feet, the elevation halfway between the 4th and 5th floors. The apartment had three bedrooms, a kitchen (which saw little use except the refrigerator – for keeping our beer cold), and a large living room, which served as our office. The office overlooked the Cowboy Bar, the prime drinkin', dancin', and shit-kickin' spot in town. Long before adult coloring books became vogue, we sat at our drafting table in the evenings, coloring maps while keeping one eye on the comings and goings at the bar. If and when enough eligible ladies arrived (eligible meaning gals weighing less than 200 pounds – we are talking about Wyoming here), we ran down five flights of stairs, dashed across the street, and tried to hustle the babes before some unworthy cowboys lassoed them.

If no eligible girls showed up, we were known to resort to desperate acts for entertainment, like the Ton O' Woman Contest. Tom Woodard, my assistant for the summer, who hailed from Colorado's San Luis Valley, introduced this exciting event to the Cowboy State. Tom grew up on a cattle ranch, as did everyone in the valley since there is nothing else there. Lacking eligible ladies, the San Luis cowpokes invented the Ton O' Women contest to entertain themselves in between dusty roundups and sweaty calf castrations. The objective was simple: dance with enough ladies until you have danced with an aggregate of 2,000 pounds of women. Whoever reaches that lofty goal in the fewest dances is declared the winner. The rules were simple: scout the bar for big, beautiful women; select a good one; ask her to dance.

Then the hard work began. You had to wrangle your catch across the dance floor and parade her in front of the judges, who sat at their table at the edge of the dance floor. That was not easy on a spacious, crowded dance floor. Tom, who had real-life experience rounding up beef on the hoof, had a distinct advantage over city-slicker dudes. For starters, he was skilled at culling heifers from the herd. He also was keenly aware of the importance of scoring a dance partner on the proper side of the dance floor. If you hooked a whopper of a BBW on the far side (where the best swamp rhinos were hiding in the dim light), it could take two or three dances to get her all the way to the judges' table. That was not good, because you would have to waste time making small talk between the songs. More importantly, after two dances, the gals figured that you really liked them, and they got a bit starry-eyed.

My preferred method was to snag a babe of prodigious proportions and get her going in a counterclockwise spin. It took some concerted effort to transfer the potential energy of those babes to kinetic energy. Once centrifugal force took over and the law of conservation of energy set in, it was just a matter of steering the spinning mass of pulsating protoplasm toward the judges. There was one problem – arresting the momentum after passing the judges' table. Colliding with fellow dancers didn't do much for one's popularity but absorbed some of the energy, reducing the velocity and allowing partial resumption of control.

The judges sized up your catch and estimated her weight (carrying a portable scale and asking your dance partner to step on it was strictly taboo). After the dance, the brave contestant put in a bid for the judges' consideration – of say, 240 pounds. Then the negotiations began. The judges might make a counteroffer, "No way, she isn't more than 220 pounds dripping wet." Eventually, the two parties settled on an official weight, the points being added to your scorecard on the way to hitting 2,000 pounds.

We were always very polite and respectful to our dance partners. Most of them were thrilled to dance (at least early in the evening – their world improves vastly as closing time nears). It was of paramount importance to never let on to our game. The consequences could be dire – like being beaten severely around the head, ears, and groin, or even crushed to death. The rules were strict. You could only get weight points one time for each dance partner, eliminating the incentive to ask for a second dance. Doing so could put you in dire danger of having a lady take a hankering to you and want to take you home. Tom and I carefully avoided that fate, but certain other contestants, who will go unnamed, got carried away – right, Jeff?

A unique twist to the Ton O' Woman contest came when Tom, my boss Bob Woodfill, and I stayed in the quaint mountain town of Saratoga for a few days to work in the surrounding Sierra Madre Range. Our accommodations were at the Saratoga Resort, a lovely place, with the best – and only – golf course for many miles around. The inn had a decent restaurant and a well-stocked bar with an expansive dance floor. We were in luck. A local cowboy band was playing while we were there. Plus, the Rawlins Retired Ladies

Golf Club was staying at the inn. It was a nearly perfect setting for a Ton O' Women contest. There was one small fly in the ointment – the golf ladies ranged in age from 65 to 90. Ladies of such vintage, who are active enough to play golf, tend not to come in extra-large sizes. They were all as skinny as rails. It would be a challenge to score 2,000 pounds of those sprites. Thus, we held the first – and last – Bicentennial Ton O' Women contest. First place would go to the contestant who scored at least 200 years and came closest to racking up 2,000 pounds of women. The combined weight and age would be tallied for the score. The bicentennial part was easy. Heck, I exceeded that in the first three dances but barely scored 270 pounds. Being the suave and debonair cowboy he is, Tom took top honors with a full ton and over 1,200 years of little old ladies, a record that has stood for decades.

We spent the summer alternating between work on the Three Mile drilling project in the Medicine Bow Mountains, near Laramie, and reconnaissance throughout Wyoming – from Cheyenne to Sheridan to Rock Springs and pretty much everywhere in between. The latter work primarily consisted of visiting uranium occurrences identified by the Atomic Energy Commission in the 1950s, when every local yokel was running around with a Geiger counter looking for fissionable material to sell to the government to use in annihilating the next evil empire. Armed with a binder of Preliminary Reconnaissance Reports (PRRs) compiled by the AEC, we set out on wild goose chases to find uranium showings based on sketchy metes-and-bounds descriptions. My favorite one read something like this: "Follow the Forest Service road to its end. Proceed up the hill on foot until you have finished smoking two cigarettes; turn left at a deadfall; proceed for one more cigarette. The prospect is a small digging next to a big pine tree." Well, I don't smoke, so I had no idea just how many feet or meters are contained in a cigarette unit of measurement. Furthermore, the description didn't specify what brand of cigarette to smoke. If you would walk a mile for a Camel, did it take a mile to smoke one? Did it take longer to smoke a 100 mm Pall Mall Long ("Outstanding and they are mild"*)* than a standard-length Lucky Strike? So many unspecified variables! Hell, there were large pine trees everywhere. It was in a bloody *forest*. Needless to say, we never found that prospect.

Tom and I spent a lot of time driving from one end of Wyoming to the other, checking those flaky PRRs. A typical day consisted of rising at 5 AM for an early breakfast at a greasy spoon café, followed by several hours of driving to the end of some dirt road in the middle of nowhere, a full day hiking, searching in vain for the equivalent of the Lost Dutchman Mine, then another long drive to the next tiny cowboy town for a late dinner at another greasy spoon café, and off to bed. After doing this for a couple of weeks, Tom came to breakfast looking tired and exasperated after a particularly long day; I think we arrived at the motel at 10 PM. He peered at me over his bowl of Wheaties and said, "Bob, I can work all day; I can drive all night. But I can't do both!" Point taken, I cut the workday back to 12 hours.

Things got interesting when we drilled the Three Mile prospect. Exxon contracted Tonto Drilling to drill all of its projects in Wyoming that year. Tonto was the best we could come up with; the Lone Ranger was booked for the summer. The drillers spent the first part of the summer drilling the Pheasant Draw project near Douglas. Carolyn Kirchner, Exxon's sole female geologist at the time, was the project geologist. Carolyn was one tough cookie, which came in very handy.

Drillers, especially diamond core drillers, are a special breed, a degree lower on the evolutionary scale than modern humans (*Homo sapiens*), whom anthropologists claim migrated out of Africa and spread across Europe and Asia about 100,000 years ago. The Neanderthals (*Homo neanderhtalensis*), who arrived in Eurasia 100,000 years earlier, supposedly went extinct 40,000 years ago. Not so. Drillers are living proof Neanderthals are alive and well. Drilling requires a combination of dirty, greasy, hard physical labor – playing with great big wrenches and lifting heavy drill rods, plus some mechanical savvy. The profession doesn't attract many people with Ph.Ds. in literature or philosophy. The pay is good for tough rednecks with little education or social skills. That is what you get.

Tonto's lead driller for Exxon in Wyoming was Donny Anderson, a strapping young Canadian of definite Neanderthalian genetic stock with all of an 8th-grade education. Donny was about a quarter-pound meat patty short of a Happy Meal. To say he was a little rough around the edges would be a glaring understatement. He spoke with elaborate profanity, never completing a sentence without throwing in a few four-letter words, including at least one f-bomb. Despite an IQ of a speed bump, Donny was not without his linguistic skills. Although he was challenged by polysyllabic words, he had a unique command of the English language. He could employ the F-word as a noun, verb, adjective, and adverb (not that he knew the difference), sometimes all in the same sentence.

In between failed attempts to get into Carolyn's pants, Donny spent his free time looking for fights. He was the undisputed heavyweight boxing champion of southern Wyoming, the cowboy version of Apollo Creed. His record was 8-0 for the summer, including two knockouts. Donny had a nasty habit of getting drunk and starting fights. If he couldn't find someone at the bar to fight, he would beat the crap out of his own helper. He was on his fourth helper when he came to drill the Three Mile project.

Tom and I split our time between geological fieldwork, watching the drill rig, and logging the drill core. One of our daily tasks was reviewing the daily drill reports prepared by each drill shift. We checked the hourly and footage charges against the work done, then approved the drill report or noted any discrepancies. Donny's drill reports were full of the latter. He was fudging his reports to bill high hourly rates when he was actually drilling on a footage rate but not making much progress. He was covering up his incompetence while charging us extra for it. Donny didn't know that we kept pretty close tabs on him, spying on the drill rig with binoculars from adjacent ridges. I busted him several times for lying on his reports. He didn't like it.

In Donny's pea brain, the solution to the problem was to switch to night shift so we couldn't see what he was doing. He further complicated that by not filling out daily drill reports. Each morning, I showed up at the rig and asked the day driller if Donny left his drill report. "No," was always the answer. This went on for more than three weeks. No daily drilling report meant that neither the day's work nor Tonto Drilling's invoice was approved for payment. The net result was Donny was not getting paid. That royally pissed him off.

Tom was going back to graduate school in New Mexico at the end of August. On his last day of work, we had a nice dinner, then retired to the historic Buckhorn Bar for a couple of celebratory beers. I was starting my second beer when the saloon door swung open. In swaggered Donny and his helper. Donny came straight over to us. I smiled and said, "Hi Donny, how is it goi-." Wham, a big fist smacked me in the face. Shocked, I jumped up and dodged a second blow from Donny as a string of expletives exploded from his mouth. The bouncer came over and intervened. He led us to the back door, with Tom and Donny's helper following. The bouncer opened the door, pushed Donny and me outside, and closed the door behind us.

Donny and I were alone in a dark alley behind the bar. This was not good. Donny was madder than an ant in the sun under a magnifying glass. He stood 6' 3" and weighed about 215 pounds. I was three inches shorter and weighed a whopping 170 pounds, dripping wet. Not only did Donny have a significantly longer reach than I and more mass with which to deliver a debilitating blow, he also had a lot more fighting experience.

We danced around in the back alley, Donny ranting that he was going to kill me. I considered replying, "Yeah, but you have to catch me first!" I bit my tongue. Hell, the last thing I wanted to do at the point was to piss him off! Fueled by adrenalin and a strong desire to survive, I peddled backward in a circle, remembering Muhammad Ali's famous words, "Dance like a butterfly, sting like a bee." I was dancing like a caterpillar and was worried about being swatted like a fly.

The last real fight I had was in 7th grade. Walking back from Catechism class one Monday afternoon, I got into it with Gary Kingston, a kid I hardly knew but didn't like because he was a "greaser." At that point, my sum total of fighting involved wrasslin' matches with my two brothers. Few punches were thrown. The objective was to take your opponent down, thump on his chest a few times, then bend his limbs into various forms of pretzels until he yelled, "Uncle!" After that, the winner was pronounced; all was good, and life returned to normal.

It didn't take long to get Gary in a headlock, then take him down and pin him until he squealed. I stood up to the cheers of my many new fans. As I started to walk away in triumph, I got whacked hard in the back of the head. I have no idea what that little shit, Gary, hit me with, but it was hard, maybe a rock, a 2x4, possibly a sledgehammer. The blow gave me a big goose egg. Stunned and bleeding from the back of my head, I stumbled back to school, wondering why anyone would violate the sacred rules of street fighting. I learned an important lesson – there are cheaters out there, and Gary Kingston was one of them. I decided then and there that I would never end a fight with my opponent standing upright.

It might also be worth noting I was an undefeated wrestler in 9th grade. Sadly, that has less to do with my wrestling prowess and a lot more to do with the fact I wrestled a grand total of three matches before quitting the team. It wasn't that wrestling was difficult or wasn't fun. Tying other kids in knots had the potential to be lots of fun. But my coach expected me to wrestle with sweaty kids I didn't know. That differed greatly from friendly matches with my brothers, whose armpit aroma I knew well and learned to love. After meeting kids from places like Depew and Cheektowaga (who may have been fine upstanding lads – or not) for all of one second as we shook hands to begin the match, I was expected to stick my nose in their stinky, sweaty, hairy armpits, and roll around on the wrestling mat like impassioned lovers trying out different kinky, contorted Kama Sutra positions. That was too much for me. I much preferred to spend my time skiing at Emery Park, a local county park that had all of one rope tow and a whopping 300 feet of vertical. The ski area was open on weekday afternoons and evenings, and it was free. All my non-wrestling buddies and I needed was to convince one of our mothers to drive us there after school and another mom to pick us up at closing time. It was a lot more fun than wrestling, and the crisp pine-scented air smelled a far sight better than the armpits of those Cheektowaga boys.

Fortunately, I learned a few good moves from my brief but admirably successful wrestling career. I also learned a thing or two from old western movies, like the most excellent gunfighter's advice, "Keep your back to the sun." As Donny and I danced around in the alley, I noted the spotlight above the backdoor. It was in my eyes when I faced the door, blinding me and making it difficult to dodge the punches that my cerebrally challenged friend was throwing. I knew I could not out-box Donny; he would eventually land a debilitating blow or two. My only hope was to take him down and wrestle him into submission. I decided to use the spotlight to my advantage by maneuvering until it was behind me and in his eyes. When he was partially blinded, I could make my move. Time was of the essence. This was a do-or-die moment. I was certain of that because Donny told me repeatedly that he was going to f'ing KILL my f'ing ass! I considered taking him down with a Tilt-a-whirl Mat Slam or a Crossbody Overhead Chop like I had seen Andre the Giant do so eloquently many times in All-Star Wrestling. I settled on a simple Double Blast tripping takedown, one that served me well during my abbreviated wrestling career.

I danced half a revolution around the ring until the light was in Donny's eyes, then I went for the takedown. I charged forward, head down, slamming my shoulder into his ribs, grabbing him with both arms – pinning his arms to his sides, and throwing my right leg behind his legs. The momentum knocked Donny off his feet and threw him down and backward, landing hard, flat on his back. The impact knocked the stinking wind out of him. That was the break I was hoping for. I sat on Donny's chest with my legs pinning his arms to his sides. At that point, with memories of getting nailed from behind by Gary some 16 years prior, something inside of me snapped. With renewed vigor and determination, I pummeled Donny's face and shoulders (to paralyze his arms in case he got them loose) with heavy blows – left, right, left, right, left. I let loose a whirlwind of wicked punches with the fury of a wounded wildcat – an enraged madman on a mission of survival. I hit him hard so many times my hands ached, and my knuckles were raw. Donny cussed and sputtered f-bombs the whole time. I was going for the KO, but the champ had a head as hard as a rock, surely something he inherited from his Neanderthal kin. After a while, my hands were too sore to do much punching. For a few extra flourishes, I rolled his head to the left, grabbed his hair, and slammed the side of his face repeatedly into the gravel. Then I rolled him to the right and repeated the procedure, just for the sake of bilateral symmetry. Half exhausted and somewhat resigned to the fact that I was not going to win by a knockout, I ended up sitting on his chest and whacking him in the face whenever he was foolish enough to sputter another expletive.

The backdoor of the Buckhorn opened. Out stepped Tom and Donny's helper. I could tell by the look on Tom's face that he was shocked to find me still among the living. He undoubtedly expected to see my dismembered limbs scattered across the alley and Donny dancing in celebration of his 9th straight win. Instead, I was sitting on top of the bloodied, defeated driller, smacking him every time he opened his mouth. The guys pulled me off Donny, who just lay there, bleeding from his mouth and his bent nose, his eyes just about swollen shut, with gravel stuck to the sides of his head and face. The former champion of the Cowboy State was not happy to be dethroned, and by a lowly geologist at that. As his helper picked his semi-limp corpse off the ground and held him up, Donny had the audacity to demand a rematch. "Meet me tomorrow morning at 7:00, and I'll teach your f'ing ass a f'ing lesson," he muttered, spitting blood from his swollen mouth.

Tom and I didn't go back to finish our beers. We returned to our humble penthouse abode above the old folks' home to clean up and get some sleep. I somehow survived the altercation with a slight black eye from Donny's first sucker punch. Tom needed to shove off early in the morning. I needed to get some rest and brush up on some new-fangled moves to impress Donny with in the morning.

I was at the designated spot bright and early, but as expected, my challenger was a no-show.

It was Saturday morning. I called my boss, Bob Woodfill, at his home to let him know what had transpired. He was not terribly surprised since he was well aware of the problems Carolyn had with Donny. Still, I was worried about losing my job. Although I had not committed the entire Employee Handbook to memory, I was reasonably sure beating the crap out of the company's contractors, even in self-defense, was on the naughty list.

I spent the rest of the day hiking to the summit of 12,013-foot Medicine Bow Peak in the Snowy Range. There is nothing like fresh mountain air to clear one's head and restore faith in humanity – not that Donny Anderson qualifies as a member.

I called the office on Monday morning to talk to my boss and learn the repercussions of the boxing match. Bob told me the drilling project was being shut down. Tonto's top manager had been in contact with Exxon's drill supervisor, and they made that decision. Besides, Tonto was one driller short of a full crew. I was concerned only one side of the story had been told – Donny's expletive-ridden one. Bob told me to retrieve any core from the drill site and bring it back to our core warehouse in Laramie, then return to Denver for debriefing on the matter.

I arrived at the office mid-day on Tuesday, expecting a big hassle and a drawn-out investigation into the matter. I walked down the hall to my office and almost went past it. The usual brown name plaque on the door with "R.G. Cuffney" in white letters was gone. In its place was a gold one with "ROCKY" in big black letters. My colleagues were waiting to congratulate me on my victory and call me "Champ." Carolyn was especially pleased since she hated Donny's guts.

Donny showed up at Tonto's office in Salt Lake City with two very black eyes nearly swollen shut, a badly split lip, a nose that looked like it belonged on a proboscis monkey, a couple of chipped teeth (damn, I had hoped to knock out at least one of them), and gravel embedded in his head, which he was digging out for days. The head of Exxon's drilling department stopped by my office with a question from the bosses at Tonto Drilling, "Who was the guy who did that to Donny?" He was not used to being bested, let alone beaten to a pulp. But he had never tangled with anyone who got nailed by Gary Kingston back in 7th grade and was determined to not repeat the experience.

"Yo, I did it!"

Epilogue

It is indeed a small world. A mere 29 years later, I was in northern Arizona working with a Canadian core driller, Mickey Clarke, who mentioned he worked for Tonto Drilling years ago. I told him I had worked with Tonto drillers. In fact, the only fistfight I had since Junior High was with a Tonto driller. He immediately knew who it was. In fact, Mickey had more than one altercation with Donny and was one of the few people who had bested him. Mickey told me Donny almost met his Maker, who I imagine was disappointed in what He (or is it She these days?) made, while working in Australia. Donny went to an Outback bar, got seriously drunk on Victoria Bitter beer, and picked a fight with the wrong bar patrons. The Aussies beat the crap out of him, then drove him into the Outback and literally threw him off a cliff, leaving him for dead. Even that didn't do Donny in. He regained consciousness the next day and stumbled back to town with a bashed-in head, broken ribs, and a punctured lung – one tough, stupid hombre.

Not unexpectedly, Donny never changed his ways. He remained a pugilistic prick until his dying day. Mickey said decades of hard living, fighting, and drinking finally took a toll on his liver. Ole foul-mouthed Donny passed away down in Mexico just a couple of months prior to my conversation with Mickey. I don't normally take pleasure in the misfortune of fellow souls, but I admit the news elicited a degree of heartwarming satisfaction in having outlived the bastard. The thought of Donny descending to his well-deserved new home put a sublime smile on my face. Then a dark shadow of gloom and despair fell over me. Was I feeling guilty for taking satisfaction in Donny's demise? Hell no, I was just disappointed that, after waiting nearly thirty years, I finally knew for certain – there would never be a rematch.

F'ing rest in peace down there, you mother f'er!

Chapter 19

WHITE KNIGHT

Noble geologists in search of the Holy Grail of gold deposits.

The best job I ever had – at least the most fun one – was working for a small Vancouver company, White Knight Resources. The company was run by two brothers, John and Gord Leask. They treated their employees like family – a rather dysfunctional one – nevertheless, family, eh?

Pat Cavanaugh, the President of White Knight, convinced me to leave the Indonesian jungle in 1996 and join White Knight to help explore for gold in my old stomping grounds of Nevada. I was more than glad to trade the heat, humidity, mud, and bugs of the steamy jungle for the civilization and relatively benign field conditions of the Great Basin.

The company acquired several good gold prospects and joint-ventured most of them with larger companies, who had the money to spend on the expensive part of exploration – drilling. White Knight was on a roll for a couple of years, then disaster struck.

The price of gold fell precipitously when central banks around the world took turns selling gold bullion. It sure seemed like a diabolical plot to drive down the price of gold. Whether or not that was true, the result was a drop in the price of gold from $400 per ounce to $250 per ounce.

For companies mining gold, the price drop was disastrous. Many marginal mines became uneconomic and shut down. Even the better-performing mines had a hard time breaking even. Miners and geologists were laid off to keep the companies afloat. Small exploration companies, such as White Knight, lost their financing and saw their stock prices plummet, making future fundraising difficult, if not impossible. Many exploration companies laid off their entire staff. Some went out of business.

It was only a matter of time before the dreaded call came from the bosses in Vancouver. John and Gord asked for my input on what to do. I said I could survive with being laid off, but we should keep our best projects at all costs and have them ready when things turn around. John and Gord disagreed. They said properties could be replaced, but good people could not. The company only had a few office personnel in Vancouver. The Nevada staff comprised Tracy Guinand, who did our land work, and me. Pat Cavanaugh decided to retire and close the Boise office.

The bosses decided we should essentially go into hibernation, stop work on the properties, keep expenses to a minimum, and decide what to do with our properties in September, when annual claim rental payments would be due to the BLM.

Tracy and I took a pay cut but remained on the payroll. Hibernation was the difficult part. I was expected to do nothing for several months, maybe a year, until the market turned around. That was unacceptable. Several of our properties needed geological mapping. I decided I could do that inexpensively by camping at the sites. My only expenses would be for peanut butter, gas, and beer (in increasing order).

I started working on the New Pass property near Austin, Nevada. I spent a week there, having fun hiking the hills and pondering the complexities of the geology. On my second trip, I was sitting around the campfire listening to the local coyotes sing while enjoying a cold Mickey's bigmouth beer when I had an epiphany. All our competitors, from small Vancouver juniors to the major mining companies, were in equal or worse financial shape. It was a given that by September 1, when fees were due to the BLM ($125 per claim), a lot of good properties would be dropped. If we did proper research, we could position the company to acquire an excellent group of properties by staking claims starting September 1 (initially inexpensive but costly when payment of $174 for each claim filed would be due to the BLM).

I pitched the idea to Gord and John, calling it a once-in-a-lifetime opportunity to acquire good properties and make the company a smashing success. The plan came with risk, but the guys agreed. Timing was the tricky part. The price of gold needed to rise, and stock prices of junior mining companies needed to recover before we needed to pay the government fees, or the effort would be for naught.

John and Gord came down from Vancouver to spend a week touring Nevada, discussing gold deposits, and speculating which companies were likely to drop properties of interest. At the end of the week, we had a plan to monitor the activities (or lack thereof) and financial status of several companies.

The plan paid off. Come September 1, claims were dropped by the thousands. One of our primary targets was Atlas Precious Metals' claims surrounding the former Gold Bar mine in the Roberts Mountains. Atlas dropped nearly all of its 50 square miles of claims. The claim-staking candy store was open for business.

Mineral rights on Federal land are acquired by physically staking claims on the ground, filing the claims with the BLM, and paying the filing and annual rental fees. The claims give the locator the exclusive right to explore for minerals on the land, but no other rights, with the specific exclusion of surface rights. Typical claims are rectangles measuring 1,500 feet by 600 feet. Five 2x2 wood posts define a claim: one on each corner and one – the location monument– in the middle of one of the two short sides, which has a notice detailing the claimant, date of staking, etc. It would take a lot of posts to locate claims over the extensive area we wanted to stake.

I became a one-man staking crew. For the next several months, I drove the company truck along marginal trails and cross-country through the sagebrush to reach places where I could drop off piles of posts. Then I carried the posts up and down hills and across valleys to claim the mineral rights abandoned by Atlas. I was always looking over my shoulder for competitors, who might also try to claim the good ground. Only one small company, Tone Resources, showed up to stake a small prospect. The rest of the time, the mountain range was mine.

After a while, I was getting behind on finishing the staking. Vancouver approved hiring a professional claim-staker, Kip Tonking, to help with the work. Kip had an ATV, perfect for off-road travel across the lower, flatter ground in the sagebrush. So, I concentrated on placing posts in the pinion-juniper forest in the steeper, high country.

One day at breakfast, Kip turned to me and said, "I have been following some of your tracks along the claim lines. I can't believe the places you've taken your truck!".

"Not my truck," was my reply.

After we acquired the land position in the Roberts Mountains and several other areas by claim staking, White Knight was able to raise enough money to keep operating until the price of gold finally rose. Then, the company was in an excellent competitive position. We assembled a team of hard-working, fun-loving geologists and went about exploring our properties and looking for more projects in northeast Nevada. We based most of our work from the tiny town of Eureka, where we rented a big house on the hillside above town to serve as an office and lodging. Evenings were for BBQs on the back deck and games – like redneck horseshoes, which are played like regular horseshoes, but using toilet seats as horseshoes (we called them "toilet shoes") and toilet plungers as stakes. Another favorite pastime was playing downhill crochet. Think of regular crochet, except the yard was on a steep hillside, making it hard to stop a ball once it passed through a wicket. Hitting the ball back up the hill, so you could then hit it through the next wicket was a helluva challenge. The balls rolled up the hill, then rolled back down. It was OK to take a swing at the ball as it rolled past, hoping to get it through the hoop on the second try. Eventually, all the crochet balls rolled down the hill and hit the retaining wall at the end of the yard. If a ball went over the edge and rolled to town, we ended the game and followed the ball to the Owl Club for drinks.

Travels Without Charlie – The Natives Ate Him

Kristen Benchley & Tim Jefferson throwing the toilet shoes.

It is unusual to see grown men standing around in 40-degree weather wearing only their skivvies. But Mario was an unusual person.

The fall of 2006 was a busy time for the mining industry. Drill rigs and crews were difficult to find. The best we could come up with was Harris Drilling, a company that was "unbusy" – for a good reason. Harris was often the low bidder on jobs, mainly because the company was cheap. They had a reputation for using old derelict equipment and cutting corners wherever possible. Broken-down drill rigs and trucks came dirt cheap from auctions, better yet from junkyards. The old junk received a new paint job and got fixed up just enough to make it to the job site. Maintenance was optional. It was preferable to wait until equipment broke down and needed repairs. Usually, the wait wasn't long.

We were drilling high up in the Roberts Mountains, accessing the drill site from the old Gold Bar Mine haul road. Driving up the dirt road one afternoon, I caught a glint of light coming from the middle of the road. As I approached, I noticed it came from the back window of the driller's water truck. It had fallen out and was lying there, abandoned. Apparently, it was one of those optional accessories, which was no longer needed. Several days later, it was still there, with a few tire tracks across its shattered glass to confirm its fall from graces.

Early one chilly morning, I stopped to talk to Mario, the driller, at the water well at the bottom of the road. It was late fall, and there was frost on the pumpkins. Mario, a little Mexican fellow with a broad grin filled with bad teeth, was standing on top of the water truck holding onto a hose pouring water into the 1,000-gallon tank through a big open hole on the top of the tank. We chatted for a while. Mario filled the tank to the brim. When he finished, I told him I would stop by later to see how the drilling was going. He jumped in the truck and drove merrily up the bumpy road toward the drill rig.

Normally, the filler hole on top of the tank has a cover, for obvious reasons. This one did not. A back window is also normal on water trucks for more obvious reasons. But this was Harris Drilling. Nothing was normal.

The drill site was high up the hill, at about 7,000 feet elevation. The three-mile-long access road was steep, bumpy, and crossed several gullies.

Half an hour later, I drove up to check on the drilling progress. There were splashes of water at every bump and sharp turn in the road, more in the dips, and a large splash at one gully about halfway to the drill site.

As I pulled up to the drill rig, I noticed a pair of overalls hanging next to the hot muffler of the engine. At the controls stood Mario, looking rather absurd in his underwear but focused on drilling, as if his attire were perfectly acceptable.

"Mario, what's going on?" I asked.

"Sheeeet man, I going down dee big hill and hit dee brakes at dee bottom of dee gully. Den a beeeg wave of water come through dee window and hit me like a beeeg douche. I freeze my ass off all the way here, man."

"Mario, did you ever think you should have replaced that back window?"

Mario laughed – or was he just shivering?

The bosses occasionally enlisted me to accompany them on marketing tours and to trade shows such as the Prospectors and Developers Association Convention (PDAC) in Toronto, the Exploration Roundup in Vancouver, and the annual Gold Investment show in San Francisco. It was fun meeting investors and other exploration geologists and learning about our competitors' activities. But spending hours standing in a booth playing "booth babe" was tedious. Having to wear a coat and tie all day added to the torture.

White Knight and Mansfield booth babes at PDAC convention in Toronto:
l to r: Jim McDonald, the author, Gord Leask, John Leask.

The English language can be confusing. Homonyms, words that are spelled alike and sound alike, but have different meanings, make for some of the biggest problems. For instance, the word "fly" could mean to travel through the air, to flee, the zipper on a pair of pants, or an obnoxious insect. Add geological jargon, and the problem is further compounded. For example, "cleavage" refers to the way minerals fracture, or cleave, along crystal planes. It is rated as distinct, good, or perfect.

Back in the heyday of White Knight Gold, we held a two-day conference in Elko to present the company's properties and exploration potential to investors and stockbrokers. After the meetings, John and Gord Leask wanted to visit the Slaven Canyon project near Battle Mountain. Karene, the company's Public Relations officer, who organized the meetings, tagged along to see what it was like working in the Nevada mountains.

It was a warm, sunny summer day. Karene, a 30-something, amply endowed blonde, showed up at the truck in the morning wearing tight jeans and a rather revealing low-cut T-shirt.

We left Elko and drove west over Emigrant Pass, then followed the mighty Humboldt River to Argenta, where we left the highway and drove south on a dusty gravel road along the west flank of the Shoshone Range. Battle Mountain – once designated as "The Armpit of America" – was visible to the west at the foot of massive 9,678-foot Mount Lewis.

We turned the truck onto a 4x4 trail and worked our way up a ridge. Karene had just finished an introductory geology course at a junior college in Vancouver. She wanted to experience geology in the field instead of from textbooks. We looked at some gold-bearing outcrops and explained the rock types and alteration and how they related to the gold mineralization.

Further down the ridge, we came upon an interesting outcrop, only about two feet high. A fault passed through there and left its tracks as parallel striations and polish on the rock face, a feature called "slickensides."

I bent down on one knee and pointed out the slick, polished rock surface on the small outcrop. Before I could begin to explain, Karene, keen to apply her newfound geological knowledge, said, "Wait, don't tell me. I know what this is."

She paused for a moment, then proudly announced, as she bent over the outcrop in front of me, "I know, we are looking at cleavage."

Karene was confused.

"No," I replied (not exactly looking her in the eye). "You are looking at slickensides. I, however, am looking at cleavage *(pretty close to the perfect rating)* – and rather enjoying it."

Karene blushed and stood up, ending my lovely valley view. Gord and John were laughing so hard I thought they would roll all the way down the hill.

Through no small effort, the White Knight team gradually acquired a large land package, including some defined gold resources and former mines, in the Battle Mountain-Eureka Mineral Belt, one of Nevada's premier gold belts. The land position caught the attention of a larger Canadian company, US Gold, which we dubbed USeless Gold, who announced a hostile takeover of White Knight. The CEO of the company, an egocentric butthead I will call, "He Who Shall Not Be Mentioned" – shortened to "He Who" – which has a nice poetic ring to it, expected the regulatory agencies to bend over backward for him in the takeover process. They were not so inclined. The process took a little over a year to complete. During that time, stock exchange regulations prevented any communication between employees of the two companies. Ann Carpenter, the President of USeless Gold, wanted to talk to me about our projects and even implied that there would be a job for me with the company after the merger was completed. She called about once a month. Every time, I reminded her that our conversation had to be limited to subjects such as the weather, skiing, or the latest craft beer, but not company matters.

On the day the takeover was consummated, we threw a big party to thank our contractors, suppliers, and supporters. The party was dubbed the "Wake of the White Knight." I called Ann and invited her to the party, also suggesting we get together afterward to discuss the things which had been on her mind for the past year.

The warehouse in the back of the office was decked out for the party. A dummy dressed as a medieval knight with a broadsword at his side and a bloody wound in his chest was laid out on a table serving hors d'oeuvres. A sign explained the evil Black Knight of the North had vanquished the White Knight in a long, drawn-out battle. The members of the White Knight staff were all dressed in medieval costumes. John and Gord Leask unexpectedly showed up from Vancouver. Gord had one arm inside his shirt and walked around, saying, "It's just a scratch. I've had worse." Throughout the evening, we occasionally paraded through the crowd clapping coconut shell halves together as we rode our invisible steeds in search of the Holy Grail of gold deposits or at least another beer.

The highlight of the evening was the lynching of the Black Knight of the North. A dummy dressed in a fancy suit (the nicest I could find at Goodwill for under $10) with a picture of He Who's face glued onto its Styrofoam head, stood on the second-floor railing with a hangman's noose around its neck. At the end of a countdown, the effigy of He Who dropped. Given this was our first lynching, the hanging did not go as expected. The dummy's head popped off. Its decapitated body fell to the floor with a thud. We all cheered anyway.

I never heard from Ann and never worked for USeless Gold. I have no idea why…fired before hired?

The White Knights: l to r: Kristen Benchley, John Leask, Gord Leask, the author, Ruth Buffa.

Chapter 20

THE HIGHWAY TO HELL

"If you're going through Hell, keep going."
— Sir Winston Churchill

June 19, 2016, Father's Day – most dads are in their backyard barbecuing steaks and enjoying a cold brew or two while proudly wearing a pair of gaudy Bermuda shorts and a T-shirt emblazoned with "World's Best Dad," the kids gave them for Fathers' Day. Not this dad, I am roaring down US Highway 95, making the 540-mile trip from Reno to Bullhead City, Arizona. One of my clients has requested I map and sample a gold property in the Mohave Desert near Oatman, Arizona. Of course, they couldn't have planned ahead and asked me to do the work in the winter or spring or wait until the fall. Nope, it has to be done now – in the heat of the summer.

It is a pleasant 66° F at 7:30 AM, as I drive east on Interstate 80 from Reno, enjoying the scenic drive through the Truckee River canyon and watching the last of the melt from the Sierra snowpack rushing to its landlocked destination at Pyramid Lake. But the top-of-the-hour news on KOH 780 AM mentions a heat wave hitting the Southwest. Something tells me the next three weeks are going to be hell.

Most motorists who have made the long drive south on Highway 95 consider it boring, with nothing to see between Reno and the glitzy lights of the Vegas strip. I, however, find it a fascinating trip full of beauty, mystery, and grandeur. There are sweeping desert vistas to behold, geological wonders to ponder, odd attractions to visit, and rich history to investigate. Granted, the towns don't offer a lot of amenities and can be far apart (like 100 miles), but they are all unique places with fascinating stories worth investigating. Abandoned mining camps and ghost towns (complete with ghosts, of course), cat houses (active and abandoned), quirky museums, off-limits secret military installations, rustic saloons, eccentric residents, bizarre outdoor art, and even an Alien Center await the curious traveler who isn't intent on getting to Sin City as fast as possible to gamble away his paycheck. I consider it an insult to one's flattened derriere not to stop at some of these attractions along the way.

Nevada is one of the last bastions of the Wild West, a rough-and-tumble place where pretty much anything short of murder is legal (Harry Reid is living proof that even larceny on a grand scale is acceptable – at least for people with "connections"). In 1861, Nevada Territory was carved out of the Territory of Utah because the freedom-loving settlers in Nevada (mostly tough miners who enjoyed their saloons and camp-followers) wanted nothing to do with the strict ways of the Utah Mormons. Nevadans are known for their rugged individualism and independent spirit, with a don't-mess-with-me attitude handed down from generations of hardy settlers of the last American frontier – my kind of place.

Paradoxically, there is nowhere in the U.S.A. more under the control of the Federal government than Nevada. The Feds, through the Bureau of Land Mismanagement, the Department of Lack of Energy, the Department of Deaf-ense, the National Park Serv-us, and the US Forest Circus, own fully 86% of the landmass of Nevada. Locally, the situation is much worse, with the federal government owning 97.7% of Nye County and 97.9% of Lincoln County, counties through which Highway 95 passes. Most, but certainly not all, BLM and USFS land is somewhat accessible to the public (although restrictions increase annually), but much of the government land is not. Nevada has long been a playground for the military and its secret weapons programs, strictly off-limits to the general public.

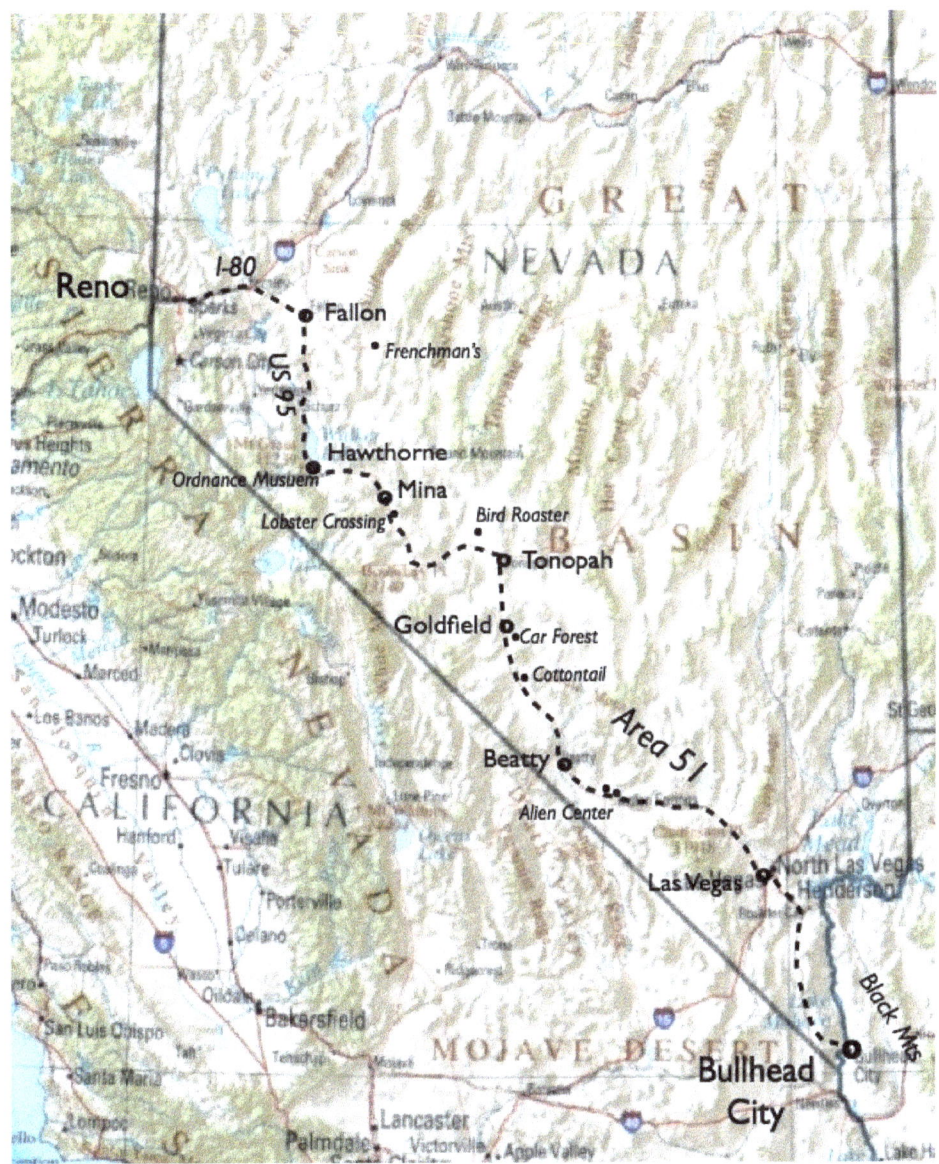

Nevada has always been a boom-and-bust state, owing to its roots in mining and ranching, both unpredictable cyclical industries. Mining is subject to ups and downs created by swings in commodity prices, technological innovations, and government policies. Ranching is a victim of similar external influences with the bonus or curse of weather fluctuations. Being a historically poor state, Nevada looked for innovative solutions and adopted more reliable businesses – gambling and prostitution – as its primary industries. Both businesses were well established early in the state's history. Although they were technically illegal, they were tolerated. Gambling was legalized in 1931. Reno led the way, but Las Vegas added scantily clad tall girls with big feathers on their heads to generate more interest in the gambling scene. It worked. The gambling boom roared until the Indian Gaming Regulatory Act of 1988 was passed, and Indian casinos popped up like mushrooms in nearly all states, especially California, cutting into Nevada's gambling business.

Brothels were a prominent part of life in early mining camps, which were populated by rugged young men and very few pioneer women, many of whom could probably whoop the tough miners. The elegant imported ladies of the night provided a necessary service to the men and a much-needed break for their burros. One by one, the mining camps eventually went bust, but the brothels stayed. Lawmakers, finally waking up to the fact that brothels were here to stay, and the state was losing out on potential tax revenue, allowed legal licensing and taxation of brothels on a county-by-county basis in 1971, thus instituting the first de facto "peter tax."

Today, Reno and Las Vegas are prosperous towns. Sadly, the rest of the state is in bust mode due to low metals prices, onerous government regulations and land grabs, and an overall downturn in the economy. Nowhere is this more evident than along Highway 95 with its old and new ghost towns and abandoned cat houses.

My route takes me east on I-80 to Fernley, where I turn onto US Highway 50 and travel further east to Fallon, a farming and ranching community along the Carson River. Not far past Fallon, what remains of the river after it has been sucked nearly dry for residential and agricultural usage, trickles into Carson Sink, where, like all rivers trapped in the Great Basin with no outlet to the sea, it evaporates in the heat of the high desert.

I pick up US Highway 95 in Fallon and head south on what will be 325 miles of twisting, hilly two-lane road before I see a 4-lane highway again. It doesn't take long for the military's presence to be felt. Two low-flying F-18 Hornets streak by as I leave Fallon, home to the Naval Air Station Fallon and Top Gun training. One has to wonder why the Navy would establish a major naval base more than 200 miles from the nearest seawater. The answer lies in the need for a lot of restricted air space where pretty much nobody lives – to reduce complaints from sonic booms as well as property destruction and loss of life from errant bombs. Unfortunately, the Navy also needs (or at least wants) a lot of restricted ground space and constantly lobbies Congress for more, creating conflicts with local ranchers, miners, hunters, and recreationists. The latest demand from the Navy called for quadrupling the size of the base to 1,262 square miles – just 300 acres shy of the size of Rhode Island.

I am still pissed that the Navy forced out the ranchers in Dixie Valley and demolished Frenchman's Station in 1987. Frenchman's, situated 30-some miles east of Fallon on US Highway 50, was a historic stage stop dating back to 1904 and was the only decent watering hole along the 111 miles of the Loneliest Highway in America between Fallon and Austin. Frenchman's saved my butt on a hot summer's day in 1984. I was driving back to Reno from some forgotten drilling project in the Shoshone Range with a full load (OK, make that more than a full load – a severe overload) of drill samples when I had my third flat tire of the day a couple miles east of Frenchman's. The Bronco limped to the station, which had the only pay phone for many miles around. I rummaged through my jeans pocket for a dime to call the office in Reno and asked rude, crude Bill Klud, our expeditor, to bring out a few more spare tires. We changed that flat, and I had one more on the way back, setting a personal record of four flats in one day. A minimum of two spare tires is a requisite for fieldwork in the backcountry of the high desert, but more than two flats in one day is unusual. I think the combination of worn tires, hot asphalt, and a weight overload may have had something to do with that record.

In addition to a pay phone, Frenchman's had a bar, a small restaurant, and a few rather Spartan motel rooms. Its demise derived from its location along the final approach of jets attacking the Navy's Bravo 17 bombing range. It was unique entertainment to watch the jets roar across the highway barely above the ground, then see the puff of their bombs somewhere near the targets in the alkali flats to the south, or sometimes nowhere near the targets. In the late 1980s, the Navy wanted to expand its real estate holdings and offered to buy out the Dixie Valley ranchers and the owners of Frenchman's Station. The last holdouts, Frenchman's being one, gave up when the Navy convinced insurance companies to deny insurance due to the risk of wayward bombs. Having watched some of the bombing runs, I think they had a point.

Being the target of Navy and Air Force jets is one of the more exciting perks of working in rural Nevada. Nevada is part of the Basin and Range geological province, a vast uplifted plateau that was stretched and pulled apart, torn into blocks that slid like a giant deck of cards, producing a series of north-south mountain ranges separated by broad gravel-filled valleys. Billowing plumes of gray dust issue forth from vehicles driving dirt roads that follow the long valleys. The dust trails can be seen for miles, make that tens-of-miles for someone 10,000 feet in the air. Driving along dirt roads at 50 mph in Radio-free Nevada, in your own little world in the middle of nowhere with nobody and nothing around, the only sounds being the humming of tires and the occasional thump of a jackrabbit that zigged when it should have zagged;

it's a shorts-filling experience to suddenly hear the roar of jet engines 20 feet above you and milliseconds later to be looking into the afterburners of an F-whatever. Jet jockeys take great joy in strafing innocent civilians and scaring the living crap out of them. I swear there were times when there would have been tire marks on the roof of my truck if the jets had their landing gear down.

The tiny town of Schurz on the Walker River Paiute Reservation makes its appearance as a strip of green standing out from the otherwise barren desert. Giant cottonwood trees flanking the Walker River as it meanders its way through the desert to its terminus at Walker Lake are the only notable features of the town, whose economy appears to be rather explosive – consisting entirely of Bad Jacks Fireworks, Tee Pee Fireworks, and the Four Seasons Smoke Shop and Fireworks.

As I turn the corner a few miles south of Schurz, the brilliant blue waters of Walker Lake come into view. Flanked by tree-shrouded 11,285-foot Mt. Grant, towering more than 7,000 feet above the lake's west shore, Walker Lake congers up images of a slightly smaller Lake Tahoe. However, Walker Lake's stark beauty disguises a sad hidden secret – it is slowly dying, a condition hinted at by the white rim of salt along the shoreline, making the lake look like a giant margarita glass *con sal*. Walker Lake is one of the last vestiges of the once great Lake Lahontan, which covered about one-third of western Nevada during the Pleistocene ice age. Greater than 12,000 years of global warming (at least 11,800 years of that being natural) and evaporation have taken a toll on Lake Lahontan. Now, only Pyramid Lake and Walker Lake remain as terminal lakes with no outlets. The rest of the great lake has been reduced to playas – mud-cracked alkali flats, such as the famous Black Rock Desert, where tens of thousands of crazed "Burners" from all corners of the globe congregate each September to breathe in the alkali dust of the playa and fulfill their pyromaniacal fantasies on a grand scale. Strandlines of old beaches perched on the sides of the basin mark the positions of former shorelines of the lake as it slowly dried up.

Typical Nevada playa – minus 50,000 ecstasy-crazed friends.

Diversion of water from the Walker River for residential, industrial, and agriculture use has further shrunk Walker Lake. The level of the lake has dropped more than 180 feet since 1882. In most years, water from the river never reaches the lake. As a result, Walker Lake is a beautiful deathtrap for aquatic life. Lahontan cutthroat trout were reintroduced in the 1950s and thrived for decades, making Walker Lake a destination famed for world-class fishing. Now, the salinity of the lake is too high. The trout hatchlings go belly up shortly after hitting the salty water. The Nevada Department of Wildlife, not being fond of euthanizing fish, wisely stopped stocking the lake. Flocks of migratory water birds, including grebes, tundra swans, cormorants, and American white pelicans, used to congregate at the lake on their annual migrations. So many common loons gathered at the lake each spring that the town of Hawthorne, at the south end of the lake, held an annual Loon Festival. The loons stopped coming when the fish they depended on declined. The last festival was in 1996. No worries – despite the lack of birds, there are still plenty of loons in Hawthorne. Check Joe's Tavern or the bar at the El Capitan if you don't believe me.

Some people, who have spent far too much time in the searing desert sun, claim there is a crack in the bottom of Walker Lake, and it is connected to the Pacific Ocean via an underground waterway. I once met a local who claimed, quite seriously, that a woman drowned in Walker Lake, and the search for her body came up empty-handed. A week later, her body surfaced in Lake Tahoe, having made the 70-plus–mile trip under four mountain ranges via an underground passageway. Her reason for taking the long posthumous journey was not mentioned. Local nut jobs claim the US Navy uses a subterranean channel to move submarines between the Pacific coast and a secret submarine base at Walker Lake (a mere 200 miles or so of solid rock between), although nobody has ever seen a submarine in or near Walker Lake, and it would be hard to hide one in a lake which is a maximum of 83 feet deep and getting shallower annually. That crazy rumor seems to be rooted in the presence of the Naval Undersea Warfare Center (NUWC) in Hawthorne. Looney Tunes locals have also informed me the reason Walker Lake is so salty is that the Navy screwed it up with their connection to the sea. That makes sense, sort of. I mean, how could all those submarines come and go without dragging along a bunch of saltwater, seaweed, and a shark or two?

Having a Naval warfare center in the middle of the desert makes no sense at all unless you consider Hawthorne's historical military role as the Hawthorne Army Depot, which, as the sign entering town proudly proclaims, is the "World's Largest Ammunition Dump." I have stayed in Hawthorne many times and can testify the word "Ammunition" could be dropped, and the sign would still be accurate. The NUWC actually has little to do with submarines. It is a storage and test center for Hawthorne's specialty – things that go boom – in this case, torpedoes, mines, and depth charges.

Hawthorne is surrounded by rows upon rows of gray concrete warehouses and thousands of earthen-covered bunkers that stretch as far as the eye can see. My curiosity got the best of me during one of my many trips through Hawthorne. I stopped at the Hawthorne Ordnance Museum to learn more about the history of the depot. It is hard to miss; just look for the giant M47 Patton tank sitting among a bunch of bombs, missiles, and anti-aircraft guns around the corner from the humongous American flag. My guide was Bob, a lively 80-something retired veteran. Bob informed me the ammunition dump is used to store, test, refurbish, and dispose of munitions, including mustard gas and other delightful forms of chemical and explosive mass destruction. Saddam Hussein would feel right at home here (had we not strung him up). Established as the Hawthorne Naval Ammunition Depot in 1930, the depot, along with all its deadly contents, was transferred to the Army in 1977. I think that was the result of the Navy losing a bet on the 1977 Army-Navy game, but I would have to verify that factoid. Eventually, the military got bored with the depot and decided the storage of enough explosives to blow up half the state is not a terribly sensitive issue, so the Army turned over operations at the depot to private contractors.

The museum is a must-see attraction, hosting an amazing variety of killing machines, projectiles, bombs, and missiles: everything from 50-caliber bullets to 2,000-pound shells and from grenades to cluster bombs – even a nuclear missile (without the active warhead, I was told). I was stunned by all the mean, nasty, ugly, and horrible ways our military has conceived to break things and maim and kill anyone who might make the mistake of wearing the wrong uniform. Good stuff!

I asked Bob if there had ever been any serious accidents at the depot. He told me two bunkers full of black powder once caught on fire but did not explode. The Hawthorne fire department arrived on the scene, ready to extinguish the fire until they were informed that spraying water on the fire would make the whole thing go "boom" big time. The worst accident was when eight Marines were killed when a mortar exploded during a training exercise. Bob chalked that up to stupid people doing stupid things (he obviously was not an ex-jarhead). He described some of the dumb things the contractors did, concluding with, "I have two cats that are smarter than some of those guys." Supplied with that tidbit of confidence-building information, I am now just a little more nervous driving past all those explosive-packed bunkers.

Traveling south-by-east from Hawthorne, past the sign warning "Next gas 100 miles," I soon approach the tiny towns of Luning and Mina. Once mining and railroad boomtowns, there is little left of them, save for dilapidated buildings and a couple of rock shops that are closed more often than are open. The towns have literally dried up and blown away. As I ratchet down from 80 to 35 mph approaching Luning, a sign proclaiming, "No explosive laden vehicles allowed to park in the town of Luning," greets me. I wonder *Why? What difference would it make if one parked and exploded?* Two minutes later, I am through town and accelerating towards Mina without seeing a single living thing in Luning.

The metropolis of Mina, population 104, has a little more character. The town even has one decent eatery, S'Socorro's Burger Shack, well named because it is a tiny red-and-white shack with a smiling lady, Socorro, inside. Mina's main attraction is (or was) the Desert Lobster Café. A local rancher, Bob Eddy, grew weary of herding cantankerous cattle and came up with a better idea – a strange one, at that. He dragged a derelict yacht, a white Navigator 5300, up from Texas and parked it high and dry along the main (and only) drag in Mina. Bob's brilliant idea was to start a lobster ranch at Sodaville, along Highway 95 just south of town, where 80° F water from natural warm springs offered the perfect medium for raising "desert lobsters," big Australian freshwater crayfish. The desert lobsters were safely housed inside a large greenhouse in tanks fed by the warm springs. Bob deeply cared for his shelled friends. He erected a sign reading "Lobster Crossing" along the highway to help protect them in case any of the crustaceans from Down Under escaped and went walkabout.

Bob had a thriving business until he butted heads with the bureaucrats at the Nevada Division of Wildlife. The biologists were worried that some of Bob's desert lobsters might crawl out of their tanks, open the door of the greenhouse, successfully cross the road at the official lobster crossing laid out for them, then somehow make a beeline to crawl more than 130 miles across the barren desert over six mountain ranges to reach Railroad Valley, where, being seriously parched and famished from their long journey, they would somehow find the six warm springs in which the endangered Railroad Valley springfish (whatever the hell they are) live, plunge in, and ravage them all. Bob fought the regulators for years, even appealing to the state legislature, until that fateful day in 2003 when a dozen armed Nevada Division of Wildlife officers descended upon the lobster ranch hell-bent on protecting the people and fragile environment of Nevada by putting Bob's tasty crustaceans to the sword. I am not sure if the heavy arms were needed to protect the agents from the lobsters' claws or to fend off Bob's many dedicated customers, who dearly loved Bob's culinary offerings. Yes, indeed, just about anything is legal in Nevada, including murder – as long as government agents are doing the killing. Building a successful business raising crawdads in the desert – now that is highly illegal. Bob and his wife continued running their restaurant out of the 53-foot white yacht sitting along Highway 95, sans lobsters, for a few years before it finally closed its doors. It was then converted to Captain Jim's Golden Treasures trading post and rock shop. The captain didn't make it in Mina either. The big boat is empty again, and the lobsters are only memories for those lucky enough to have tasted such a delicious desert delicacy.

Another 60-some-miles down the road, as I crest the divide between the Monte Cristo Range and Lone Mountain and enter Big Smoky Valley, I am blinded by a white light off to the northeast, suspended on a tall pedestal, invoking images of the Eye of Sauron flashing from Barad-dur, the Dark Tower in the Lord of Rings. What the heck? Turns out it is the Crescent Dunes solar power plant, aka the Tonopah Avian Frying Pan. Crescent Dunes is a supposedly state-of-the-art "concentrating solar power plant" (CSP), which uses more than 10,000 large curved mirrors (heliostats) arranged in a circle to concentrate sunlight on a 640-foot-high salt-filled tower in the center of the array.

The power plant lives in infamy for its initial test run on January 14, 2015, when the searing heat from the mirrors turned at least 130 birds into overdone versions of Kentucky Fried Chicken. The light from the mirrors attracts insects, which attract birds, which fly through the death rays of concentrated sunlight and are severely burned, thoroughly roasted, or completely incinerated into smoking pieces of aerial charcoal plummeting to the ground – "streamers" as power plant workers affectionately call them, or "collateral damage" as green power advocates refer to them. This carnage is nothing new to CSPs. Another plant, Ivanpah, in southern California, has been frying an average of 30 birds per hour since it began operating in 2014. That's a lot of feathered friends going up in smoke (and adding deadly bird CO_2 to the atmosphere in the process). I am thrilled that Crescent Dunes, much like the highly successful Solyndra solar panel plant, was funded through a $737 million federal loan guarantee. There is hope for Tweety Bird and his friends. Solar Reserve, the owner of Crescent Dunes power plant, and recipient of that generous loan from we-the-people, was quick to announce they solved the fried bird problem with a simple adjustment to the "standby algorithm" for aiming the heliostats. I assume they either stopped counting scorched carcasses or adjusted the mirrors to completely vaporize the birds, leaving no trace of the avian casualties. Make mine extra crispy with barbecue sauce, please.

The next stop on Highway 95 is Tonopah, wedged between 8,515-foot Mount Seibert and Ararat Mountain. Tonopah's moniker as the Queen of the Silver Camps is evident before entering town. Bright white mill tailings fill the wash alongside the highway (appropriately named "Slime Wash"); yellow-and-red-streaked mine dumps and rusty skeletons of mine headframes are sprinkled across the hills.

Tonopah has gone through many booms and busts, with perhaps one too many of the latter. Rich silver veins were discovered in 1900 by Jim Butler (actually by his burro, but Jim got the credit, the money, and the babes). Town sprang up almost instantly. The mining boom was short-lived; the mines pretty much wound down by 1920.

Tonopah also boomed in the late 1970s with the government's infamous MX missile program. Like a giant and deadly version of the three-shells-and-a-pea game, the plan was to hide hundreds of intercontinental ballistic missiles in thousands of shelters mounted on railroad cars that would move in giant loops within Nevada's big intermontane valleys, creating moving targets, most of them decoys, to frustrate the nasty Soviets in case they decided to attack the country. That crazy idea made for a brief boom followed by yet another bust for Tonopah.

The military came to the rescue again in the 1990s with development and testing of the Stealth Fighter, the F-117 Nighthawk, at Groom Lake in the nearby Tonopah Testing Range. Tonopah residents not only got a financial benefit from the program but also got the first glimpses of the mysterious planes, which looked like black obsidian arrowheads streaking across the sky.

More recently, Tonopah boomed with construction of the Crescent Dunes bird roaster in 2011-2014, which employed several hundred workers during the construction phase. For a couple of years, you could not get a motel room in town. People were living in RVs. Not anymore, Tonopah has reverted to the bust half of the cycle. Today, Tonopah ekes out an existence as a stopover for people traveling between Las Vegas and Reno, simply because it is about halfway between the two cities and a long way from anywhere else. There are a few decent motels, including the Clown Motel on the west side of town. The lobby is stuffed with hundreds of clown curios and memorabilia – not a good place if you have coulrophobia (fear of clowns, in case you didn't know), are creeped out by Batman's nemesis, the Joker; or are still traumatized from watching Killer Klowns from Outer Space. But never fear; only good clowns hang out there.

As I enter town, on my left I spy the abandoned Tonopah Garage with its faded sign and empty service bays. The garage was formerly the Exxon station run by Sarge, a classic gasoline alley character with a round beer belly, greasy overalls, an old Army ballcap, and a friendly smile, minus a tooth or two. Working out of Tonopah in 1980, I immediately bonded with Sarge, partly because of his amicable personality, partly because I didn't know anybody else in town, and partly because we were fellow employees of sorts. I worked for Exxon Minerals, and he sold Exxon gas. We remained good friends until the day he ran out of unleaded gas, which had recently been mandated for new vehicles. He filled up the company Ford Bronco with leaded gas, saying, "It won't be a problem. Leaded gas works just as good as unleaded. The oil companies just want to charge you more for their new-fangled gas." That sort of almost made sense to an ignorant young lad since unleaded gas didn't really have the extra cost of removing lead; lead just wasn't added to begin with. I went through a couple tanks of leaded gas with no problem.

Dang it; Sarge was wrong. When I was driving back to Reno on Highway 50 to catch a flight to Denver, the Bronco started bucking like its namesake and finally came to a coughing and sputtering halt on the east side of Bob Scott summit (turns out the leaded gas plugged up the catalytic convertor). I ended up spending the night sleeping in the Bronco; then I decided to hitchhike to Reno. The Loneliest Highway lived up to its name. The few people who drove by did not want to pick up a grubby geologist with a backpack – until Larry showed up. At first a godsend, ole Larry turned into the opposite. He was a derelict and stopped at every bar along the 111 miles of Highway 50 between Austin and Reno. Fortunately, they are few and far between. The last straw was when he wanted to go to the Mustang Ranch for a beer (and probably more). I pleaded with him to just take me to the airport, which he reluctantly did.

Leaving Tonopah, I crest 6,256-foot Tonopah Summit and descend toward Goldfield. Although the pass is as high as Lake Tahoe, the temperature at 11:00 am is 92° F, an ill omen.

I dial my radio to 89.1 FM – KGFN Radio Goldfield, "the voice of the old west," broadcasting from the tiny town of its namesake, in time to catch the Tumbleweed Show featuring Leo the Weather Burro. Unfortunately, I missed the Jackalope Show and the Desert Rat Show. It is fun to listen to a small local station without the annoying commercial breaks of mainstream radio. The programming is not bad, either: lots of classic western songs, cowboy poetry, bluegrass, and folk music – often from obscure but very talented artists. Where else could you listen to a 1936 recording of Gene Autry singing *Yodeling Cowboy*, or catch an episode of Tony's Place, a KGFN original – a really corny, 1940s-style radio drama featuring the Sheriff of Goldfield and his hapless deputy, Blue Balls (yes, really …Blue Balls), as they deal with the corrupt mayor and Lisa, the madam of the local brothel, the Big Crack? Only in Nevada!

Just north of Goldfield, scattered scraggly small trees join the monotonous desert foliage of low sage and rabbitbrush, the nemeses of allergy sufferers. These are Joshua trees (*Yucca brevifolia*), named by Mormon settlers who thought they looked like Joshua holding his hands up to the heavens and beckoning them to the Promised Land. The Mormons must have had a rough trip across the desert. To me, the multiple twisted spiny branches of the Joshua trees look more like a severely dehydrated Medusa after she stuck her finger into an electrical outlet. Joshua trees are considered a marker plant for the Mohave Desert. Thus, Goldfield is, at least by botanical definition, the boundary between the Great Basin Desert on the north and the Mohave Desert on the south.

Looking at Goldfield today (population about 200 plus at least four ghosts), it is hard to believe it was once the largest city in Nevada, although with an estimated peak population of only 30,000, that is not saying much. The town boomed between 1902 and 1920, was serviced by three railroad lines; and at its peak featured five banks, three newspapers, upscale restaurants, countless saloons, and 53 brothels. It even had its own stock exchange. Goldfield survived fires in 1905 and 1906 and a flash flood in 1913 that destroyed large parts of the city. But it never recovered from a devastating fire in 1923 that burned most of the town – 54 blocks in all. Shortly thereafter, the high-grade ores dwindled, and the miners and their camp followers abandoned the city for other gold camps, dealing the final death blow to Goldfield. After the fire, about all that was left of the original town were the schoolhouse, courthouse, jail, High School, the Goldfield Hotel, and the Fire Station (imagine that), which not surprisingly are made of good old fire-resistant rock and brick. I guess the architects of Goldfield never read the story of the *Three Little Pigs*.

The four-story Goldfield Hotel, built in 1907, featured 154 rooms and was billed as the most luxurious hotel between San Francisco and Chicago at the time, featuring elevators and rooms with running water and telephones. The hotel is rumored to be haunted. There are many ghost stories, most of them entirely fictional and debunked. One story has some merit. Shirley Porter, who lived in the hotel from 1976 to 1981 while trying to renovate it, spins a good yarn in her book, *But You Can't Leave, Shirley*. If not true, at least it is a spellbinding tale. The hotel was never successfully renovated despite several failed attempts, perhaps because of what lurks in room 109. The hotel is now boarded up. White pigeon shit streaks the windows, a sad fate for a once majestic historical landmark.

Goldfield never experienced the later booms and busts that Tonopah did, but there have been recent attempts to revive the town. To the rescue of Goldfield came two artistic dudes who started the International Car Forest of the Last Church. Mark, a long-time Goldfield resident, and Chad, who stumbled upon Goldfield while wandering between Burning Man festivals, got the idea to create an international art phenomenon that would attract visitors from around the world – maybe even from as far away as Tonopah, to see their artistic majesty. Fortunately for them, they didn't have to go far for material for their art – junk cars. In Goldfield, everyone's backyard is a junkyard; nobody ever gets rid of a junker unless some crazy artist needs it for his car forest project. The artists held their own version of a car dealer's Push, Pull, or Drag sale and rounded up about 40 derelict vehicle carcasses, including a couple buses. They rented an excavator, dug some holes, and buried the wrecks end-up in the desert. For a finishing touch, the vehicle carcasses were painted with colorful images, hairy eyeballs being a common theme. Although Chad and Mark broke up several years ago (a lover's spat?), the church stands today – a monument to whatever.

I need a break to relieve my numbed behind from the rigors of driving. After passing the Dinky Diner, I hang a left at the south edge of town, just past the boarded-up Golden Parrot saloon, bounce my way over a dirt trail, and past a few derelict mobile homes to the Car Forest. I park my pickup truck next to the first half-buried vehicle, where I am greeted by a sign reading, "World's Largest National Junk Car Forest: ARTIST'S PLAYGROUND AND ATV Park with Ultra light runways" Walking through the artist's playground, trying to discern the upended cars from the forest and keeping an eye out for ultralight aircraft on final approach, I pause next to a silver Subaru wagon buried nose down in the dirt with a big black basalt boulder balanced on its trunk. Just as I was beginning to think my boring monochromatic blue Toyota Tacoma is the only horizontal vehicle there, I spy an old GMC delivery van sitting on its wheels with a beat-up Cadillac stretch limo perched on top of it. It is good to have level company in the Car Forest.

The International Car Forest of the Last Church.

One of my favorite works of car art in the forest is a blue hatchback with this missive painted in red, *"I'm afraid I might not be succesful."* I can't help wonder if learning to spell might help alleviate that fear.

Crossing Goldfield Pass, I enter Malpais Mesa with its yucca-studded flat-topped hills. The mesa is capped by rhyolite tuffs erupted from the Timber Mountain caldera about 15 miles to the east. Eleven million years ago, a series of eruptions puked out roughly 500 cubic miles of rock and ash from the supervolcano, creating a collapse caldera 25 miles across, making all historic volcanic eruptions look like popcorn farts. Another 15 miles down the road, I pull over at Lida Junction, the intersection of Highway 95 and Route 266, and park in the gravel lot of a weather-beaten white building with a big faded yellow "For Sale" sign in front. It is time to stretch my legs and water the desert flora. I turn my back to the stiff scorching breeze (a handy trick I learned the hard way as a young lad), facing the building and its fenced yard full of trapped tumbleweeds. As I stand there holding my pride and joy, I ponder what other blokes did with theirs over the 37-year life of the Cottontail Ranch, one of Nevada's most famous brothels and supposedly a hangout of Howard Hughes, but now an abandoned shell of its former fame.

A few miles later, I pass the Shady Lady Bed and Breakfast, a couple of yellow buildings set back from the road in a little treed oasis. This was formerly the Shady Lady brothel, famous for having the first (and last) male prostitute or "prostidude," as he was known. That experiment didn't last long. I wonder if people staying at the B&B have any idea what the place was previously. And just what did the owners do to convert it? Did they keep the old beds and just add the breakfast? More importantly, have they at least changed the sheets?

Thirty minutes later, I am approaching Beatty, "The Gateway to Death Valley." To my left, a small herd of wild burros grazes lazily on a thin green strip of grass within a broad band of white alkali dust marking the course of the desiccated Amargosa River. The steep slopes of Bare Mountain, aptly named for its lack of vegetation, rise from the river valley and surround the town of Beatty. A familiar sight greets me along the right side of the road – the carcass of a twin-engine plane just off the highway at the entrance to Angels Ladies brothel (formerly Fran's Star Ranch). The temperature has soared to 104° F, but I stop to investigate. The fuselage is covered in graffiti and stickers left behind by nearly 40 years of visitors. An odd one is an anonymous message scrawled on the inside of the cabin: "*Hugh and Patricia are psycho. Don't ever date them. You have been warned.*" Who would want to date a couple, even if they were not psychos? Oh yeah, Californians; I should have thought of that.

The story of the wrecked plane is far more interesting than the graffiti that covers it. The Pahrump Valley Times reported:

> In 1978 Fran York announced that she was going to put a mattress in the center of a large star painted in the parking lot of her establishment. She notified pilots and parachute

jumpers across the west that anyone who could jump from an airplane and hit the mattress would win a free night at the brothel with the lady of his choice.

The day of the big event, Fran's women lined up along the runway dressed in spiked heels and fishnet stockings, g-strings, black lace bras, rhinestone chokers and long capes blowing in the breeze.

A number of planes were in the air that afternoon. One was a twin-engine Beechcraft D-18 with seven passengers piloted by a man named Buster. Buster attempted to land his aircraft on the small airstrip located near the brothel. In preparing to land he did all the right things – adjusted the throttle, trimmed the propellers, lowered the landing gear, flaps down.

But as he made his final approach, instead of watching the windsock, which indicated strong crosswinds, his eyes understandably became focused on the beautiful women on the ground and his plane hit a crosswind. Buster attempted to make a correction. As he tried to regain control, he realized too late that in his distraction he'd forgotten to lock his tail wheel down.

As the ladies looked on, the plane touched down, the landing gear gave out and the craft spun toward the edge of the runway where it hit a high post and came to a halt in the sagebrush. Fortunately no one was injured, but the aircraft was a wreck. Buster was later able to salvage the engines and some other parts, but that was all.

Meanwhile Fran noticed an increase in traffic past her brothel to view the wreck and decided to leave the fuselage where it sat. It was good for business.

Today, it remains a Nye County landmark.

Sadly, Fran's Star Ranch does not. Fran sold the business in 1997 and it was renamed Angels Ladies by the new owners. It was sold again in 2010, but the purchaser did not keep up his payments; the brothel went bust in 2014. Meanwhile, Bikinis Gentlemen's Club opened its doors near the airport at the south end of town in 2011. Advertised as a strip club and with "Nude Girls" written in big letters across the side of the building, it was actually a brothel in disguise – perhaps too well disguised. The club had a brief run and closed its doors, dance floor, and beds in 2013.

Beatty offers a few decent motels and a couple of bars and restaurants for weary travelers. The town used to hold an annual burro race but canceled the event several years ago, supposedly because the mule skinners were too tough for the town. As rough and tumble as Beatty is, that is hard to believe. There must be some awfully poorly behaved mule skinners out there.

I leave Beatty and drive past dry playas off to the west, beckoning me to take a cool dip in their nonexistent deep blue waters conjured up by the shimmering afternoon heat. I am not buying it, but I suspect many thirsty settlers were fooled by mirages on their long wagon-train journeys west. To the left is the infamous Area 51, site of Top-Secret military installations, alien landings, and countless conspiracy theories, including pickled aliens and their spacecraft kept there by the federal government. President Eisenhower established Area 51 as a training center and home for the AEC's nuclear tests. Rumors about the secret facility ran rampant, mostly just because it was top secret. The US Government denied the existence of Area 5 until 2013, when the CIA officially acknowledged it and declassified some related documents.

The next attraction along the way is the Area 51 Alien Center in Amargosa Valley. This is a standard gas stop on my Highway 95 travels, partly because of its cheap gas, but also for its entertainment factor. The center features a gas station, a diner, and a convenience store selling every conceivable Area 51 and alien souvenir (from Alien T-shirts and panties to Alien hot sauce and tequila).

At the north edge of the center stands a tall water tank painted red, with big white letters announcing, "M 800, Alamo Fireworks – World's Biggest Firecracker" and a perpetual sale – "EVERYTHING 50% OFF" (doesn't that mean the price is set to twice normal to begin with?). Just a stone's throw from the gas pumps is Daddy D's, a 1950s-themed diner, which features red vinyl chairs and a big red Coca-Cola sign. Attached to the restaurant is the offensively conspicuous, boldly advertised Alien Cathouse Brothel, where you can get "Hot girls and cold beer." One time when I stopped there, a big sign on the restaurant door proclaimed, "HIRING TODAY– ALL POSITIONS." I was not sure if the restaurant was hiring or if the brothel was looking for specialists. Usually, brothels are tucked away out of view of most travelers, but not here. Signs on the gas pumps invite everyone from grannies to little kids to come in for a free tour. I can just see it now. Mom and pop pull in to fill up with gas. Little Sally sees the signs and asks, "Mommy, can we take a tour to see the kitties from outer space?" Yikes!

To the east lies the recently renamed Nevada National Security Site (NNSS), a fancy politically correct moniker for the infamous Nuclear Test Center, part of Nellis Air Force Gunnery and Bombing Range, previously known as the Nevada Proving Grounds or Nevada Test Site. According to the government's official website, the NNSS is *"A large geographically diverse outdoor testing, training, and evaluation complex."* Large it is – at 1,360 square miles, it is slightly bigger than the state of Rhode Island. The government men must think changing the name now and then will confuse the ignorant masses, who will forget what went on there, back in the 1950s and 1960s when 828 underground nuclear explosions rocked the area and the deadly radioactive fallout from 100 above-ground nuclear explosions drifted eastward to poison the unsuspecting residents of Utah. One thing is for sure, the facility is tightly secured. Even thinking about trying to get in there will get you arrested or filled with lead.

Highway 95 continues for another 20-some miles of twisted two-lane road until the Mercury exit, where the road finally becomes four lanes to accommodate the heavy traffic between Creech Air Force Base and Las Vegas. Behind the security gates of Creech, over 500 Air force computer geeks hold on to joysticks and stare at computer screens, remotely flying drones to terrorize terrorists half a world away – pretty amazing stuff.

Las Vegas lies a mere 40 miles ahead. The desert is really cooking now – at a toasty 113° F.

Unlike most travelers, who would be eager to descend upon the casinos of Las Vegas, I am hell-bent on seeing Sin City and forever construction and traffic jams in my rearview mirror. I negotiate my way through the maze of traffic and finally escape east of Henderson, which has grown to be indistinguishable from Las Vegas. Just before Boulder City, Highway 95 turns south and heads to the old mining town of Searchlight, home of Senator Harry Reid, "Dirty Harry," as Nevadans call him. There is not much left of the once-prosperous mining town, just a small casino, a couple of gas stations, and a McDonalds. I slow down to the mandatory 25 mph speed limit, knowing the town is an infamous speed trap.

South of Searchlight, I am entertained by a pair of dust devils, miniature white tornadoes carrying tumbleweeds and other desert flotsam aloft as they do their counterclockwise dance across the desert flats.

Twenty miles south of Searchlight, I turn east off Highway 95 onto Route 163 and work my way up and over the Newberry Mountains. From the pass, it is a 2,400-foot drop in elevation to Laughlin on the Colorado River. With every foot of elevation drop comes a corresponding increase in temperature.

My travels for the day near an end as I cross the Colorado River into Arizona and pull into Bullhead City just after 6:00 PM. I glance at the thermometer on my truck console. It reads an impressive 122° F. *No way*, I think, *that can't be correct*. I pass a digital sign on the local bank. It also reads 122 degrees. The highway has taken me smack dab into the inferno of hell, where I will spend the next three weeks frying in the hot desert sun. Actually, hell may be better. Local lore has it that when residents along the lower Colorado River, from Bullhead City south to Yuma, die and go to hell, they send word back to please send blankets.

Chapter 21

LIVING HELL

...Shoulda Kept Going.

After arriving in sizzling Bullhead City and checking into the Days Inn motel, I go online to read emails and the latest news. I am greeted by welcoming headlines like Deadly Heatwave Hits Southwest and Intense Heatwave Kills Four, Feeds Southwest Wildfires. The details are disturbing. From NBC News:

> At least four people have died in Arizona from separate heat-related emergencies, authorities said.
> The first, a 25-year-old man, died while hiking the Peralita Trail in Pinal County on Saturday, Sheriff Paul Babeu said.
> Sunday, a 25-year-old woman who worked as a personal trainer died during a morning hike along the Desert Vista Trail in Maricopa County, Phoenix fire officials told NBC News. The heat overcame her so swiftly that she died despite having taken along plenty of water and being immediately treated by a doctor in her hiking group.
> Last Sunday two hikers, a man and a woman in their early 20s hiking in Pima County left the trail without taking along water and had to be rescued by helicopter, authorities said. The woman died before deputies arrived, and the man was being treated at a hospital, the Pima County Sheriff's office said Sunday night.
> And one man died and another remained unaccounted for Sunday night after their hiking group was overcome in Ventana Canyon near Tucson, the Pima County sheriff's office told NBC News. The man who died was believed to be from Germany.

Oh dear, all the people who perished were out hiking. They were healthy youngsters in their 20s. Just what is this 66-year-old going to be doing for the next three weeks? Why, hiking, of course – doing most of it solo in rugged terrain. Official recommendations from authorities are especially helpful:

> Federal, state, county and local health officials issued a reminder that the best defenses against the heat is (sic) to stay indoors if possible, limit strenuous activity to early morning or late-night hours when temperatures are cooler and stay hydrated by drinking plenty of water, avoiding both alcohol and caffeine, which can make dehydration worse.

Good advice, but there is no way I can abide by those rules, especially the staying inside part – the not drinking alcohol part, too. Proper rehydration is essential, you know.

Before sunrise, I am in my truck driving to the edge of the Black Mountains east of town to start my work. By Bullhead standards, it is a chilly 88° F at 4:30 AM. I drive a few miles up the sometimes-graded bone-jarring wash-boarded Silver Creek Road, trying to keep the fillings in my teeth. I turn north and bounce my way over a 4x4 trail, doing my best to avoid huge rocks and holes that could swallow a Volkswagen, for a little over two miles to the start of my first day of fieldwork. The sun has yet to rise over the silhouetted crest of the Black Range, but it is light enough to start work. I slather on sunscreen; don my wide-brimmed hat; grab my rock hammer, map clipboard, and my pack with several bottles of cold water; and set out across the desert in search of gold.

An all-day hike could be disastrous; it is out of the question. My plan is to do short looping hikes and return to the truck every two hours to restock water and cool down. The first traverse is no problem. It is even quite pleasant with the early morning light casting soft shadows across the landscape – barren but for creosote bush, a few small cacti, and the occasional yucca – nothing big enough to provide shade for even a rabbit. The sun gains altitude far too rapidly during the second hike. It is not long before I am sweating heavily. The dry air evaporates the perspiration instantly, leaving white crusts of salt on my shirt. My water supply dwindles quickly. I return to the truck to down cold water and Gatorade. Before departing on another traverse, I drench myself in water and become a walking swamp cooler with temporarily renewed vigor. That lasts for a while until my clothes dry and evaporative cooling ceases. I use water at a ratio of one bottle for drinking and one for drenching to fuel the portable air conditioner.

There is not a cloud in the intensely blue sky. El sol beats down relentlessly. By noon the air temperature has shot up to 112 degrees. It is getting too hot to pick up rocks. At 1 PM, the temperature is 118° F. I use my handy infrared thermometer gun and zap the handle of my hammer. It reads a toasty 144 degrees. That is exceeded by a 154° F measurement on the blistering sand in the dry wash. I am singing a tune by UFO – "Too hot to handle". It is time to quit work and live for another day.

The week goes by in similar fashion, although the heatwave backs off a little with daily highs of only 114-116° F. I am accompanied on my hikes by a few desert denizens. Roadrunners occasionally dash between bushes in the sandy washes, trying to outrun ole Wile E. Coyote with his bag of tricks from Acme Corporation. Small lizards and an occasional chuckwalla peer at me from rock perches in the cool morning. They are nowhere to be found in the scorching afternoons. I don't worry about rattlesnakes. The serpents are smart enough to avoid the searing sun; they stay in their dens until nightfall. Nothing moves after about 10 AM, except one crazy old geologist and wild burros, descendants of the pack animals used by turn-of-the-century prospectors and miners. The burros are beyond tough and seem impervious to the heat. I see them or hear them braying to each other several times a day. Burros have an uncanny ability to smell water from miles away and several feet beneath the ground. They dig holes in the bottoms of dry sandy washes, which magically fill with water. When the holes dry up, the burros sniff out new places to dig. Their abandoned water holes create a veritable minefield in the washes, making driving up them treacherous.

Geologists are accustomed to spending long periods of time away from home, staying in Podunk little towns with little or no entertainment. Bullhead City is different, offering plenty of entertainment. To put it mildly, the people there are "different." I am not sure if that is because of something in the water or just too much desert sun and heat. Either way, there are a lot of very different people wandering around providing free entertainment, whether or not you want it. Many of them are pushing shopping carts down the sidewalk while talking on their Obama phones or just chatting with themselves.

The allure of Bullhead City, other than lots of meth labs, is the Colorado River. I am stuck in Butthead City, as we affectionately call the place, over the Fourth of July weekend. Coming back to town in the afternoon of the Fourth, I look to where a placid river flowed alongside giant casinos on the Nevada shore in the morning. What I see is more like the inside of a giant cocktail blender at happy hour. For some odd reason, people flock here from California in the summer to roast in the simmering heat, then jump on "personal watercraft," aka damn annoying jets skis, to roar up and down the river, attempting to lower their overheated bodies from 110 degrees to 98.6 degrees, only to drop below 90 degrees in the cold waters of the river. There are hundreds, if not thousands, of jet skis zooming up and down the river in a wild and wacky watery version of a game of Chicken. There are no requirements for renting and piloting jet skis other than a credit card to pay for the rental and any damages incurred after playing bumper watercraft with other boats and miscellaneous floating and/or stationary objects. Too many of the watercraft "captains" are drunks or just idiots. Every year there are many incidents of jet skis colliding with each other, with swimmers (even hitting sunbathers on the beach), or with large, clearly visible immovable objects – like bridge abutments. All too often, there are fatalities. There were three in 2016 alone. I am surprised there were not more deaths.

Jet ski mayhem on the Colorado River – 4th of July madness.

On week two, I am joined by two good friends, Tom and Mike Kennedy from Kimberly, British Columbia, who will spend five days frying in the desert sun with me. Tom and Mike are part of the famous Kennedy clan, the best dang prospectors since the sourdoughs of the Klondike gold rush. We work as a team – the Kennedy brothers smashing outcrops with their 10-pound sledgehammers and taking samples while I map the veins and other geological curiosities. They are much better company than the burros and roadrunners. Mike is a strapping lad with a shiny baldhead and full beard – like an odd conglomeration of ZZ Top and Mr. Clean.

As we return to the motel from an early dinner two days into the work, we notice big puffy white clouds to the south over the Black Mountains. Although clouds are a novelty in the clear blue summer skies, we pay little attention to them. In the middle of the night, I am awakened by loud claps of thunder and the sound of wind and heavy rain. *Just a quick thunderstorm*, I think, and go back to sleep. My alarm clock goes off at 4:30 AM. I reach for the light next to my bed to flip it on, but nothing. Crap, the power is out. I bang around in the dark, searching for my clothes, running into a few pieces of motel furniture, and cursing. Eventually, I find the door and open it. Black clouds are pissing down rain while lightning provides a spectacular light show against the blackness of a town without power. It will be a late start this morning.

Mike collecting rock samples at Grapevine Canyon.

The monsoon season is upon us. Each summer, moist Gulf air gets sucked into Arizona and wreaks havoc on the desert. Powerful thunderstorms come out of nowhere and instantly dump inches of rain on bare rock and compacted soils that can't soak up the water. This results in flash floods that wash out roads and occasionally take unsuspecting travelers with them. This storm is no exception. When the rain finally subsides, we drive to the worksite. The graded county road is now a barely passable 4x4 trail where the flash floods and heavy sheet wash crossed it and eroded it into gullies. We enjoy the temporary respite in temperature provided by the rains, but that is largely negated by the accompanying humidity, making for muggy Florida-style conditions. Seems you just can't get a break in these parts during July.

On the Kennedys' last day of work, we escape the noon heat and head to the old mining town of Oatman, just a few miles up the road on historic Route 66, to have lunch at the Oatman hotel. The hotel, built in 1902, was destroyed by fire (like most buildings in old mining camps) but was rebuilt in 1920. It is famous for two reasons: Clark Gable and Carole Lombard honeymooned there in 1939, and today the walls and ceilings are covered with tens of thousands of dollar bills autographed by the tourists who donate them. I order a "Dynamite Burger" – smothered with jalapeno peppers, and a side order of "Burros' Ears" (the local variety of potato chips) washed down with a cold Gold Road beer, while we listen to a local entertainer sing a few Willie Nelson songs in between enlightening the patrons about the colorful history of Oatman. Besides being a heck of a prospector, Mike is also a wicked harmonica player. It just so happens that he has his harps with him. It is not long before Mike has struck up a conversation with the guitarist, Mike West, and has joined him on the stage to jam a few songs.

Mike and Mike jamming.

Oatman burros (and part-time town council members).

After lunch, we greet the burros that wander the street begging for handouts, all the while leaving steaming gifts for unsuspecting and inattentive tourists to step in. The burros are a big attraction for the town and bring in a lot of business. They are not residents. They commute to work. I have seen them marching single file out of town to their desert homes at the end of the workday. They must be clock-punching union workers.

The next day, I drive Mike and Tom to Las Vegas to catch their flight back to the cool pines of Canada. I am on my own again for the last week of work. I saved the best for the last, the dreaded Cliffs of Mordor, a nasty combination of cliffs and ledges of unstable rotten rock interspersed with steep slopes covered in loose scree. I approach the dreaded cliffs with trepidation and extreme caution. I plan to tackle the area early in the morning when it is cool, and I will be able to use handholds to scale the rocks. It is not long before the rocks heat up. It is extremely difficult to get a good handhold on rocks that are too hot to touch. I back off and retreat to lower ground. Mordor will have to wait until cooler weather arrives before giving up its geological secrets.

The Cliffs of Mordor…bad enough under normal conditions, living hell when the rocks are >120° F.

This is terrifying terrain. The blazing desert sun bakes the rocks until they disintegrate into sharp marble-size pieces that sit on top of outcrops at the angle of repose, waiting for an unsuspecting geologist to step on them. When you take a step, the loose marbles slide and a balancing act begins. Going uphill is a classic case of one-step-forward-two-steps back. Downhill travel is more problematic – more like one step forward, three slides forward – on your ass. When the ball bearings under your feet slide out from under you, you tend to pitch backward. Many falls ensue. The trick to survival is to pull off a "controlled" fall. The first things that go are my hammer and map clipboard – tossed to the wind as I try to recover and free my hands to break the fall. Rolling with the fall to absorb the impact can also help, unless you are in the Cliffs of Mordor, where that approach could prove fatal. The other primary rule is not to fall into a forest of Teddy Bear cholla. I have violated that rule a time or two, much to my detriment and pain.

According to the National Council on Aging, 25% of people over 65 fall at least once a year. The consequences are shocking: "A growing number of older adults fear falling and, as a result, limit their activities and social engagements. This can result in further physical decline, depression, social isolation, and feelings of helplessness." Most such falls happen at home. I, however, prefer away games. As much as I have been falling, I should be totally depressed. Actually, I am, but I attribute that to being stuck in Bullhead City. I am also socially isolated, but that is normal. There must be something to this falling thing. The good news for seniors is that if I am included in those statistics, then at least 100 seniors are off the hook for a fall this year. I average three or four crashes per day while mapping and sampling, not counting three-point landings.

Yea, though I walk through the Valley of the Shadow of Teddy Bears, I fear no evil, except the Teddys; Ouch! Why you don't fall near Teddy Bears - time to get the scissors and pliers, Moss, AZ, 2016.

The end of my sentence in hell approaches. While I am getting dressed for the last day in the field, I look for the dirty work shirt I wore for the past three weeks. It is leaning against the wall, stiff from all the salt that encrusts it. The shirt crackles as I put it on. It is also getting more than a little ripe. Perhaps I should have taken the time to wash it. Nah, it would just get sweaty and dirty again, so why bother?

The next day I make the trek back to Reno without incident, reversing course on Highway 95 and ending my adventure along the Highway to Hell.

January finds me back in Bullhead City, managing a drilling program and following up on the work the Kennedy brothers and I did in July, including traversing the foreboding Cliffs of Mordor.

Upper Cliffs of Mordor on a cool January day.

While gingerly traversing the upper cliffs, I come across the skeleton of a desert bighorn ram lying between two sections of cliffs. He apparently missed his footing and fell to his death. Good grief, if sure-footed sheep can't handle this terrain, what is this wobbly old geologist doing here?

King of the hill – Cliffs of Mordor.

Bullhead City and its neighbor across the Colorado River, Laughlin, are very different in the winter than they are in the summer. Gone are the crazed jet skiers and boaters. A horde of Snowbirds replaces them. Every winter, from mid-November until April, Bullhead City and Laughlin are invaded by swarms of senior citizens, who, following the course of Canadian geese, migrate south to escape the frozen wastelands of places like Minnesota, Wisconsin, and Alberta to spend the winter in sunny Arizona.

Laughlin has grown from Don Laughlin's original one-casino town into a mini-Las Vegas strip stretched along the west shore of the river. Third-rate bands play in the casinos during the week. Occasionally, some top-notch acts and bands show up on weekends. In addition to robbing the old folks of their retirement savings at the slot machines and gambling tables, the Laughlin casinos offer senior citizens several unique forms of entertainment (try to find these anywhere else), such as Meat Bingo. That is just what it sounds like. Be the first to fill out your bingo card, jump up excitedly (trying not to have a heart attack), and yell "Bingo," and you get – you guessed it – a hunk of cholesterol-laden, artery-choking red meat. On the same theme (those seniors must be starved for protein), there is also the Slab-O-Rama, a barbeque contest sponsored by the Rotary Club.

My favorite event is the Senior Games. Amongst manly (OK, and womanly, too) athletic competitions such as swimming, running, bicycling, and powerlifting, the Senior Games feature more mellow sports like slots, poker, and some real zingers like the "Cornhole Contest." I am not making this up. It is printed in the January 2017 issue of the Laughlin, Nevada Entertainer magazine. No explanation is provided for what goes on in a good Cornhole match, so one is left to one's imagination, which can be dangerous for individuals with sick minds. I am not sure how many players are involved in a match or what positions there are, but I'm guessing two: a pitcher and a catcher. The event is held at the Edgewater Hotel Casino in Laughlin and is open to the public. Somehow, I would think that any old farts frisky enough to enter that event would want a bit more privacy for the contest. But hey, what happens in Laughlin apparently doesn't stay in Laughlin. I can just see it now: grandma and grandpa return to Minneapolis in April. Little Johnny asks,

"Grandma, what did you and grandpa do in Arizona this winter?"

Grannie replies, "Lots of things, Johnnie. We saw the London Bridge. I played slot machines and won $100, and we did a lot of cornholing. In fact, your grandpa won a gold medal in the Cornhole Contest at the Senior Games."

This gets Johnnie's attention. He exclaims with excitement, "Really, can I see it, grandpa?" (I am hoping he is referring to the medal rather than a cornholing demonstration).

Gramps pulls out the big gold medal hanging around his neck and proudly shows little Johnnie his sparkling prize, with "Top Cornholer – Bullhead/Laughlin Senior Games"– inscribed on it; no doubt something every octogenarian would be proud to show his grandkids.

Only in Nevada!

Chapter 22

RING OF FIRE

And it burned, burned, burned.

I spent a couple of days working in the Snowstorm Mountains of northwest Nevada in mid-July to design a drilling program and locate drill sites and access routes for a client. I would have spent another day there if things had worked out differently.

Record snowfall in the winter of 2016-2017 and an unusually wet spring in northern Nevada spawned heavy growth of vegetation, not the least of which was highly flammable cheatgrass. Hot, dry summer weather dried out the grasses, making for a potentially explosive situation. All it would take would be a spark from metal meeting rock, a carelessly discarded cigarette, or a zap from good old Thor, to set the countryside ablaze. By mid-summer, extensive areas of northern Nevada had burned, setting the course for a near-record wildfire season. Before it was all over, more than 600 wildfires consumed 1.2 million acres of rangeland and forest.

The project area is very remote, lying 80 road miles from Winnemucca, of which 50 miles are dirt road, and the last 15 miles are a two-track 4x4 trail. The jeep trail is the only road into the area. It ends just past the project, which borders a wilderness area on the south and east and rugged mountains and canyons on the west.

I put my faithful Toyota pickup into 4-wheel drive and bounced my way up the rough road to a saddle on the east flank of Snowstorm Peak. At 7,400 feet elevation, the temperature was about 80° F under clear blue Nevada skies, a far sight better than the 100-degree temperature I left in the valley below. I quickly set up my pup tent, organized my gear, and went to work.

The heavy winter snowpack finished melting less than a month before. Lush spring vegetation covered the hillsides and subalpine meadows. The sweet aroma of wildflowers was in the air. Pink and purple lupine, crimson Indian paintbrush, and purple horsemint poked out of the tall grass and sagebrush, while bumblebees and black-and-yellow butterflies busied themselves gathering nectar and pollinating the next generation of flowers. I was alone in nature's glory, miles from any form of civilization, annoying noises, people, or pollution. And I was getting paid to be there. Heck, it just doesn't get much better than that.

I was hiking back to camp at 6 PM on the second day when dark clouds moved in from the east. I felt a few sprinkles of rain. *Oops, I left my sleeping bag outside to air out.* I hurried back to camp before it started to rain hard. Luckily, the storm passed to the north. Just before arriving at camp at 6:30 PM, I heard a couple of thunderclaps from up north of camp.

I didn't think much about it.

Back at camp, I set about preparing yet another gourmet camp dinner. I was finishing my appetizer of Geisha smoked oysters on whole-grain crackers and about to begin the main course – a can of Hormel chili (I consumed the beanie weenies the previous evening) – when I caught a whiff of smoke. *Uh oh, that is not good*, I thought, *but there are lots of fires burning in Northern Nevada; it is probably blowing in from a far-off fire*. To be safe, I decided to check.

I put dinner on hold, hopped into my truck, and drove up the steep 4x4 trail that leads to the high ridge to the north, where I would have a sweeping view to the north, east, and south.

Sure enough, there was a brush fire burning to the northeast, across the rocky canyon of the South Fork of the Little Humboldt River. The claps of thunder I heard came from lightning that struck there.

Ribbons of yellow flame formed a ring of fire surrounding an area of black devastation that moments before was tall grass and sagebrush. A line of fire was expanding on the southeast side of the fire. A smaller, nearly stationary ribbon of flame marked the northwest border of the blaze, near where the lightning strike started the fire.

The wind was out of the northwest, so the fire was moving to the southeast. I watched for about 20 minutes as the fire advanced rapidly downwind but had largely died out on the northwest side.

That was good enough for me; the fire was heading away from me. It looked like it would burn out when it hit the cliffs along the northeast side of the canyon, preventing it from coming my way.

I drove back down the hill, fired up the Coleman stove, and cooked the best dang dinner I had all day. Then, to ensure I would have a good night's sleep and would not need to worry about waking up to an advancing fire, I decided to make a trip back up the ridge at dusk to verify the fire had moved on.

As I crested the top of the ridge at 8:15 PM, I saw the billowing smoke and thin ribbon of flame at the front of the fire, about a mile east of where it had been. *Good*. I looked to the left and – *HOLY CRAP!* What had been the dead windward side of the fire was now a roaring wall of flame. The wind shifted, and the fire took off to the west, *NOT good!* It was heading for the 4x4 trail leading to the project. If uncontrolled, it would eventually come toward camp.

Being at the dead-end of 15 miles of rough trail with no way out other than the one threatened by a raging fire is not a desirable situation. I had to weigh my options carefully. I could stick it out and hope the wind would shift again and stop the fire's westward advance or that the BLM Hot Shots would arrive and extinguish the blaze. The latter was unlikely because the fire was burning in a wilderness area, where the government has a policy of not fighting wildfires. Darkness was approaching, and no aircraft had flown over the area yet. So, it was possible that I was the only person even aware of the fire in this remote location. Not having cell phone reception didn't help either (thank you very much, ATT). Plus, the BLM was busy fighting at least a dozen more pressing fires threatening people and property elsewhere.

If I waited until the morning, the fire could be somewhere between where it was now and where I was camped – if not IN camp. If it reached the road, I would be cut off with no escape. What to do then – race through the front of the fire and hope for the best? Why not? Back when the Barnum & Bailey Circus used to come to town, I saw tigers jump through hoops of flame. Why couldn't a pickup truck do the same with this ring of fire? After all, the active flames at the front of the fire are probably only 50-100 feet wide. A quick calculation was in order. Let's see – at 30 mph, it would take less than two seconds to travel 100 feet. That calculation didn't factor in the heat wave from the superheated air in front of the advancing fire. It turns out the wall of hot air cooks along at a whopping 1,470° F (I looked up that little fact later. I knew it was hot, but not THAT hot!). The idea of trying to drive through the wall of heat and the flames with a half-tank of highly flammable gas and a somewhat flammable driver was not very appealing. And what if something went wrong? What if I lost the road in all the smoke and crashed or got stuck? Roasted geologist, that's what.

My die was cast. I would go for an immediate escape and attempt to beat the fire before it gets to the road. If I failed, I would turn around and head back to the end of the road, monitor the fire from the high ridge, and be prepared to move out on foot if it came in my direction. Needless to say, it would be a sleepless night. If the fire did come close, I planned to abandon the truck on an outcrop with as little vegetation as possible around it – where it might survive. Then I would hike west over the shoulder of Snowstorm Mountains down to Kelly Creek. The fire would move more slowly downhill. I could probably outrun it, given a decent head start of a few hours.

The Toyota launched down the hill. I set a new world record for breaking camp. Organization be damned; I threw everything into the back of the truck in a pile. At 8:35 PM, I began the long bumpy trip down the hill – toward the fire.

It took over three hours to make the uphill trip to camp. That was in the daytime when I could see the ruts and rocks in the trail and negotiate a way around them. The downhill trip in the darkness was reckless abandon. The truck lurched from rut to rut; boulders banged against the skid plates. I flew over a big mud hole I carefully worked my way through two days before. No worries, this was a mad race to beat the flames to the bottom of the mountain.

After a few miles of tooth-jarring high-speed travel and almost getting high-centered when I didn't line up with the proper ruts, my headlights flushed a group of Black Angus from the tall grass, fat and happy from eating the lush grasses and forbs in their summer range. I rarely give range cattle a thought, but I felt sorry for these bovines, who were unaware their worry-free existence could end suddenly with them becoming well-done steaks long before the arranged time.

Over every rise in the road, I looked to the right and saw the flames lighting up the night sky to the northeast. The first several miles of the road headed north and northeast, bringing me closer to the fire. That only made me more anxious. I drove faster and more recklessly.

After several miles and a nervous hour of driving, the trail turned to the northwest. I started to put some real estate between the fire and me. Still, the specter of the fire catching up to me pressed me on, and I drove posthaste with no looking back.

About two miles before the end of the trail, I noticed flashing lights, which appeared to be near the intersection of the jeep trail with the county road. Once I reached the graded county road, I figured I was home free. Not! Someone else noticed the fire. Just when I thought my night of anxious flight was over, I found myself out of the fire and into the frying pan of a convoy of BLM fire-fighting vehicles. At the first corner, I was almost run over by a big BLM fire truck, which kicked up enough dust that I could barely see. Out of the dust came a huge water tanker. I veered off into the sagebrush to avoid a collision with that monster. Behind that were several more vehicles full of firefighters and equipment. It would be ironic to escape the fire but end up squashed like a bug under some giant piece of government metal.

Another ten miles down the road, I came to Chimney Reservoir, where there is a small campground. There was only one other site occupied, so I had my pick. I pulled into a campsite and parked. Rummaging through the disorganized mess in the back of my truck, I found my camp cot and set it up next to the truck. I threw my sleeping on top of the cot and lay down just before 11 PM. The Milky Way arched overhead and dazzled me with its zillion stars. To the north lay the Big Dipper and my old friends Cassiopeia and Draco, reminding me of stargazing on family camping trips when I was a young lad. Orion's shield stood out in the southern sky; a bright satellite, perhaps one providing GPS service, moved steadily across the Milky Way. I lay there in silence, alternating between absorbing the tranquility of the night with millions of stars shining brightly in the clear, thin Nevada air and reflecting on the events of the evening and what could have been, as words to Johnny Cash's *Ring of Fire* danced in my head. A stark reminder lay off to the east, where an eerie luminous cloud from the fire I escaped blocked out the stars. To the west, a bright glow washed out the bottom of the Milky Way. It was like looking back at the lights of LA, but there was no city there, only open range and another massive brush fire. Such is wildfire season in Nevada.

I eventually drifted off to sleep, content in having successfully put miles between me and the fire, and knowing I camped in an open area with no vegetation that could burn.

Saturday morning, I drove back to Winnemucca, then east to Battle Mountain and south to Austin to start another job. Looking to the north between Winnemucca and Battle Mountain, I could see the smoke from the fire. My thoughts went out to the firefighters, who would have a challenge fighting that blaze due to lack of access and rugged terrain.

The fire, which the BLM appropriately named the Snowstorm fire, raged for nearly a week and ended up burning 158,000 acres…that's 247 square miles! The Elko Daily News quoted the US Forest Service as stating: "The fire expanded rapidly on Saturday and fire-retardant drops were used to slow the spread." Bulldozers were also used to build firebreaks. The Forest Service added, "significant infrastructure and wildlife resources are at risk. Resources include habitat for threatened and game species such as sage grouse, big horn sheep, mule deer and elk; live waters including trout streams," – and one threatened geologist. Ok, so the Forest Service didn't really say the last part, but they should have.

The fire didn't reach the road to the Snowstorm project. I never found out if the BLM firefighters stopped its westward advance or if Mother Nature intervened and sent it off to the east. I guess I could have stuck it out safely in camp, but with age comes a certain degree of discretion and caution. For once I actually used both.

Chapter 23

Odd Quest

*If you think our liquor laws are strange,
you should see our underwear.*

Utah bumper sticker

"Can I help you find something?" the cute young blonde asked.

"Uh, well…ummm…sure…uh, belts, I need a new belt." I was lying to the saleslady at the Deseret Industries store. My mission was to create a diversion while Glenn rummaged through the underwear section.

During my time working for Newmont's Elko exploration group, I spent the field season of 1990 partnered with Glenn King, an entertaining young geologist with a novel and bizarre perversion. Glenn was a wiry little fellow with a lot of energy – most of it misdirected. Our assignment was to explore for gold in southern Idaho and Utah, principally the Oquirrh Range west of Salt Lake City. Before starting fieldwork, I compiled maps highlighting host rocks, gold occurrences, mineralized structures, et cetera – favorable elements that could guide our work and lead to discovering the next big gold deposit. Meanwhile, Glenn was making his own compilation: thrift stores, laundromats, public swimming pools, saunas, and gyms throughout the state of Utah. What the heck?

For reasons I never understood, Glenn had something against the Mormons. Maybe he was jealous of the polygamists since they each had a bevy of young blonde babes, and Glenn had none. Foolish young fellow! Glen didn't realize it, but the good Lord was punishing the Mormons by giving them polygamy. I had one wife and couldn't keep her happy. Lord knows what hell on earth the polygamists were going through. Perhaps Glenn was annoyed by the smartly dressed missionaries, who kept knocking on his door and trying to convert him. Or maybe he was a distant relative of one of the 120 emigrants whom Mormon militia disguised as Indians slaughtered for their wealth of livestock and supplies in the infamous Mountain Meadows Massacre near Cedar City. Then again, Glenn served as a tank driver in the Army Reserve. Banging around in one of those oversized tin cans may have scrambled his brain a bit too much.

Glenn claimed he read the *Book of Mormon*. He was more than happy to educate this ignorant geologist with his vast knowledge of the good book: gems of wisdom like the original Mormons traveled from Saudi Arabia to Central America some 2,700 years ago (via submarine, no less), upon their arrival, they established the great Olmec and Mayan cultures (a factoid I am sure would be news to the Mayans), and the original book of Mormon was written on gold plates which Joseph Smith mysteriously found under a rock in Ohio (left there by a wayward Mayan, of course). Glenn could even quote from the Holy Scriptures, usually starting with, "And it came to pass..." I had no reason to doubt Glenn's stories but found them bizarre, to say the least. If true, the bizarreness rests squarely with Joseph Smith and the Church of Latter-Day Saints.

Glenn's particular peculiarity was his underwear obsession. Now, I could understand if he were obsessed with ladies' panties, having been on an exciting panty raid or two back in my college days. But no, Glenn had a strange proclivity toward men's underwear – and not your typical BVDs. He had a penchant for Mormon underwear, Temple Undergarments, "magical underwear," full-length union suits worn by devout Mormon men. The underwear is supposed to have great religious significance and is considered sacred. Some people even say it has magical powers and keeps the wearer safe from all kinds of mishaps and evil powers.

Glenn was on his unique version of a Mormon mission. He desperately wanted a pair of Mormon underwear. The size didn't matter. He just wanted a pair, any pair, as a kinky souvenir, perhaps. I had no idea what he would do if and when he succeeded in his quest – frame them, run them up a flagpole like Old Glory? Regardless, his perverted passion led us to a summer of seriously depraved activities, or at least a lot of fun planning them. Partners stick together, so I was committed to assist in the quest.

The simplest approach to accomplish Glenn's goal would have been to go to the Mormon underwear store and buy a set. That required actually being a Mormon and presenting your official Mormon membership registration number and your Magic Moroni Decoder Ring, both of which Glenn lacked. The backup plan was to find a used pair for sale, a set of holey holy underwear – hence Glenn's list of thrift stores. We visited every Goodwill, Salvation Army, and Deseret Industries store from Bonneville to Provo with high hopes but found no used magic undies. I did, however, score a field hat, a sturdy Army surplus shirt, and a belt from our visits.

The curious thing about Mormon temple garments is you are not supposed to take them off. Not unless absolutely necessary, that is. And then, they should be separated from your body for only a brief time, just long enough to accomplish the task that required removing them. Oddly enough, procreation does not seem to qualify as one of those tasks. The underwear has flaps in all the right places to allow for whoopee without undressing – for expediency, I guess. I mean, think of the lost time if you had to dress and undress for each of your dozen wives. Joseph Smith thought of just about everything. Valid reasons for temporarily ditching the undies include showering and swimming. That explained Glenn's list of swimming pools and gyms. Glenn wanted to join health clubs just to have access to the locker rooms. Equipped with a pair of bolt cutters cleverly hidden in his gym bag, Glenn planned to hang out nonchalantly in the locker room, trying to look casual and not too much like the pervert he was, waiting for someone to strip down, revealing his special undergarments. Once the object of his desire went into a locker, and the owner left to work out or shower, Glenn's ploy was to pounce upon the locker, cut the lock if necessary, and make off with the booty. There was a fatal flaw in the scheme. The guys working out were much bigger and buffer than Glenn. He reconsidered that idea lest they catch him in the act of divine underwear theft and beat the holy crap out of him.

Laundromats were a brilliant idea. Nobody could wear the same underwear every day, at least not without stinking to high heaven. We assumed people must own at least two pairs. Whenever we did our laundry, we kept an eye out for a load that contained Glenn's quarry. The plan, if we came across a pair of holy underwear in the wash, was for me to distract the launderer with lively conversation about some pressing subject, such as the latest vision of the Elders or news from the missionaries in Botswana, while Glenn raided the washer and ran out the door with his soggy prize. Alas, we didn't see a single pair of the elusive underwear during our laundry scouting forays. Maybe they never wash them.

And it came to pass; Glenn never found any Mormon underwear. We didn't find much gold either. Defeated, we retreated to Nevada, the land of world-class gold deposits and normal BVDs.

Escaping Utah did not mean escaping all things Mormon. Back in Nevada, we soon ran into an invasion of Mormons, ones that would never give up their undies to Glenn – because they have none. The summer of 1990 saw a bumper crop of Mormon crickets, the Jolly Green Giants of the insect world, migrating in seething swarms across the state.

Mormon crickets (*Anabrus simplex*), which technically are katydids, not true crickets (in case anyone actually cares), are three-inch-long disgustingly ugly bugs with even more disgusting habits. In certain years, the crickets hatch in profusion. When they mature, they form giant armies, numbering in the millions, which move as a mass across the landscape, eating anything and everything in their path – grasses, sagebrush, other insects – and each other.

Mormon crickets may disagree, but to me, female crickets are uglier than their male counterparts. The ladies are armed with an evil-looking long spike sticking out of their butts. It looks like a giant stinger but is an ovipositor, a harmless tube the cricket sticks into the ground to lay its eggs.

We ran into swarms of crickets while walking through the sagebrush. Suddenly, odd squeaks emanated from the bushes, which shook under the mass of bugs chewing on their leaves. Panicked crickets hopped and ran all over the place. The march of crickets continued across the valleys and mountains, unchecked until they ran into vertical obstacles, such as the walls of prospect pits. There, the nasty bugs attempted to scale the walls, eventually falling back into a growing pile of frustrated six-legged migrants. The worst places were where the crickets ran into standing water – common in prospect pits. Mormon crickets are not proficient at either the backstroke or the six-legged breaststroke. The pits filled with the rotting carcasses of drowned crickets, creating an awful stench.

We quickly learned that if you ran over a cricket or two on a road and squished them, then returned a little while later, a pile of crickets would be on top of the dead or injured ones, happily munching on their fallen comrades. That made it delightful fun to run over the mass of cannibalistic crickets, which grew exponentially on the 4x4 trail as more hungry soldiers arrived and piled on to consume the bugs that were consuming the bugs. The crickets presented a slippery situation along the paved highways. The hordes of migrating crickets crossing highways did not zig or zag to avoid vehicles as a rabbit would. They just marched on – and piled on. After a while, there was so much cricket gore on the highway, the road became slick and hazardous to drive. Eventually, the Highway Department had to bring snowplows out of their summer mothballs to scrape the guts off the road.

Crickets dining on each other on Highway 50; a sexy Mormon lady.

Given the cannibalistic tendencies of the crickets, Glenn and I wondered what would happen if crickets ran out of food and only had each other to eat. We knew they readily devoured their wounded and dead, but were they true cannibals that killed to eat their fellow crickets? It was time to devise a scientific experiment. Glenn and I set out to find the biggest, meanest crickets in the horde. We each captured a big nasty male (the gals were too ugly to be contestants). Resurrecting memories of childhood zoological expeditions, we put them in a mason jar with holes punched in the lid to provide ventilation. Thus, began the great game of Creepy Cricket Combat.

We placed bets on which cricket – mine, dubbed Jiminy, or Glenn's, Bugsy – would defeat the other in mortal combat, consume the vanquished bug, and be declared the winner. We expected rapid results. The dang crickets just sat there and ignored each other. "OK, so they are too well fed, let's starve them for a while," Glenn suggested. A week later, Jiminy and Bugsy were looking famished but made no attempt to attack each other. I surmised they were waiting for the bell to ring, officially starting the fight-to-the-death match. I got a bell and rang it. They sat in their respective round corners and glared at each other. Another week passed with no action from our green gladiators. We wondered if we selected the wrong crickets. Maybe they were vegans, the only two out of millions of crickets.

Glenn and I went on break for a few days. I took Jiminy and Bugsy home to show my boys, who were just little buggers back then. They allowed the crickets were the grossest, creepiest things they had seen since Pee-wee Herman showed up at our door on Halloween night (young Eric was unfazed by Dracula, Frankenstein, and the Wicked Witch of the West, but freaked out when he saw Pee Wee. He has yet to recover from the trauma).

During the third week without any aggression, we considered wounding each of the crickets. Then, like sharks sensing blood in the water, they would recognize their roommate as dinner, and the attack would begin. But Glenn had become attached to Bugsy and didn't want to hurt him. I must confess I was growing fond of old Jiminy as well, if only for his stamina and Mahatma Gandhi-style determination to stick with his hunger strike. After a month with no winner in sight, we gave up. The Mormons won. We released them in the sagebrush where they could live out their last days munching on greens instead of each other. They would not have the opportunity to hook up with any ladies to add to the next generation of crickets. The swarm was a good 30 miles to the west by then.

Glenn and I went prospecting in the Monitor Range, northeast of Tonopah. We found an extensive area of alteration with prominent silicified outcrops on the east side of the range. The day was spent hiking, pounding on the outcrops with our rock hammers to collect rock samples – basically making little rocks out of big ones – and putting the chips into cloth sample bags. We carefully labeled the bags and recorded the sample locations and descriptions in our little yellow notebooks. At the end of the day, we unloaded the samples from our packs and stacked the bags in the back of the pickup.

It was a hot Indian summer day. We had a two-hour drive to Tonopah over rough, dusty roads. We were seriously parched and would miss happy hour, so we broke out a cold six-pack of Bud to quench our thirst – cheap swill for sure, but it did the job. To ward off hunger on the trip back, we opened a bag of Doritos to complete our mobile happy hour. Our route followed Little Fish Lake Valley south, thence west along a bumpy 4x4 trail over the mountains and down to East Stone Cabin Valley. We popped open our second Bud as we crossed Eagle Pass. The sun was dipping behind the Toquima Range in a blaze of orange. As we bounced down the west flank of the pass, we noticed a stream of dust behind a truck heading south on the road in the valley ahead of us. The truck slowed down as it approached the intersection with our road. It stopped…and waited.

We pulled up to the intersection to find a uniformed man standing next to a Ford F-150 with an official-looking logo on the side of the driver's door. *Uh-oh,* I thought. *What would the Sheriff be doing way out here? I hope it is not illegal to drink and drive off-road?* He turned out to be a game warden with the Nevada Department of Wildlife. No big deal to us; we were hunting gold, not big game. The bird-hunting season began the next day. It was time for brave men in camouflage vests to roam the hills with double-barreled shotguns, firing away at anything with feathers: chukker, grouse, partridge, doves, sparrows, hummingbirds, Indian chiefs, down pillows – anything that would explode in a cloud of feathers when blasted with birdshot. Jumping the gun and hunting a day early would be a big no-no.

The game warden approached the truck and came to the window. The officer could not have chosen better words for the serendipitous conversation that ensued.

"How are you guys doing?" the game warden asked in a friendly tone.

"Great," we replied

"Did you have any luck today?"

"Oh yeah, we were really lucky today."

"Really? Did you bag any?" There's the key word.

"You bet. We bagged a bunch."

"How many?

"Let's see; I think I bagged 18. How about you, Glenn?"

"Twenty-two," Glenn replied with a satisfied grin on his face.

"Really? Where are they?"

"Oh, they're all stacked up in the back of the camper shell."

"Can I see them?"

"Sure, go knock yourself out."

I watched in the side mirror as the game warden walked to the back of the truck with a spring in his step, no doubt contemplating the fabulous accolades he was about to receive – maybe even a promotion and big raise – for making the biggest bust of bird poachers in Nevada history. He lifted the tailgate window and peered in, certainly not prepared for what greeted him – a pile of rocks in white cloth bags, samples we "bagged," After a couple of minutes, he closed the window and walked back along the side of the truck – a changed man, shoulders slumped, and head hung low. He came to the window and said, "Damn, here I thought I made a big bust of bird poachers, and all I got was a couple of geologists drinking beer and eating chips."

We laughed all the way to Tonopah

Chapter 24

KAY MINE DISASTER

Not a creature was stirring, the rods were stuck tight.

The number one rule in the military is, "Don't volunteer." Being that I never served my country in a killing capacity, I didn't get the message.

It was early December 1983 when my boss in Exxon Minerals Reno office, Tom Irwin, asked the staff if anyone was willing to help the Tucson office by managing a drill rig in Arizona for a week. My fieldwork was done for the season. I had nothing pressing to do. There wasn't enough snow for good skiing yet. So, I raised my hand. I packed my gear and was off to Black Canyon City, Arizona, to work with the core drillers.

The drilling was at the old Kay Mine, a massive-sulfide deposit that produced high-grade copper-lead-zinc-silver ore from 1916 until it closed in 1956. My assignment was to supervise the drilling and log core while the project geologist was on vacation. The drillers, who worked for Longyear Drilling, needed to complete the current drill hole, then take their holiday break. The drilling was expected to take about a week, leaving a cushion of a week to finish the hole before Christmas in case the drilling slowed down.

The drill hole was down about 1,100 feet on its way to a final depth of 1,400 feet. When I arrived at the drill site, things were not going well. The hole was caving; the drillers were worried about getting stuck. I took a quick look at the core, walked around the area looking at outcrops, and realized the problem. The drill had penetrated a major fault zone – a badly broken zone of loose rock, that was falling apart and blocking the hole as the drillers tried to advance it. Worse yet, it appeared the hole had steepened and was drilling right down the fault and would not get out of it.

A fast-moving cold front moved through that evening, bringing a taste of winter to the normally balmy desert. The following morning, I made a dawn visit to the drill rig, then went for a hike – partly to check out the geology of the area and partly just to stay warm in the crisp 40-degree morning air. I walked along briskly, occasionally stopping to inspect outcrops for signs of alteration and mineralization. As I climbed a steep ridge, I spied something odd, a chuckwalla sitting on a small outcrop. Seeing a chuckwalla in the Arizona desert is not unusual; seeing one that doesn't run away at warp speed is. Chuckwallas are large (up to 18 inches long) bulky lizards with smooth wrinkly skin. Their defense mechanism is to run to the nearest outcrop or pile of rocks, where they flatten their bodies and slip into a crack in the rocks. Then they inflate their lungs and wedge themselves in the rocks, where it is nearly impossible to extract them. This chuckwalla didn't move as I approached it. Fearing it was dead, I picked it up to inspect it. The poor bewildered fellow was alive, just numbed and paralyzed by the rapid drop in temperature. The sun would not clear the surrounding mountains for another hour. I opened my jacket and put the lizard inside, hooking his claws into my shirt so he could hang on. I zipped up my coat and decided to take Chuck, as I named him, with me on my hike.

Chuck and I spent a fun hour looking at outcrops and inspecting old mine workings. The low December sun finally crested over the ridge to the southeast, casting long shadows behind the saguaros. I unzipped my jacket and pried Chuck off my shirt. He was not so stiff and cold. I placed him on my left shoulder so he could get some sunshine to charge his batteries.

I took him to visit Coati Cave, an adit where I saw a coatimundi the day before. Coatis are strange beasts, members of the raccoon family, but the stretch-limo version on stilts. They are long-legged, thin critters with a long fluffy tail and a stretched snout, looking more like a cross between a raccoon and an anteater. Mr. Coati was not home, so we moved on.

Half an hour later, Chuck was starting to stir. It was about time to check on the drillers' progress. I took Chuck to the entrance of a small adit, where the sun would soon heat the rocks and cure his stupor. I placed him there and bid my cold-blooded friend goodbye. Then I walked back to the drill to check on the progress, only to learn there was none.

Chuck thawing out.

After a few of days of very little progress, the inevitable happened. The hole caved, and the rods and drill bit were hopelessly stuck in the hole. You can't drill through a diamond-studded carbide drill bit. The only solution was to lower a charge of dynamite down the hole to shoot off the stuck bottom portion of the rods. After blasting the rods free, the drillers cemented a hardened steel wedge in the hole above the stuck drill bit to use the wedge to deviate the hole past the obstruction.

The drillers were scheduled to complete the hole by the end of the week, when I should have been on my way back home. Come Friday, the drill bit was 20 feet higher in the hole than when I arrived. I wrote my weekly drilling report and recommended stopping the hole and moving to the other side of the target to drill back across the foliation and the fault. My boss insisted on continuing the hole.

The drillers got the hole back on its new course and began making slow progress. Thirty feet down the hole, the hole once again caved in. At least the rods were not stuck. The drillers drilled through the obstruction, but 20 feet later, they were stuck again. The rest of the week went by in a similar one-step-forward, two-steps-backwards style. The drillers would make 20 to 30 feet of progress; then, the hole would cave again. They moved higher up the hole, set another wedge, and drilled again.

The drillers were scheduled to take their Christmas break on Saturday, December 22, unless the hole was finished earlier. That looked unlikely. Sure enough, on Thursday, they once again got stuck in the hole. On Friday, when my weekly drilling report was due, we were stuck 50 feet higher than when I arrived two weeks before. The hole would not be finished before the drillers' Christmas break. I think we set a record for backward drilling.

Feeling frustrated and somewhat embarrassed, I sat in my pickup truck and composed my drilling report. I drove to the pay phone at the 7-11 in Black Canyon City to read the report to our secretary, Ellie. To make matters worse, it was the night of the company Christmas party. I would not be there. The smart geologists, who know better than to volunteer, would be having a merry time and I would be in Black Rock City as stuck as the drill rods.

The following is what Ellie got over the phone. It took quite a while to dictate the report due to the laughing and occasional restroom breaks, she took to avoid peeing her panties.

FOUR NIGHTS BEFORE CHRISTMAS

T'was four nights before Christmas, while at the drill site,
Not a creature was stirring, the rods were stuck tight.
The core was logged at the warehouse with care
In hopes an orebody soon would be there.
The managers were nestled all snug in their beds,
While visions of budget cuts danced in their heads.
The drillers in their hardhats and I in my cap,
Had just given up on drilling this crap.
When up on the hill there arose such a clatter,
I sprang from my truck to see what was the matter.
Away to the drill rig I flew like a flash,
Tore a hole in my jeans and tripped on some trash.
When what to my wondering eyes should appear
But a miniature sleigh and eight cases of beer!
With a little old driver who smelled like a skunk;
I knew in a moment he must be half drunk.
More rapid than eagles his beer cans they came,
And he whistled and shouted and called them by name:
"Now Bud, now Coors, now Henry's and Heineken,
On Oly, on Molson, on Millers and Schlitzen!
To the top of the rig, to the top of the mast
Now dash away, now dash away, now dash away fast!
So up to the drill rig the beer cans they flew
With a sleigh full of tools and St. Longyear, too.
And then in a twinkling I heard on the kelly,
The gurgling and sloshing of beer in his belly.
As I drew in my head and was turning around,
Down the drill stem St. Longyear came with a bound.
He was dressed all in rags from his head to his foot,
And his clothes were all tarnished with drill grease and soot.
A bundle of tools he had flung on his back,
And he looked like a derelict just opening his pack.
His eyes were dilated, his dimples were hairy!
His cheeks were like roses, his nose was real scary.
His droll little mouth was drawn up like a bow,
And the beard on his chin was like old yellow snow.
The end of a wrench he held tight in his teeth,
And his hardhat encircled his head like a wreath.
He had a broad face and a large, round belly

That shook when he laughed like a bowl of Quik-gelly.
He was chubby and plump, a right grubby old elf.
And I gagged when I smelled him in spite of myself.
A wink of his eye and a twist of his wrench
Soon gave me to know the source of that stench.
He spoke not a word, but went straight to his work.
He filled the hole with Low-loss; turned the rods with a jerk.
And laying his finger aside of his nose,
He sprang to his sleigh, to his beers gave a whistle.
And away they all flew like the down of a thistle.
But I heard him exclaim as he drove in the night,
"Happy Christmas to all, the orebody's in sight!"

Ellie read the report to the holiday revelers in my absence. I am sure that was the only time a weekly drilling report was read at a company Christmas party, let alone enjoyed by all.

Chapter 25

RADIOACTIVE

Ban the bomb!

    During the early days of my career working as a geologist exploring for uranium, I spent the better part of ten years walking over uranium-enriched outcrops, carrying highly radioactive rock samples, and breathing in radon gas. Given the cumulative doses of gamma rays and alpha and beta particles that bombarded my body, I should have been dead years ago, or at least should have mutated into some hideous evil creature. But, wait, my ex claimed the latter did happen.

    Geologists must be related to pack rats. We delight in carrying home pretty rocks, which we proudly place on our desks, bookshelves, kitchen tables, and any other flat surfaces we can find, where they sit for years, collecting dust. For uranium exploration geologists, that means having lots of radioactive rocks hanging around. One day at Exxon Minerals' office in Denver, one of the geologists walked down the hall with his scintillometer turned on. The instrument clicked enthusiastically as he passed the rock collection in each geologist's office and screamed loudly when he passed the storage room, where lots of hot rocks were kept. The secretary freaked out. We were ordered to remove all our precious rock samples and store them in the basement. Bummer, I had to restock my entire rock collection.

Radiation emanating from rocks was problematic, but not really of great concern. Alpha particles are stopped by clothing; beta particles travel only a few feet. Gamma rays were a bigger concern since they are high-energy radiation and penetrate just about everything. Fortunately, they attenuate quickly with distance from the source. The real problem was radon, the "invisible killer", a colorless, odorless, tasteless gas. Despite the best attempts to ventilate underground uranium mines, levels of radon gas were difficult to control. Many old miners eventually succumbed to lung cancer, caused by long-term exposure to elevated levels of radon gas. There was an interesting correlation. Miners who never smoked remained healthy. Heavy smokers died from lung cancer. Needless to say, armed with that knowledge, cigarettes never touched my lips.

During my initiation as a temporary geologist, my boss, Steve Collings, and I visited an occurrence of radioactivity mentioned in an old government report. It was located in an isolated area along the border between Arizona and Mexico. To our surprise, the site had turned into a hippie commune. A group of longhaired, freaky people dressed in dirty, baggy clothes was camped in tents and old VW buses decorated in an array of psychedelic patterns. The camp was centered on a sparkling spring that filled a small pond and created a little oasis in the otherwise barren desert.

The hippies were as surprised to see us as we were to see them. Sage, the head hippie, a tall bearded character wearing sandals and a long, tattered robe, who could have passed for JC himself, came to greet us. A hippie chick, Sunbeam, who was barefoot, dressed in a tie-dyed flowing granny dress, bedecked in beads, joined him. They wanted to know what we were doing there. We explained our business. They were relieved to learn we were harmless geologists and not narcotics agents.

Accompanied by Sage and a group of grubby hippie offspring, who thought we were the strangest characters they had ever seen, we walked around the camp using our scintillometers to locate the source of the anomalous radiation. The few rock outcroppings in the area produced nothing. When we walked up to the spring, our scintillometers went crazy; the needles zoomed off-scale. *Uh, oh.* This was not what we expected. Radon gas, which is soluble in water and can travel long distances in groundwater until it reaches the surface, was bubbling up through the water.

The spring and pond were the commune's source of drinking water. Sage was shocked at what we found. Flower children are not fond of nuclear bombs, radioactive fallout, or even nuclear power. Learning that beloved Mother Nature was polluting their desert utopia with radioactive gas did not make them happy campers. We advised them not to hang out for long periods next to the spring and to let their drinking water sit for a day or so in pots before drinking it or using it for cooking, so any remaining radon would dissipate. We were going to advise them not to bathe in the water. But after seeing and getting a good whiff of the unkempt people, we decided the advice was unnecessary since they were clearly not into such hygienic activities.

I had my own run-in with the invisible killer a few years later while investigating a prospect in the Front Range, west of Boulder, Colorado. I was immediately drawn to a small shaft and mine dump. A wooden ladder led to a drift 20 feet down the shaft. Being a curious cat and forgetting what curiosity does to kitties, I climbed down the ladder.

At the bottom of the shaft, I noticed the needle of my scintillometer climbing. So, I switched to a higher range. As I walked down the drift, it went off-scale. *Wow*, I thought, *there is some high-grade uranium here*. Excited by the discovery, I spent an hour mapping the underground workings and collecting samples of the "ore."

Back on the surface, I walked around, checking the level of radioactivity. Everywhere I went, the readings were three to four times normal background. Beaming with excitement, I took notes postulating I found a high-grade deposit surrounded by a giant low-grade disseminated uranium deposit.

My excitement soon turned to utter terror. I tripped over a rock. As I pitched forward, the scintillometer that hung at hip level from a shoulder strap swung around and pointed at my belly. The buzzer went off, indicating high radiation. I stopped and ran the scintillometer up and down my body. When I pointed it at my chest, the reading was off-scale. Holy shit, my lungs were full of radon that built up in the confined space of the mine workings.

As radon 222 decays to its daughter product, polonium 218, it emits alpha particles. Alpha particles are large, low-energy atomic particles that can be stopped by a single sheet of paper. However, lungs are not coated in paper. Having radon and her offspring in direct contact with my lungs was definitely not good. Radon has frightening effects on the human body; a dose of radon being equivalent to 20 times that of a dose of high-energy gamma-ray radiation. The half-life of radon, at which point half of the original amount has decayed, is 3.8 days. As far as radioactive isotopes go, that is short but not short enough. I was not about to wait for it to slowly disintegrate. I wanted that crap, all of it – not just half, out of my lungs right then and there.

I had a scintillating conversation with myself, consisting of an audible, "Oh, Shit!" I dropped my pack, hammer and scintillometer and took off at a sprint. I ran as fast as I could up the nearest hill and reached the top, panting like a greyhound after yet another futile attempt to catch that dang mechanical rabbit. Fagged from the effort, I ambled down the hill to where I left my equipment.

I picked up the scintillometer and ran it over my chest. Hallelujah – normal readings! I successfully cleared my lungs of the invisible killer. Sadly, without a chest full of radon, all the rocks now gave low readings. Denied a big discovery but spared to live another day or two, I packed up my gear and went home.

Despite extensive radiation exposure, I have managed to survive to a ripe old age and somehow even sired two sons, who turned out to be relatively normal individuals. Well, maybe not quite normal, but their abnormalities could be due to their DNA being sourced from the shallow end of the gene pool rather than any nuclear influences. I am delighted they at least have the appropriate number of digits on their hands and feet. On a personal level, I did receive an unusual but useful benefit from the many years of being zapped by deadly radiation during my uranium exploration days. I now save money on nightlights since I no longer need them. I just have to drop my drawers; the soft blue glow from my private parts lights the way.

Chapter 26

POKING THE BEAR

"You're fired!

— Donald J. Trump

Mineral exploration is a risky business – one that is expensive, requires a long time to yield results, and has low odds of success. The risks are offset by the giant prize that awaits success – ore deposits worth hundreds of millions to billions of dollars, making the risk-to-reward ratio palatable to companies with long-term business plans and financial stability. Unfortunately, the business attracts far too many companies with short-term/get-rich schemes and minimal financing.

Reconnaissance exploration, looking in places where little is known, requires years of groundwork: prospecting; geological mapping; regional geophysical surveying, sampling of rocks, soils, and stream sediments; and a lot of luck to find mineralization worthy of further work.

Follow-up exploration of promising prospects is also expensive, requiring more detailed mapping and sampling, possibly some geophysical surveys, and expensive drilling to sample the subsurface. Despite all the hard work involved in finding enough ore to justify spending tens or hundreds of millions of dollars to develop a mine, only about one in one thousand prospects is advanced to be a mine.

Patience and money are required for success. Sadly, management tends to be short on both. As a result, many exploration programs are terminated before they have a chance to be successful.

Exploration geologists find the deposits that keep mine engineers and metallurgists employed throughout their careers. Sadly, we get little credit for it. As soon as a mine goes into production, the exploration geologists are laid off. Mines have a defined life – based on the size of the deposit and the rate of mining. The mine life is finite unless economics change or exploration finds more ore. There is a classic story, surely based in fact, about a mine that was in the last few months of its planned life. The mine manager wakes up and realizes the mine is about to run out of ore and will have to shut down. He calls up the Chief Geologist and demands, "Send the exploration geologists out to find more ore."

"I can't," replies the Chief Geologist.

"Why not?"

"You fired them three years ago."

My career has had more ups and downs than a basketball during March Madness – not only because mining is a cyclical boom-or-bust business, but also because I often can't resist antagonizing the boss.

Exploration geologists are at the bottom of the mining industry's totem pole as well as on the bottom floor of the industry's two-floor outhouse. Funding for exploration, the very foundation of the industry, without which it could not exist, is subject to even more cycles than mining is, including commodity cycles, political cycles, economic cycles, investment cycles, icicles, and tricycles. To make matters worse, besides the cyclical pressures, I have somehow managed to piss off just about every boss or manager that I ever had – especially at the big companies. It took me a while to learn I am not corporate material. Good corporate employees are "yes men"– brown-nosers, who kowtow to the boss and whose aspiration is to climb the corporate ladder. Like the boy who yelled, "The Emperor has no clothes," I have too often challenged the bosses – not that I necessarily thought I was right. I just wanted them to know they were wrong. That effectively tore the rungs off the corporate ladder as I slowly reached for them. Despite that annoying habit, I don't think I have ever been fired, although I am not quite certain about that minor detail. All it takes is a downturn in one of the many cycles, the passage of some new regulation, diversion of corporate funds to some pet project of the CEO, or the arrival of a cold front or a brand-new Monday for management to lay off dispensable personnel, such as lowly geologists.

Another factor in my corporate downfall relates to diplomacy, which is best defined by Sir Winston Churchill: "Diplomacy is the art of telling people to go to hell in such a way that they ask for directions." I had a decent grasp of diplomacy. The problem was I focused entirely on the first half of the definition.

The US Exploration Manager for Exxon Minerals was an infamous character named Dick Weaver. His parents must have been prescient because his first name was appropriate. It would have been even better if his middle name had been Head because the two names together described him perfectly. Dick was a formidable antagonist who loved to argue, just for the sake of arguing. He also loved to fire people. After Dick left Exxon Minerals, he was the CEO of Atlas Precious Metals. He routinely fired people for fun or to flex his corporate muscles and keep the employees in fear of his wrath. In classic Dick fashion, he fired a geologist for fidgeting with his dinner fork while Dick gave a long-winded speech at the company's Christmas party. It is no wonder the response when his name was mentioned was often, "Dick Weaver — before he dicks you!" or "If the sun still rises in the east, Dick Weaver is still an asshole."

I had a few run-ins with Dick. He came to visit the Stone Rock project on Prince of Wales Island in Alaska shortly before the summer's drilling program began. Alaska Drilling Company won the drilling contract. Two drill rigs were being barged to the site the following week. The plan called for using a lightweight man-portable rig that could drill shallow reconnaissance holes without pad preparation or permitting and a larger rig to drill deeper holes, which required building drill pads, permitting the sites, and moving the rig with a helicopter. My boss, Bob Woodfill, and I agreed this was the best approach for the project. Dick, in his classic butthead style, argued we should only use the big drill rig. I explained the need for the small one. Then he argued we should only use the small rig, not the big one. I explained the need for the larger drill; he switched his argument once again. After going around in circles for about 20 minutes, I turned to him; in my best statesmanlike manner, I said, "It doesn't really matter. The two rigs are on the way, and that is how we are going to proceed." At that point, Bob held up his clipboard, on which he wrote, *Dick is tired. Let's go back to camp.* Tired and ornery would be a better description! We walked back to camp, gave Dick a cocktail, and sent him off fishing for the remainder of the afternoon. All seemed fine.

Two years later, while having a beer with Bob Woodfill, I learned Dick went back to the Denver headquarters after his trip to Alaska and did everything in his power to fire me or transfer me to the company's worst hellhole office. It took intervention from the VP to block him. Yes, the sun still rises in the east, and Dick Weaver (RIP) is still a flaming asshole.

Part of my employment (or lack thereof) problem is due to the fact that geologists and mining engineers don't get along. It is not the geologists' fault. We are sociable types and will drink beer with anyone. I drank a lot of beer with future mining engineers when I was in college. They were fun fellows back then, when we were equals, studying hard, getting lots of wrong answers on tests, and being verbally abused by our professors.

Once they become professional engineers, mining engineers think they have to be correct all the time. They convince themselves that they *are* right all the time. They are not risk-takers, being sure to dot every "i," cross every "t," double or triple check every calculation, and avoid making decisions until they are 110% sure they are right. Geologists, on the other hand, are risk-takers and gamblers. We are not afraid of making decisions, be they right or wrong – something that is much more fun and easier to do if you are using someone else's money.

Exploration geologists are rarely ever right – and we know it. Come to think of it, the only times in my career when I have been right were when I finally admitted I was wrong. Never fear; that does not preclude success. As Sir Winston Churchill so eloquently noted, "Success is the ability to go from one failure to another with no loss of enthusiasm."

The difference between engineers and geologists was summed up succinctly by the famous author, Anonymous, in his insightful essay, "Observations".

"Observations"
Anonymous

An Engineer is said to be a man knows a great deal about very little and who goes along knowing more and more about less and less until finally he knows practically everything about nothing.

Whereas, a geologist, on the other hand, is a man who knows very little about a great deal and keeps knowing less and less about more and more until he knows practically nothing about everything.

A landman starts out knowing practically everything, but ends up knowing nothing about anything, due to his association with engineers and geologists.

After I was laid off by Exxon Minerals in 1985 (along with some other colleagues – I dragged a few victims down with me that time), I took a job as Exploration Manager for Rio Algom, a division of the mining giant, Rio Tinto, managing the company's Reno exploration office. It sounded like a great job, pursuing acquisitions of advanced exploration and development projects for a respected mining company, one with a successful history spanning more than 200 years. I was wrong. I reported to a freaking mining engineer, an unprincipled rascal who shall go by the pseudonym, Chuck, since he really has a big tattooed cousin in Philly, who runs a pizza parlor and has a side business involving a baseball bat and people's kneecaps. Mining engineers belong in mines, directing big Tonka Toy shovels and trucks, and calculating ways to shave a few pennies off production costs. They should not be in charge of exploration.

Chuck was a real piece of work, a piece of stinking excrement, that is – a spineless butthead, bereft of any redeeming qualities. He had a curiously inert personality and had no balls at all (enough of the sugarcoating – remind me sometime to tell you how I really feel about him). Chuck was terrified of making decisions since each one carried a 50% chance of being wrong. Suffering from paralysis by analysis, he either tried to avoid decisions or passed them off to the company's attorney, who has happy to charge excessive fees for making Chuck's decisions for him.

My partner in crime, Bruce Harvey, and I were tasked with finding exploration opportunities for the company. Nothing was good enough for Chuck, including properties that eventually became profitable mines. One project was not acceptable because it was too close to a highway, another because it was too far from a highway. Every property we recommended had some sort of fatal flaw, which allowed Chuck to kill it. We joked that Chuck would not acquire any property where water flowed downhill, a definite project killer.

The final straw came when Bruce found an outcrop of high-grade gold mineralization near Ely, Nevada, on open land adjacent to a group of claims held by an attorney in Reno. We planned to stake claims on the open ground, but we also needed to acquire the adjoining land to have a big enough property for a mine. I knew the attorney and initiated negotiations with him. Chuck wouldn't approve staking the open ground until the deal was concluded. We were close to having a signed agreement when Bruce noticed a competitor snooping around the area where he found the mineralized outcrop. Bruce and I agreed we needed to stake the ground right away. I tried to call Chuck to convince him of that. He was off hunting ducks or pursuing some other diversion that relieved him of business decisions. I couldn't reach him. I called the owner of the adjacent claims to notify him we were staking the claims, which would be included in the joint venture, as per the draft agreement. He agreed with the plan. We went ahead and staked the claims. Bruce and I drove back to Reno.

Friday morning, I called Chuck at the Denver office and told him we staked the claims to protect the critical ground in case some competitor found it. Good grief, you would think I peeked up the dress of the Virgin Mary in the Holy Loo or committed some other mortal sin. Chuck, acting like the pile of fecal matter that he was, went ballistic, telling me how we had exposed the company to potential lawsuits, stock fraud, insider trading, various venereal diseases, and a host of other mean, nasty, ugly things that would cause the entire company to implode within the next 24 hours. Chuck demanded that Bruce and I drive back to the site and remove the claim posts that very day. I explained I obtained permission from the owners of the adjacent claims, so there was no conflict of interest. Besides, the claims were not legal entities until we recorded the paperwork. I further explained that Bruce worked for three weeks straight and was taking time off with his family. Likewise, I worked for ten days and had plans for the weekend.

"Can't we go back and remove the claim posts next week?" I asked.

The answer was a firm, "No, they have to come out today."

I lost it and replied, "Well, in that case, you can go pull out the f'ing posts yourself!"

Well, ole Chuckie was not happy. Nor did he take me up on my gracious offer. Bruce went out the next week and "un-staked" the claims, a first in exploration history. Yet another property was cursed with a fatal flaw or two (water did indeed run downhill there). The deal was terminated.

Needless to say, my relationship with the boss, which was already rocky, commenced a steady decline down the slippery slope to a final showdown. A few months later, the company decided to consolidate offices. Chuck was transferred from Denver to Reno so we could be one happy family together. A week after he arrived, he called me into his office and began the conversation with:

"You have an hour to clear out your office."

"Oh, I see. Does that mean I am being fired?"

"You can resign or be fired. The choice is yours."

"Is there a difference in severance?"

"No."

So, without giving Chuck the satisfaction of selecting the prize from Door #1 or Door #2, and without kicking him in his hanging down parts (which I dearly wanted to do, but I suspected there was nothing there to kick), I stood up and in departing, stated casually, "Well, then, I'd better get busy."

To this day, I don't know if I was fired or if I quit. Nor do I really care. After two frustrating years of getting nothing done, I was free of the ball-less mining engineer.

After the experience with Rio Algom, I decided working as an employee of big companies was not for me. Working as an independent contractor was a much better fit. Sometimes such work is hard to find. I was in between consulting jobs for a while in early 1991. I was looking for consulting work but the only thing I could find was a full-time job as a Senior Geologist for Newmont Mining based out of the Reno exploration office. Out of desperation, I took the job. Given my propensity for burning the rungs on the corporate ladder, I knew working for a big company would be a mistake. I would get in trouble at some point. I just didn't realize how soon that would be. Nor did I have a clue I would get into more trouble in the three years I worked for the company than in the 30 preceding years.

My initial assignment was to study Newmont's Nevada gold deposits and compare them to similar deposits worldwide to develop exploration criteria. It sounded like a fun research project. I arranged visits to the company's mines and talked to each of the mine geologists to get their input. In appreciation of their efforts, we invited the mine geologists to a dinner at the Pine Lodge in Lamoille. My partner for the project, Pat Mallette, the mine geologists, a couple of geologists from the Reno exploration office, my boss, and a few other high-level dignitaries, attended.

The event commenced with cocktail hour; wine flowed freely with dinner; after-dinner drinks followed. Pat and I sat at the far end of the long dinner table, across from two young blonde lady geologists. We were having a great time, telling blonde jokes and otherwise being politically and socially unacceptable. The big wigs were at the other end of the table, engrossed in serious business chatter. As the evening wound down, someone who ate trout for dinner stuck the fish head in a shot glass and passed it down the table as a joke. When it got to me, I stuck my fingers in it and turned it into a puppet. We were laughing and having a good old time until the trout's lower jaw dropped off. At that point, I screwed up royally. The talking trout became Doug Wood, the former manager of the Reno office, who had been replaced by Rick shortly before I hired on with the company.

A little background is needed here. Doug broke his jaw in a car wreck a few months earlier. His jaw was wired shut for a month, much to the delight of everyone in the office. Doug was a bit of an oddball. He had some really strange ideas, the latest of which was a "holistic" approach to exploration. Science be damned, Doug was going to find gold with incense, a handful of bones, and a crystal ball. Nobody believed in what Doug was doing. His program was the brunt of many a joke around the office. The trout head with half a jaw repeated some of those jokes, much to the delight of the gals sitting across from Pat and me.

After dinner, I rode back to Elko with my boss, Rick. He was strangely silent all the way back.

The next day, I went into the wilds of northeast Nevada to do a week of fieldwork. When I returned to Reno, Rick came by my office to announce it was time for my three-month personnel review. I was to be in his office at 10:00 AM. I walked in with a smile and sat down, expecting to hear what a great job I had been doing. Rick turned to me and said, "You are not fit to be a Senior Geologist at Newmont."

I was stunned. That was not quite what I expected. Dispirited but not intimidated, I used the most of a pregnant pause to call upon my keen sense of diplomatic style and fired back, "That's OK. I am not hung up on titles. Call me a Junior Geologist if you want, as long as the pay is the same." That was true. I have never been big on titles. I learned early in my career that one company's Project Geologist could be a high-level position, whereas it could be an entry-level one at another company. I registered to attend a conference on uranium exploration organized by the Atomic Energy Commission in Grand Junction, Colorado, back in 1978. The pre-registration form asked for my title. Given that job titles are often a joke, I wanted to leave the line blank but finally decided to invent my own title. The line on the form for my job title thus read, "Imperial Geowizard." At the conference, I got in the A-C line to pick up my registration packet. The gal searched for it, but it wasn't there. She went to get her supervisor, who returned with her and said, "Oh, we have it over here. We've been waiting for you." Everyone's badge had their name in large letters, under which their title and company were in smaller type. Mine was switched. My name and company were in small type under the job title, IMPERIAL GEOWIZARD, in bold capital letters. I wore that badge proudly for the entire conference.

I was taken back by Rick's comment. I had been busting my butt at work and thought I was doing an exemplary job. Rick went on a rant about how disgusting I was and how he had never been so embarrassed in his life. I was confused and didn't put two and two together for a while until he mentioned, "that night in Lamoille." Then it dawned on me. He was either upset about the blonde jokes or about the talking fish head. I apologized and said it would never happen again, whatever it was. My plan going forward was to avoid restaurants that serve trout and to never dine with Rick again.

After my ass-chewing, I talked to a couple of my colleagues and learned the dirty secret behind the story. Well, bugger me, it turned out that Doug, the former manager of broken jaw fame, was no longer the manager because he was involved in a sexual harassment lawsuit with one of the company's female geologists – specifically, one of the blondes that the fish head imitating Doug had been entertaining. Because it was an ongoing lawsuit, it was supposed to be a big hush-hush subject. If only I had known!

The harassment lawsuit was settled by the yearend. Doug left Newmont to pursue a career as the one and only holistic consulting geologist. The other half of the lawsuit drove west to California in her shiny new sports car. I somehow managed to keep my job and my title, but my annual bonus was a big fat zero that year. I would have preferred to trade the title for the money.

A year later, I got into even more hot water with Rick. He had a grandiose plan for exploring the western United States. Hardly holistic, it was based on geophysics and geochemistry. Geology was pretty much ignored. The entire office: the geophysicist, geochemist, draftsman, all the geologists (except Pat and me), and the computer geeks, spent full time working on the project for nearly two years. They compiled geophysical and geochemical data from company files and every published source imaginable. The data were then plotted on regional maps of the western United States: straight, curved, and squiggly lines for magnetic and gravity linears; dots for anomalous values of various elements detected in the National Uranium Resource Evaluation (NURE) program of the 1970s; and bigger dots for gold deposits. The final product was a series of maps that overlaid magnetic linears and NURE gold anomalies on top of known gold deposits. Linears that passed through NURE gold anomalies were deemed the chosen ones. Those magnetic linears were given a 10-kilometer area of influence. Low and behold, 60% of the known gold deposits in Nevada lay within their rather broad area of influence. We were to focus all our efforts for the next three years within those zones.

I got to thinking – there were so many lines and points that just about any set of data with a similar density would likely produce similar results. I set out to test my hypothesis, but which data sets to use? I tried a few things before settling on paved highways for the linears (the geophysical lines were not necessarily straight ones) and something unique to Nevada for the geochemical data points – brothels, using Robert Engles' essential book *Brothels of Nevada*, supplemented by my own field observations.

I plotted my non-biased data on a map of the state of Nevada. The draftsman, who was finally available to do some work outside of the boss's pet project, drafted the map and used AutoCad to calculate the percentage of gold deposits that fell within the area of influence of my linears – 69%. I did not make that number up; it came from his computer. The map hung proudly on the wall of my office. Fellow geologists came by to view it and get a chuckle.

Word about the map, dubbed "Trails and Tails," spread rapidly. Bruce Harvey, my former colleague at Rio Algom, who was in charge of exploration around Newmont's mines, called and invited me to present my findings to the mine geologists at their weekly meeting in Elko. Bruce offered to buy me a beer after the meeting. With that as an incentive, I gladly accepted. The whole idea was to add a little levity to the meeting.

I began with a background on how the compilation study was conducted. I showed a few humorous examples of its evolution, like when there were so many geophysical lines on the map it went black (some sorting and prioritization was in order). Without meaning to pick on geophysicists (although they deserve it), I proceeded to tell the audience how to ascertain the difference between geologists and geophysicists using a simple math quiz:

Ask a mathematician what 2+2 equals; the answer is, "Obviously, the sum of two plus two is the integer 4."

Ask the same question to an engineer and the answer is, "Precisely 4.000."

Pose the question to a geologist, and after a bit of head-scratching, the reply is, "Umm, somewhere between 3 and 5?"

But ask a geophysicist what 2+2 is, and the answer is, "Depends. What do you want it to be?"

Geology is an inexact science (but it *is* a real science). Although geophysics uses a lot of fancy black boxes and advanced math, it is also an inexact science. There are always multiple interpretations of the data.

The presentation finished with my "Trails and Tails" map. The geologists were chuckling. I noticed there was not much laughter from the back row of the audience.

The next day I went to the field in search of gold in the hot zone, being sure to stay within 10 kilometers of a brothel.

I returned to the Reno office nine days later. A note from the boss was on my desk: "See me. Right away." *Uh. Oh. What is it this time?*

Well, it seems the proverbial poop hit the fan over my presentation in Elko. The people in the back row, who were not amused, were the geophysicists. Jim Wright, the leader of the pack, was so upset with my abuse of geophysical data that he could not sleep. He called Newmont's Chief Geophysicist in the Denver headquarters the next morning to complain. Of course, the Chief Geophysicist went down the hall to notify the CEO of the horrible things I had done. Following upon ole Dougie's sexual harassment lawsuit, which cost the company a fair amount of money (at least as much as the price of a new Corvette), the top dog asked, "Were there any women present?" Of course, there were. Now the problem expanded from insulting geophysicists to potential sexual harassment since words like "cat house" and "tails" were used in the presence of *females*.

Rick informed me there would be an inquisition the following morning at nine o'clock in the conference room. All the personnel who participated in the compilation project were there to take turns raking me over the coals. Bruce Ferneyhough, the geophysicist, was the first to attack. "This is an insult to my profession. You have destroyed all the hard work we have done to improve the relationship between geophysicists and geologists," he said in a trembling voice. I tried to explain I thought the work they did to compile the geophysical data was good. I only disagreed with the interpretation. Scott, the geochemist, was next in line. He berated me for making a joke about his work. I noted the NURE geochemical data was seriously flawed (a well-known fact in the industry – the samples were collected by poorly supervised incompetents and analyzed by unqualified laboratories – it was a government program after all), so putting emphasis on that database was my biggest problem with the study. Scott admitted he knew the data were flawed, but NURE was the only regional geochemical data set available to use. He felt he needed to contribute something to the study, so he used it. He then backed off a bit and said he was not really upset because "Geochemists are used to being abused." He was the only one with half a sense of humor. One by one, each of the staff took their shots at me.

After the prosecution rested its case, I finally got a chance to defend my actions. I began by telling everyone to lighten up, that the "Trails and Tails" study was done as a joke, and asking where the hell their sense of humor was. I also noted there was an underlying moral to the story. One of the critical data sets was flawed, making the conclusions shaky. Furthermore, my hypothesis that just about any similar data set could find an equal number of gold deposits was proven correct. Maybe they should put their emotions aside and take a second look at the study. Did we really want to devote three years and millions of dollars of budget to exclusively exploring the areas defined by those flawed criteria?

We never followed up on the recommendations from the compilation project. It quietly faded away. Instead, we went about our normal exploration. There was no fallout from sexual harassment since the ladies present in Elko were geologists who had a sense of humor. I once again survived being fired, but my annual bonus was more than slim. Somehow, I think it was worth it.

Another conflict with Rick involved safety, or at least his concept of it. Rick had worked in the desolate Outback of Australia and was paranoid about geologists getting lost in the wilds of Nevada, especially during the winter, an odd season he did not understand. He instituted a system by which each geologist had to complete a form stating where he would be and how he could be reached each night when working in the field. We were to call the office at the end of the field day (long after the office closed) and leave a message on the answering machine saying we returned safely. This was before the advent of cell phones. Finding a working pay phone and a dime or two was often a challenge. Supposedly, the secretary would check the calls in the morning. If someone failed to call in, a search and rescue operation would commence (although perhaps a bit late). Gene "the machine" Urie and I worked out of Austin, a tiny town in the center of Nevada, over the Fourth of July weekend. We came back from the field and went straight to join in the festivities, forgetting to call the office. We didn't think it would be a big deal since nobody would check the messages for three days anyway (a small flaw in the system).

Once again, upon my return to the office, I was in trouble. There was no message from Gene and me over the long weekend. One would think that should have initiated some sort of search for our bodies, which by then likely would have assumed ambient temperature on an outcrop somewhere in the mountains of Lander County. But no, the reaction was for the boss and the secretary to argue over whose fault it was the system didn't work. Nobody was concerned about us. That was fine by me. I had worked on my own for 30-some years without a problem and didn't see any need for the call-in system.

That problem never arose again. At times we worked in places too remote to commute from the nearest town and its phone booth, in which case we camped. Rick acknowledged we could not be expected to call the office from a nonexistent pay phone at a campsite in the middle of nowhere. That was the loophole I was looking for. Every time Gene, Pat, and I went to the field, we filled out our form stating we were camping – regardless of where we were staying. It was a ruse, but the boss accepted it; I think because he was tired of dealing with my insubordination.

Eventually, Rick moved on to bigger and better things and was replaced by Bill Oriel. Bill and I got along great. However, that did not exempt me from doing stupid things and getting into trouble.

Bill and I were part of a team evaluating a group of gold prospects south of Las Vegas. Richard Gorton joined us from the Denver office. John Squyres, a local consulting geologist, led the field trip. We spent the better part of a day examining outcrops and inspecting old mine workings. Late in the afternoon, we drove down a dry wash and stopped at the portal of an old mine tunnel. Bill, Richard, and John went in. I was getting bored from hearing the same old stories, so, I decided to walk down the wash and look at the outcrops instead of joining the group in the mine.

The wash took a sharp bend to the right. Around the bend was another small portal. I turned on my flashlight and went in. About 50 feet inside, I heard voices. How could that be – unless the two workings joined? I continued walking. The voices got louder. I could see a faint light coming from the right side of the tunnel. When I reached the intersection, I looked to the right, down about 100 feet of tunnel, and saw the silhouettes of Bill, Richard, and John standing about 20 feet inside the mine. My sick sense of humor kicked in. I backed up a couple feet from the intersection, worked up my best imitation of a mountain lion scream, and let it rip. It must have been pretty damn convincing. I poked my head around the corner to see three scared-shitless geologists running ass-over-teakettle out of the tunnel.

I strolled down the tunnel, chuckling at the practical joke I pulled. As I walked out of the mine, I saw the guys frantically trying to get into the Bronco parked in the wash. It was funny until I noticed Bill was standing on one leg with a pained expression on his face. He took off running so fast he tore his calf muscle. He was hurting and spent the remaining two days of the trip sitting in the Bronco. We brought him offerings of rocks to keep him occupied. When it was time to fly back to Reno, I pushed him around the Las Vegas airport in a luggage cart. I probably should have been fired for that infraction. Practical jokes, as well as impractical ones, were a big "no-no." Fortunately, Bill had a great sense of humor and didn't make a big deal out of my indiscretion, but I felt bad for his pain and suffering.

The gig with Newmont came to an end in 1994, when the company closed its Reno office. I went back to working as an independent consulting geologist, a much better fit for me, especially since there was no corporate ladder to attempt to climb and fall off. Most of my clients were small companies with minimal bureaucracy, a corporate stepstool instead of a ladder, and few corporate politics, making it harder to get into trouble – harder, but not impossible. The challenge of consulting was not so much in keeping a job as in finding one. One good way of making contacts was to attend field trips organized by the Geological Society of Nevada. I attended many such trips and was the leader of a few. I would have made a lot more contacts and landed more jobs from those efforts if it were not once again for my unacceptable behavior.

Early on the first day of one of the trips I led, I met an older chap who managed a small exploration company. We chatted for a while. He was impressed with my knowledge of the area, its geology, and gold deposits. Our conversation concluded with him asking me to get back to him before the trip was over. He had some work for which he thought I would be a perfect fit. *Great*, I thought. But that was on the quiet bus ride out from Reno. I caught up with the gentleman on the bus ride back and asked about the work. He seemed put off. For some strange reason, the job evaporated. He no longer needed my services.

Was it something I said, something I did? I pondered the events of the past two days. Maybe it was the bawdy rugby song session Tim Jefferson and I led, or the rude and crude joke session Dave Shaddrick and I provided as in-flight entertainment. Or he could have been upset about the foaming tide of spilled beer that flowed up and down the aisle every time the bus ascended or descended a hill. Perhaps he forgot to lift his feet as one of the waves passed, and he got his expensive patent leather shoes wet. That would be his fault. There was plenty of warning of the ensuing hops-and-malt-infused tsunami from the shouts of the people in the adjacent rows. Then again, maybe he was offended by the Ton O' Woman contest or the attack of the Lahontan Land Shark on the dance floor of the tavern.

Land sharks are another invention of those crazy San Luis Valley cowpokes, acting out Jimmy Buffet's hit song, *Fins*. The shark is composed of three people: two serve as fins and one as the body with its head and teeth. The fins are constructed by two guys facing each other and holding arms out at waist level – hands interlocked. The shark lies prone in the cradle formed by the fins. The fins cruise the dance floor searching for shark bait – cute gals' butts. When the fins, which also serve as the shark's eyes (the shark has a limited field of vision), spot a good meal, they direct the shark to the quarry, whereupon the shark's jaws snap wildly in an attempt to bite the gal on the behind, which is about all the shark can see (no problem, Land Sharks are butt men). The most desirable targets have tight jeans, making it difficult to get a firm attachment. Most gals are cool about the shark attack and just laugh or run away giggling.

Sometimes the shark attack does not go so well. Things went seriously wrong on the crowded dance floor of a cowboy bar in Tucson when I was the shark. My fins spied a tall thin blonde with a nice tight butt and directed the jaws to the target. As they closed in, I craned my neck to see beautiful blonde hair flowing down to mid-back level above a shapely ass. I opened wide and clamped down, catching my quarry by the Levi tag on the jeans. The shark held on tight. The catch struggled, broke loose, and turned around. To our shock, we were looking at a bearded face with a big scowl. That was no cute chick; it was a pissed-off long-haired dude! The fins spun 180 degrees and carried the shark out of the bar post haste.

Another time the shark attack did not go well was at the 2005 Geological Society of Nevada Symposium. The big event of the last night was a dance in the Rose Ballroom of John Ascuaga's Nugget. There was a lot of shark bait on the dance floor. A shark team was hastily assembled. My fins were Eric Seedorf, who lives in the clouds – standing 6'6', and Tim Jefferson, who is all of 5'9". No matter how much Eric bent over and Tim tried to lift his fin, the cradle was tilted toward Tim. The shark kept rolling downhill toward him. That made chasing gals difficult and scoring a kill even harder since a belly-up shark has a helluva hard time biting accurately.

Despite being known as a pain-in-the-butt for managers and being labeled a bit of a rabble-rouser, I somehow managed to eke out an existence as a consulting geologist. That is in part because, over the years, I gained a reputation as a no-BS, straight-shooter. There is a need for that in the mining and exploration businesses. Exploration geologists are optimists. They need to be. But all too often, they fall in love with their projects. They hate to see them die. They lobby management to provide more money to do more work, including expensive drilling, despite a lack of encouraging results. In part, that is because they start to believe their own bullshit, a dangerous mental malady. In part, it is because, when their project dies, they have to do the hard work of coming up with another one.

Property owners are even worse. Needless to say, they have a vested interest in the dogs they promote. Most owners are honest, but some unscrupulous scoundrels will do anything to promote their property: making gross exaggerations, falsifying data, withholding negative data, even salting samples. Mark Twain said it succinctly, "A mine is a hole in the ground with a liar next to it." *Caveat emptor* is the rule for dealing with property acquisitions. Companies that don't conduct proper due diligence get taken to the cleaner.

As a consulting geologist, I have spent a lot of time reviewing projects and potential property acquisitions for clients. They appreciate the ability to critically analyze data, sort the weed from the chaff, and not hesitate to call, "Bullshit", when it arises – except when it is called on them.

Chapter 27

FOUR-WHEEL MISADVENTURES

"You take the high road; I'll take the low road,"
...and we'll see who gets stuck first.

      One of the challenges of being an exploration geologist is the simple fact the rocks don't come to you; you must go to them. Getting there can be a challenge, involving various means of transportation: airplanes, helicopters, boats, canoes, rafts, ATVs, dirt bikes, ox carts, rickshaws, horses, mountain bikes, skis, snowshoes, and most of all – 4-wheel-drive vehicles. During my career, I have driven the equivalent of several circumnavigations of the globe, much of it off-road, getting to and from my "office". I have had far too many challenging off-road experiences. I could write an entire book about the bazillion times I have been mired in mud, sunk in sand, or stuck in snow; and spent hours sweating, digging, jacking up vehicles, and carrying tons of rocks and logs to get unstuck. Most of those situations are far too commonplace to write about, but a few notable tales of 4x4 travels gone awry are worth mentioning.

Left-handed Threads

Whether it is opening a jar of peanut butter, a bottle of wine (whatever happened to corks?), or loosening a bolt or nut, *righty tighty, lefty loosey* was always the rule to follow. It turns out that is not always the case.

I learned everything I know (which I admit is not much) about auto mechanics the hard way from owning two old beater vehicles while I was a poor college student. The first was a 1955 Buick Century, a two-ton behemoth I bought from a frat brother's sister for all of $40. It came with a bald tire on the right front wheel, a knobby snow tire on the left, two mismatched tires on the rear wheels, a lovely coating of house paint, and a peace sign on the driver's door. The Beast, as it was named, ran surprisingly well but needed a lot of work to be roadworthy (like brakes, seat belts, windshield wipers, working lights, etc.). I finally got it up and running properly. Sadly, the old Buick eventually committed suicide at an intersection in Denver when it failed to stop at a red light. I would like to blame that on the Beast, but I think operator error may have been a factor. Let's move on, shall we?

The Beast was replaced with a 1949 CJ2A Willys Jeep. It barely ran, but I just had to buy it for off-road fun. Over the course of a year, I became very familiar with the JC Whitney parts catalog. I rewired it (fortunately, few wires involved); added turn signals and a horn; replaced the shocks, wheel bearings, and u-joints; added locking hubs; rebuilt the brakes, the carburetor, and the transmission. The latter task was quite an undertaking that was done on the living room floor of my second-floor apartment, which I can proudly say I accomplished without getting a drop of oil or grease on the carpet.

I learned a lot doing that mechanical work, and had no major problems with those projects. My nemesis proved to be the simplest task of all – changing tires. I bought a new set of tires and started taking the wheels off the jeep. The ole '49 was a bit of a rust bucket. Loosening rusted nuts was always a challenge. I got the two tires on the right side off relatively easily after applying WD 40 to loosen the rust. I went to the left side and tried to take off the front wheel. Man, those lug nuts were stuck tight. I cranked as hard as I could with the lug wrench. They would not budge. Having recently taken and passed (barely) my Physics 401 exam, I knew the importance of leverage. Off I went to get a cheater bar. Still no luck. I got a bigger cheater bar and cranked as hard as I could. The lug nut did not budge, but the wheel lug did. It snapped off. *Damn*, I thought, *these lugs must really be rusted.* I proceeded to do the same with the remaining lugs, resigning myself to replacing all of them. The left rear wheel presented the same problem. When the last lug snapped, I held it in my hand and pondered why the left-side wheels were so much more rusted than the right-side ones. I looked at the lug nut, still firmly attached to the sheared lug, and noticed a big "L" engraved on the top of the nut. I grabbed a lug nut from the right side. It had a big "R" engraved on it. Whoever heard of right- and left-handed lug nuts? Apparently, the engineers at Willys did.

I learned that lesson the hard way. I should have thought of left-handed threads long before breaking off the lugs. After rugby games, we routinely sang a verse of a limerick about a man named Dick, "who was cursed with a corkscrew prick." Poor Dick set out to find a match for his abnormality. After a lifelong search, he finally succeeded.

> He found one and took her to bed,
> Then in chagrin he dropped dead.
> For that spiraling snatch
> Just would not match,
> The damn thing had a left-handed thread.

Fast forward to more recent times and the true story of a similar fateful romance. As reported by NPR, it involved a terrestrial gastropod, a snail, specifically a British brown garden snail (*Cornu asperum*) named Jeremy. Jeremy would have been an ordinary snail, but he had a unique twist. Like Dick of limerick fame, Jeremy was born with a rare genetic deformity – a shell with a left-spiraling whorl, complete with a left-twisted unit. Nearly all others of his species have right-spiraling shells. Jeremy dreamed of falling in love, settling down in his shell, and raising a family. Sadly, Jeremy could not find love with normal snails. He couldn't get it "right" since his sinistral snail unit turned left. But he was in luck. Instead of being served as escargot in a left-leaning British restaurant, he was donated to the University of Nottingham, where biologists could study his exotic condition. The researchers at that esteemed institution of higher learning took pity on poor love-deprived Jeremy. They contacted the BBC to put out an all-points bulletin for anyone finding a left-twisting snail to send it to them. Lo-and-behold, two such snails were located, one in Ipswich, Britain, and one in Spain. The plan was for Jeremy to mate with the two new snails to carry on his deformed legacy. As fate would have it, the gallant Spanish snail romanced the Ipswich snail with his passionate courting. The two lovers ignored Jeremy and got it on with each other. Poor Jeremy died of a broken heart, but not before he pulled off a *menage a tois* and mated with the debonaire Spanish snail, who, like Jeremy and all snails, was a hermaphrodite.

The old '49 Willys was a bucket of bolts and needed lots of fixing and maintenance. Hence, it was dubbed the "Piece of Shit" or POS, out of frustration, particularly with parts that really were frozen with rust. The saving grace was that it was as simple a vehicle as was ever produced and could be worked on easily. Despite its mechanical challenges, the ole POS proved to be an unstoppable off-road beast. With its flathead 4-banger under the hood and low gears, it would lug down and creep up and over just about anything. During its lifetime, we bounced our way over just about every 4x4 pass in the state of Colorado. The only problem was getting there. It could hit a maximum speed of 50 mph – going downhill.

The San Juan Mountains of southwest Colorado offer some of the best 4-wheeling in the country. My buddy, Duncan, and I set out to jeep into Bullion King Basin along the old Brooklyn Road, which starts at the top of Red Mountain Pass on Highway 550 between Ouray and Silverton. We made it a couple of miles up the rocky road when suddenly the steering wheel went loose, spinning freely. We stopped and inspected the steering linkage to see what the problem was. The bolt in the bellcrank that holds the linkage together had gone missing. We searched for it on the rocky trail but came up empty-handed. The backup position was to go to the obligatory bucket of nuts, bolts, screws, washers, baling wire, and duct tape that accompanied the POS wherever it went. The best we could find was a bolt that was longer and had a smaller diameter than the original one. It sort of fit – loosely, so we installed it. It took a revolution or two of the steering wheel to get the front wheels to turn, but that was a significant improvement.

We had to abort our 4x4 adventure and limp to the town of Ouray, 12 miles down the infamous Million Dollar Highway. And I mean infamous. In 2013, USA Today (for what it is worth) included the section of Highway 550 from Ouray to the top of Red Mountain Pass in its list of the "World's most dangerous roads." That may be a stretch, but the road is indeed dangerous thanks to its tight switchbacks, narrow lanes, and lack of shoulders or guardrails to prevent vehicles from plunging hundreds of feet into the Uncompahgre Canyon below. There is no room for such luxuries to fit between the vertical wall of rock blasted into the cliffs and the drop-off into the gorge on the outside lane. In places, the shoulder of the road is less than one foot wide. That makes for some puckered-up driving. I have seen many tourists (especially ones with Texas plates) literally freeze and refuse to drive farther.

The website "dangerous roads" describes the drive as:

> This stretch through the gorge is challenging and potentially hazardous to drive; it is characterized by steep cliffs, narrow lanes, and a lack of guardrails; the ascent of Red Mountain Pass is marked with a number of hairpin "S" curves used to gain elevation, and again narrow lanes for traffic — many cut directly into the sides of mountains."

Such was the 12 miles of highway hell we needed to descend driving a jeep with almost no steering capability. I put the tranny in first gear. We slowly crept down the road. After a while, I kind of almost got the hang of steering the POS and kept us more or less in one lane. The tight switchbacks were a challenge. We were all over the road, using both lanes as I cranked the steering wheel multiple turns to the left, then the right to weave our way down the road. Fortunately, it was the offseason. There was very little traffic on the road. We finally made it to Ouray at the bottom of the pass and pulled into the Conoco station, breathing a deep sigh of relief and wringing very sweaty palms.

Al Fedel, a crusty ex-WWII marine and former mayor of Ouray, ran the gas station. I knew Al well from a few summers prior, when I worked for his son-in-law, mapping the Red Mountains. I explained our predicament to Al, who went to the back of his garage. Can you believe he came out with the exact part we needed? We were back on our way in minutes with fully functional steering.

The author with the POS at 13,114' Imogene Pass, Colorado, Sept 1974,

No Smoking Allowed

As a summer student working for Exxon Minerals in Arizona during the spring of 1975, I was assigned a leased Chevy Blazer to drive. The poor thing had been driven hard and put away wet by several geologists when I got it. One day, I was bouncing my way up the steep 4x4 trail out of Hess Canyon in the remote White Ledges area, doing my best to avoid the large rocks in the trail and navigate my way around the washouts when the Blazer came to an abrupt stop. I tried restarting it. The engine turned over, but the ignition wouldn't fire. I knew the Blazer was not out of gas since I filled it in Globe that morning. I popped the hood and inspected the starter, wiring, fuel lines, fuel pump, etc. Everything looked normal. As I turned around and looked back, I saw a large metal object lying in the middle of the trail. *What the...?* It was the gas tank, a good 20 feet behind the vehicle. Gee, maybe that explained why the truck stopped. How did that happen?

The tank was held in place by a single metal strap. It had so many big dents in it that a gap developed between the bottom of the tank and the strap. Far too many hard bumps caused the tank to bounce up and down against the strap until it fatigued and broke.

There I was halfway up a steep, gnarly 4x4 trail, 50 miles from town, in a predicament. Today, I couldn't change the spark plugs on a modern vehicle, but back then, vehicles were much simpler, although far less reliable. You could easily work on them and often needed to. I carried a complete set of tools, including a tubing cutter and flaring tool. I assessed the situation, measured the gas lines hanging under the truck, and came up with a plan. There was no way I could get the tank back into position and hold it there, but there was just enough gas line to put it in another location – the front passenger's seat. The next challenge was getting it there. It was a 31-gallon tank, which, nearly full, must have weighed close to 160 pounds. I wrestled it into place. I cut the gas line, flared the end, and rerouted it through the wing window with the gas spout coming out of the top of the window. Before moving the tank, I poured some gas into an empty beer bottle (of which there was a selection) to prime the carburetor. She fired right up.

The Blazer with its unique gas tank arrangement.

I drove the Blazer that way for a week while Exxon arranged for a replacement. Gas stations were full service. It was a hoot pulling up saying, "Fill 'er with regular," watching the attendant go to where the gas cap should be and finding a hole. While the attendant was scratching his head, I pointed to the passenger's window.

Exxon leased its vehicles from a dealership in Deming, New Mexico. I drove the Blazer to a truck stop outside of Lordsburg, along the border, to meet an employee of the leasing company who was bringing a replacement vehicle. I drove up and met him driver's-door to-driver's-door. We exchanged keys. As he got in the old Blazer, I said in parting with no further explanation, "I hope you don't smoke."

Caught between a rock and a hard spot

While doing reconnaissance for uranium in southeast Arizona way back in 1975, I drove east of the small ranching town of Wilcox to visit an old mine at Mineral Park in the Dos Cabezas Mountains. Reaching the mine site required driving several miles up a narrow 4x4 trail. The further I went, the narrower and rougher the trail became. The last two miles were incredibly sketchy since the trail was barely as wide as the wheelbase of the Chevy Blazer, with a wall of solid rock on the right side and a steep drop-off to a ravine on the left. My hope was there would be a turnaround area at my destination.

I arrived at the mine and spent a couple of hours checking outcrops and mine dumps for anomalous radiation. There was none. Unfortunately, there was also no place to turn the truck around.

I was forced to back down two miles of very narrow trail. About a mile into the exercise, as I was going around a bend. I came too close to the rock wall. The rear bumper banged the wall, causing the front end to lurch about a foot counterclockwise. With the back end against the wall, there was no way to go backward. I was uncertain about going forward.

I opened the driver-side door and peered down the steep slope – gazing at the ravine 100 feet below. *Oh boy,* I thought, *this could be interesting.* I held onto the door and carefully stepped out to see what was in front of the truck and if I could drive forward enough to pull away from the wall and take another shot at backing around the corner. When I let go of the door, the Blazer rocked back and forth. Yikes! There was nothing but air under the left front tire. Moving forward was out of the question.

I was 40 miles from the nearest town; nobody had a clue where I was; I could not go forward or backward. What to do? Walk back to town and call a tow truck, with the odds being they couldn't or wouldn't go near the place? I surveyed the situation and came up with a plan. It was time to play County Road Department and do some road building.

I jacked up the front of the Blazer to lift the front left wheel and level the vehicle. Armed with a shovel and pickaxe, I cut a swath about two feet wide and two feet deep into the steep slope, extending along the length of the vehicle and for 20 feet in front of the truck. The next step was building an extension to the road. I collected several tons of flat rocks and laid them in layers in the cut, separated by layers of dirt. Four hours later, the construction was completed. It was time to test the new road. I knew I had only one shot at it. If the road repair didn't hold, at the very least, I would remain stuck. The worst scenario was catastrophic failure of the new roadbed, resulting in the Blazer and the construction crew rolling to the bottom of the ravine.

The blazer – neither going forward nor backward.

I put the transfer case in 4-low and slowly pulled forward, watching the front wheel through the bottom of the open driver's door. So far, so good. With a slight turn of the steering wheel to the right, the truck followed the new road back onto the main highway. Breathing a deep sigh of relief, I pulled forward, then carefully backed up and around the corner, making sure not to repeat the folly of the first attempt.

Cheered by my engineering prowess, I continued backing down the trail, with thoughts of getting back to town to enjoy dinner and a celebratory beer or two before the restaurant closed. Wouldn't you know it, at the point, I got a damn flat tire, causing another 20-minute delay in happy hour. Some days you just can't win.

Doing a 180 – the hard way

Highway 395 north of Reno is a treacherous stretch of winding two-lane road infamous for head-on collisions. To improve safety, the Nevada Department of Transportation requires headlights to be on during the daytime. Not a bad idea – that is for people with functioning memories.

I drove up the highway to do some prospecting for gold in the Warner Mountains near Lakeview, Oregon, obeying all traffic rules along the way, as a good driver should.

My destination for the day was high in the mountains east of Lakeview. I followed a dusty logging road for about 12 miles into the forest, then turned onto a small spur trail that continued another quarter mile up a ridge, where I parked my 1979 Toyota pickup. From there, I went on foot to investigate my destination, a small rhyolite plug intruding the Warner basalt.

I returned to my truck at dusk, with thoughts of dinner and a cold beer once I got to town. I put the key in the ignition and gave it a turn to the right – nothing, not even a click. Looking at the headlight switch, I noticed it was in the "On" position. *Shit,* I dutifully turned my lights on for safety but stupidly left them on all day. The battery was drained, leaving me in a sticky situation. *No worries,* I thought, *this is why I have a manual transmission. It will be easy to kick start the truck. I have done this maneuver many times in the past.* Fortunately, I parked the truck facing downhill across the ridge in an area of open sagebrush. All I had to do was put the clutch in, give the truck a shove, bounce downhill through the sage, pop the clutch to start the truck, and drive back up onto the trail to be on my way. I gave her a shove and popped the clutch, but nothing. I quickly spun the wheel to the right, using the last ounce of gravity-fed momentum to reach the trail. Heading down the trail, I gained some speed and popped the clutch again – nothing. I tried two more times with the same disappointing results. I was rapidly approaching the logging road and had to decide to turn right or left. It was uphill to the right and downhill to the left – for how far I did not know. I opted for the downhill to try one more time to kickstart the truck.

Bad decision.

The road went downhill for about 100 feet to a dip over a small creek, after which it was uphill again. I realized with blinding clarity that I screwed up and turned the wrong way. I probably had enough momentum to make it over the rise if I turned to the right. After that, it was ten miles downhill. Now I was stuck in a low spot. The truck was pointed in the wrong direction, making matters worse.

I calmly assessed the situation and took stock of my tools and critical supplies. I was admittedly unprepared for such a situation. All I had was a shovel, an axe, a Hi-lift jack, and some rope. My emergency survival kit consisted of one warm six-pack of Henry Weinhard Ale. I performed a triage to prioritize my plan of attack and immediately put the six Henrys in the creek to cool down. I would need their assistance later.

If I had a winch or come-along and cable, I would have been out of there in no time, but I had neither. I was stuck with the few tools I had, plus an abundance of small logs left over from the recent logging operation.

I conferred with Necessity, the mother of my good friend, Invention, and came up with a plan. The first part of the mission was to get the truck pointed in the right direction, which required rotating it 180 degrees in the narrow road. Hi-Lift or Handyman jacks, as they are known, are magnificent, albeit dangerous, tools. I blocked the truck's left front wheel with a piece of pine, jacked up the rear bumper as far as I could – about three feet – then pushed the truck off the jack to the right. That can be a dangerous maneuver, called "casting," and is certainly not something the manufacturer would recommend. But it worked, swinging the rear end of the truck about 10 degrees counterclockwise in the road. I then blocked the rear wheel, jacked up the front bumper, and swung the front end of the truck. I repeated the process for a hard, sweaty hour until the truck had rotated 90 degrees and was now perpendicular to the road. A well-earned break featuring a refreshing conversation with one of the Henrys was in order.

Another hour later, I had successfully swung the truck a full 180 degrees. It was finally pointed in the right direction. I was tired but feeling confident, having completed the first challenge. I celebrated with a second Henry.

Now the real challenge began – pushing the truck up and over the hill. That could be a big losing battle against gravity, a worthy adversary, to say the least. The trick was not to fight gravity but to use it to my advantage. That is not easy. If you have ever tried to push a rope or a heavy truck uphill, you know what I mean. The truck was in a flat spot. My goal, the crest of the hill, lay about 100 feet away and at least 20 feet higher. I was two Henrys short of a six-pack but went straight to my work.

I decided I could turn an uphill trek into a series of short downhill ones, once again using the Hi-lift jack. It would be inch-by-inch, step-by-step work.

I jacked up the rear bumper as high as it could go, so the truck's ass was higher than its head, providing the truck with a short downward slope into an uphill grade. I split a round of pine and tied a wedge-shaped piece of wood to my ankle with a short length of rope, making me look like a prisoner with a wooden ball & chain. When the bumper was at its zenith, I charged the truck from behind, giving a mighty push. The truck fell off the jack and rolled forward. I kept pushing with all my strength. The truck rolled a few more feet forward and slowly came to a stop. As it started to roll backward, I swung my foot forward, bringing the piece of wood around and kicking it in place behind the rear wheel. That move kept the truck from losing precious forward gain and saved me from being run over. I repeated the process for five sweaty, exhausting hours, reaching the top of the hill at 11 PM. By then, I was beyond exhausted. I was absolutely beat, pooped, seriously knackered. Worse yet, I was five Henrys short of a six-pack.

I thought I had it made. From there, it was ten miles downhill. I surely could kick start the truck in that stretch.

Once again, I put the transmission in neutral and gave the truck a big push. The last surviving Henry and I jumped in; we rolled down the hill. It was a totally dark, moonless night. Bouncing down the dusty road without headlights, I lit the way with a flashlight held with my teeth. We gained speed. In the hope or expectation of the truck starting, I dropped the tranny into second gear and popped the clutch – nothing, except an enveloping cloud of dust, which obscured my view. I stopped and waited for the impenetrable murk to clear, then tried again. I must have repeated that 20 times with the same disappointing results (isn't that the definition of insanity?). Henry and I finally descended to the bottom of the hill and coasted to a stop on the flat road below. Defeated and six Henrys short of a six-pack, I got out my sleeping bag and pad, walked off into the woods, and crashed for the night.

In the morning, I walked three miles down the road to a ranch. The friendly rancher gave me a jumpstart, and I was finally on my way to Lakeview. I made it there only to learn my battery was toast and would not hold a charge. I also learned you cannot kick start a truck with a completely discharged battery. That was news to me and a lesson I will not forget.

Backward Drunk

Dark clouds were building, and a cold November wind was blowing as I tried to squeeze in a little fieldwork before the winter snows arrived. I traveled south from the town of Carlin, Nevada, turned onto the gravel road up Woodruff Gulch, and followed 4x4 trails for several miles to where I would start my work.

I stopped at the crest of a ridge at the headwaters of Woodruff Creek. From there, I went on a geological traverse, searching for signs of gold. About two hours into the effort, light snow started falling. A few minutes later, I was in a blizzard. Not wanting to spend the winter there, I opted to end my fieldwork and hiked back to my truck.

I jumped into the Bronco, happy to get out of the stiff wind and snow sifting down my neck. But then, a minor inconvenience presented itself. I stuck the key in the ignition, stepped on the clutch, and grabbed the shift lever to put the tranny in neutral. It went limp. The shift lever just flopped around loosely, like a chicken with a freshly wrung neck.

This is not good, I thought. *I am stuck in one gear, but which one is it?* Neutral would be the worst since I could go nowhere. Third or fourth gear would be problematic because I would need low gears to traverse the steep 4x4 sections of trail and the gully crossings. I slowly let out the clutch to see what would happen – and went backward. *Rats!* I forgot about reverse. Now it would be a challenge to drive back. *Well, at least reverse is a low gear that will get me up and over the 4x4 sections.*

I made my way down the hill, traveling backassward. For an hour, I slowly drove up and down steep hills, around sharp corners, and across ravines, navigating by alternately using the rearview mirror, the side mirrors, and looking over my shoulder. My neck ached from craning to see what was behind (or was that in front of) me.

I breathed a sigh of relief when I finally reached Highway 278, the Eureka Highway. A new challenge presented itself. It is one thing to drive slowly down a dirt trail in the middle of nowhere. It is entirely different to drive backward on a two-lane paved State Highway. Fortunately, it was only about a mile-and-a-half to the west side of the town of Carlin. Then the real fun began, driving through the residential area to reach the far side of town where there was a pay phone at the gas station at the I-80 exit. I must have provided quite a spectacle, as I nonchalantly backed my way through town, trying to look cool, not wanting to draw attention to my unconventional driving style.

I pulled up to the pay phone in front of the gas station to call a tow truck and parked, heading the wrong way on the street. I started to walk over to the phone booth. A Carlin Police patrol car pulled up.

"A lady called to report a drunk driver going backward through town," the officer said.

"That would be me," I replied, "except for the drinking part. I am sober but feeling like I could use a stiff drink."

I explained the situation to the officer, who got a good chuckle out of the story. He went off to calm down the little old lady who ratted on me to the cops, telling her it was not a drunk, just another backward geologist passing through town. I am sure she understood.

Nuts

I have managed to extricate myself from many vehicular near disasters and, with few exceptions, have somehow limped back to town on my own. In over 50 years of off-road driving, I have needed to be rescued only a handful of times. Most of them involved severe mechanical breakdowns (like a failed fuel pump, a bad alternator, or Sarge's dang leaded gas). Part of pulling through such incidents is the old Boy Scout motto: "Be Prepared." However, one recent routine incident went horribly wrong despite impeccable preparations. Harking back to the beginning of this story, it also involved tires and lug nuts – those pesky little things that hold the tires onto the wheel – plus a series of compounding screwups.

Back in June 2020, I left the Moss mine in Arizona late on a Friday afternoon, the last person to leave. I was staying in Kingman, a 30-mile drive via the highway, but five miles shorter, taking the (sometimes) graded Silver Creek Road east to Oatman, then joining historic Route 66 and driving over Sitgreaves Pass to reach Kingman. Forgetting what happens when taking shortcuts, I elected to take the scenic route. I drove the mile of gravel road from the mine to Silver Creek Road and turned east. Everyone else turned west to Bullhead City, a 10-mile drive on a decent gravel road.

Three miles down the road, which had degraded significantly since the last time I drove it and was a rocky, bumpy, wash-boarded, bone-jarring track, I heard a sound I dreaded – the hiss of a flat tire. *Damn*, I thought, *this will delay cocktail hour by a good 15 minutes.* Other than the time delay, I was unconcerned. Why should I worry? I was more than prepared, with not one, not two, but three spare tires (and have needed all three many times). I also had two jacks, four different tire irons, an air compressor, a tire repair kit, a spare set of lug nuts, good old WD-40, and a can of Fix-A-Flat tire inflator (as a last resort). What could possibly go wrong?

Apparently, a lot. I inspected the tire, the front left. It was indeed flat, but only on the bottom. I set the emergency brake and blocked the tires with my mine-approved wheel chocks so the truck would not roll once jacked up. Then I went to work to loosen the lug nuts. Man, they were tight. I cranked as hard as I could and got one off. The next one would not budge. I decided to tackle that later using a bigger lug wrench and went to the next lug nut. I managed to get three of the six lug nuts off. The other three would not move. *No problem, a little more leverage is all that is needed.*

Much to my surprise, more leverage did not work. The lug wrench slipped as I applied more force. *What the...?* I looked closely at the stuck lug nuts and noticed they were no longer pentagons; they were round. The corners of the lugs were worn off so that the lug wrench had no grip. The last tire monkey to rotate the tires used way too much torque and wore the corners off the lugs. Crap, there was no way I was getting those lugs, or the tire, off.

The next step, the last resort, was to use the can of Fix-A-Flat to seal the puncture and inflate the tire. Well, it sounds good in the advertisements, but it did nothing, maybe because the hole in the tire was too large. Perhaps I should have tried to patch the tire first, but too late.

That left me decidedly stuck on a dirt road to nowhere that was likely to not see any traffic for hours. Grudgingly, it was time to admit defeat and call for the cavalry. I reached for my cell phone, noticing there was almost no charge left. I looked for the phone number of Trevor, the newly hired contract geologist working at the mine, who left for Bullhead City ahead of me. *Rats*, I had not entered it. No worries, I was sure it was in my contacts on my laptop. I opened it. I was searching for Trevor's number when *blink*, the battery ran out, and the computer turned off. Still no worries, I had a charger for the laptop (but not my cell phone) and a power convertor. I fired up the ole Tacoma and ran the engine long enough to charge the laptop and find the number I needed.

I entered the number into my cell phone and dialed it. Nothing. As fate would have it, I was in an area with no cell service. That left me with little choice but to abandon Old Blue, throw some water into my pack, and start walking back to town, hoping to get cell phone reception somewhere along the way. By then, it was 5:30 PM. The temperature had cooled down to a balmy 103 degrees, perfect for a 12-mile evening walk in the desert.

I walked along the dirt road, periodically stopping to see if I had phone reception, but I had none. At one high point, I got a weak signal, called Trevor, but only got his voicemail. I left a short message letting him know of my predicament. Over the next hour, I checked for a return message every time I had any reception. I also noticed my battery rapidly going flat. Two miles down the road and with less than 1% battery left, I got a call from Trevor. *Fantastic, saved,* I thought.

Trevor came to the rescue. We drove back to the Tacoma and assessed the situation. Although he had a good selection of tools, Trevor also had nothing that would loosen the rounded lug nuts. It was off to the auto parts store in Bullhead City to look for a suitable tool and get more Fix-a-flat. We made it just before the store closed and headed back.

Of course, the tool that I spent $30 on that was supposed to fit the rounded lug nuts did not work. The next step was to attempt to fix the tire while it was still on the truck. I got out my portable compressor to inflate the tire and find the hole, but no go – the compressor would not work. Good grief, what else could go wrong?

We jacked up the truck and spun the wheel to find the rather obvious hole. We used the tire repair kit to plug the hole, then inflated the tire with a can of Fix-a-flat. That sort of worked, but there was still a slow leak around the patch. In went the second can of goo and air, and I had a partially inflated tire with about 10 psi air.

It was a long, slow, nervous trip down the rough gravel road, but we miraculously made it back to Bullhead City before the tire went totally flat.

In the morning, I took the truck to the local tire shop. The mechanic had to cut off the lug nuts with a chisel. As it turns out, those lug nuts are comprised of two parts – a solid steel interior and a shiny chromed shell of softer steel, that eventually gets worn and creates mayhem. I was not the first or only person to have this problem. In fact, there is a class-action lawsuit against the manufacturer and some of the companies that provide the lug nuts as OEM parts. Sign me up!

Chapter 28

ILLEGAL (DOCUMENTED) ALIEN

You are not welcome here.

As a descendant of Scottish and Irish immigrants, I have always believed in the importance of immigration, as long as people come legally. Contrarily, I have never been fond of illegal or "undocumented" aliens.

Then I became one.

My conversion to the wrong side of immigration came during a quick trip to British Columbia for a client, Kootenay Gold, in the fall of 2011. The job was managing a short drilling program near the tiny town of Nakusp, along the shores of beautiful Upper Arrow Lake, part of the Columbia River system. I also planned to meet the company's Vice President of Exploration to review a report I just finished on a gold property in Idaho, and to deliver several file boxes of data on the project.

It was a long drive from Reno. I arrived at the border crossing on Highway 395 between Kettle Falls, Washington, and the small town of Grand Forks, British Columbia, at about 9 PM. That was my second mistake. I was the only customer at the sleepy border station at that late hour. The officials had plenty of time to screw with me. They did just that.

Having worked in several foreign countries, including Third World ones, I knew there must be a procedure for entering Canada for short-term professional work. When I questioned the managers of Kootenay Gold, they didn't know. "Just come as a tourist," was the recommendation. And "NAFTA allows for reciprocal employment opportunities." That almost made sense. Canadian geologists work in Nevada all the time. Surely the Canucks wouldn't mind a Yankee geologist working in their backyard for a week.

My first mistake was accepting that advice.

I pulled up to the station, rolled down my window, and handed my passport to the immigration agent. The officer quizzed me about how long I would be in Canada and why I was coming. I told him I was meeting a friend to deliver some files and planned on spending a week fishing and hiking. *That should satisfy the bored bureaucrat,* I thought.

I was dead wrong.

"Pull over there," he demanded, pointing to a parking spot on the left with a wooden bench next to it. The officer ordered me to exit the truck and sit on the bench. He rummaged through everything in the cab and the bed of my truck, like he was searching for illegal drugs, wads of ill-gotten Loonies, or contraband Girl Scout cookies being smuggled to feed the deprived natives of Nakusp.

He retrieved my Daily Planner from my computer bag. Thumbing through it, he came to September and studied it. "What's this about drilling starting on Thursday?" he inquired in an accusing voice.

"I am going to help my friend for a few days before we go fishing," I said.

"Oh yeah? What is his name? I want his phone number!"

I gave him the phone number of Bob Thompson, the VP of Kootenay, who was driving from Vancouver to meet me.

"Sit there, I will be back," the officer barked.

About 45 minutes later, the agent came back. "Follow me," was all he said as he led me into the station. He opened the door to his office and directed me to sit on another bench outside the office.

An hour rolled by before the officer opened his door and led me in. I sat in a proper chair for a change while he raked me over the coals for "Attempting to enter Canada under false pretenses," a crime normally committed by terrorists or drug smugglers and punishable by caning followed by waterboarding and public hanging. Fortunately, since it was my first offense, I was offered a more lenient sentence of spending the night in a cozy jail cell. The alternative was to turn around and drive back to the U.S. with my tail between my legs.

Damn, I hate those tough decisions. I carefully weighed the options. Either way, I would be late for my meeting with Bob the next morning. It was a toss-up. It was too late at night to call a friend. There was no audience to poll. I was on my own. I decided to decline the generous offer of a free night's lodging and opted to drive the 50 miles back to Kettle Falls.

The agent agreed that was the right decision. He eased up and said, "Your boss explained the need for you to come here. Because your work helps with local employment and benefits our economy, you can come to work. You need to bring a letter from your employer detailing who you are working for, what work you will be doing, where the work will be conducted, and how long the job will last," then you can come work here.

Gee, I wish I had known that little secret earlier.

I drove back to Kettle Falls and checked into a motel a little past midnight. Before hitting the sack, I fired off an email to Bob telling him I needed a letter and what should be in it. Early in the morning, Bob's letter arrived. I printed it and drove north to the border.

I pulled up to the checkpoint; the immigration agent asked the same questions. This time I answered truthfully: I would be in Canada for a week of work, followed by a few days of R&R. The letter from the VP was sitting on the passenger's seat. Before I could grab it to show to the officer, he barked an all too familiar order, "Pull over there!" *Oh boy, here we go again.*

I parked. The obviously pissed-off immigration officer stomped over to the truck. He raised holy hell with me, "Think you're smart, eh? Coming back when there is a different agent and trying to sneak in." In a repeat of the previous evening, he ripped through everything in the truck, leaving a disheveled mess in his wake.

I finally had enough and blurted out, "Wait a minute. Last night I was told to come back with a letter from the company, and I would be allowed to enter. It is sitting on the passenger seat. You didn't give me a chance to show it before you ripped into me about trying to sneak across the border."

He looked at the letter. "Harrumph! Come with me."

We went into the office, where I was once again directed to sit on the Group W bench. A second agent took my computer bag to his office. He opened my laptop but didn't turn it on. He sat there staring at a blank screen.

When there was a break in the ranting and raving, I asked if there was a set procedure to work in the country, maybe a business visa or a temporary work permit. After all, every other country has one. What the hell is it? The company didn't know, and they couldn't find out. "There must be a proper way to apply for temporary work in advance. Shouldn't the company be able to do that from Vancouver?"

"No, there is no procedure for that." After a pause, the officer added, "But here at a Port of Entry, you can fill out the Employer Specific Work Permit application and pay $100; then you can work in the country."

You would think the agent who told me the previous evening that I should come back with a letter from my employer might have mentioned that. Nah, that would take all the fun out of it for his buddies the next day. Encouraged by the news, I responded, "I don't suppose you have an Employer Specific Work Permit application that I can fill out?"

"No, we don't." My hopes were dashed. "Headquarters should have them." It makes perfect sense not to stock the application forms at the only place where they can be used. Gotta love government bureaucracy.

"Could you maybe have them fax one over?"

"I will see about it," the grumpy agent replied.

Half an hour later, while the second agent sat glued to the black computer screen, the officer came over with a blank form. "Here, fill this out." I did so. I paid my $100. At that point, I foolishly thought I was about to be granted permission to enter our friendly neighbor to the north, Canada.

Not.

"So, you came to work as a geologist, eh? Prove it."

What? After rummaging through everything in my truck, finding my field gear, seeing the notes about drilling, and reading a letter from the VP stating that I am a geologist coming to help with, of all things – geology, he doubts I am indeed a geologist. I really wanted to say, "Give me a word, any word, and I will add '–ite' to it." Or, "Throw me a rock. I will lick it and interpret the evolution of the entire Canadian Cordillera from it." I was too stunned by the absurdity of the demand to counter it.

"Uh, what kind of proof do you want?"

"Do you have a resume with you?"

"I have one on my computer," thinking the second agent who was still staring at a blank screen could finally do something useful, like turn on the computer.

"That won't do. It has to be a paper copy. How about a business card?"

"No, I didn't bring any of those." Why would I when I was supposed to be a tourist?

"If you can't prove you are a geologist, you can't come here to work as one."

"Then can I come in as a tourist?"

All I got from the humorless agent was an evil stare. I was struggling to come up with a way to prove my credentials when I remembered I had my American Institute of Professional Geologists stamp in my computer bag. "Oh wait, in my computer bag that your associate has is my professional stamp." Agent 2 took a break from trying to decipher the meaning of a black screen and brought my computer bag over to me. I reached in, grabbed the stamp, and slammed it down on the back of the application form. The agent stared at the words, "Certified Professional Geologist #11603," circling the stamp. He reluctantly said, "Ok, you are good to go."

I made sure I didn't overstay my permitted time. I had no issues returning to the States, except for having an apple confiscated at the border. It likely originated in Washington State but was not allowed to return – maybe because the border-crossing agent was hungry.

I thought I was finished with the Canadian immigration officials. Unfortunately, Canada has a memory like an elephant, or at least their computers do. I soon learned that I am considered a *person non grata,* a nasty foreign terrorist, and am forever in the crosshairs of Canadian immigration. Every time I fly to Vancouver or Toronto, I go through the immigration gauntlet. I make it through Immigration fine. At Customs, the last step in the process, agents are eagerly waiting to ruin my day. They look at my paperwork, which must have some secret invisible mark on it. Instead of directing me to the exit, they point to the detention room on the right, where the other terrorists, spies, and smugglers are cross-examined.

On my first trip to Vancouver after the border incident, I nervously entered the detention room. An officer told me to take a seat on the bench. I plunked my ass down between a swarthy bearded man sporting a bulky, oversized stuffed vest and a jittery guy wearing a Panama hat, who had white powder all over his nose. An agent finally called me to the counter. He grilled me with the usual questions: "Why are you coming here? How long will you be here?" Then he brought up the issue at the BC border. I explained it was due to miscommunication and would never happen again. The reasonable officer accepted that. He pointed me to the exit.

The immigration people in Toronto are not nearly so accommodating. A couple of years later, I flew to Toronto for two days of meetings with a new client. Once again, I made the mistake of arriving at Immigration late at night, midnight to be exact.

I approached the door of the detention room with trepidation, not sure what to expect. As I entered the rather large room and looked around, my heart sank. I was the only customer – no need to wait on the bench. The agent at the far end waved me over. *Oh no, not another bored officer with nothing better to do.* I handed the gentleman my passport. He scanned it and flipped through the pages. The Canadian Inquisition followed. I explained I was there for two days of meetings with a client. I tried to add some details when he stopped me mid-sentence.

301

"Why should I believe you? You have been known to lie about your reason for coming in the past," he said. No amount of apologizing was good enough for him.

"You need to prove why you are coming here. Do you have a letter of invitation or an agenda for the meetings? If not, you cannot enter and will be sent back."

Oh, shit. This asshole was hardcore. I opened my laptop and tried to connect to the WI-FI.

Damn, it wouldn't connect. The agent gave me shit, saying, "Everyone else has been able to connect. What's your problem?"

I gave up on the computer and checked to see if I had any emails regarding the meetings on my phone. I turned it on, immediately remembering I forgot to charge it before leaving. The battery was just about depleted. *Crap, I have to do this quickly.* I frantically searched through emails to find one requesting I come to Toronto for the meeting. There were a lot of emails from the client. I was not finding what I needed. The battery was just about flat when I found one referring to the meetings and showed it to the agent. He reluctantly accepted it. He told me to come better prepared with an official invitation letter if I ever wanted to return to Canada.

I left the detention center and walked to the curb to hail a cab. I glanced at my cell phone – the screen was black. The battery was dead as a doornail. I came within a gnat's eyelash of being sent back to the U.S. The digital gods were screwing with me but thankfully had mercy at the last second.

I don't worry about having an official letter of invitation before going to Canada. I have not been back since the overly inhospitable reception in Toronto. It is not worth the hassle of dealing with the border bozos flexing the undeserved power bestowed upon them by even bigger bozos. When clients ask me to come north for meetings, I convince them they will save a lot of money if we do a conference call instead.

.

Chapter 29

OFFICE MATES

"Crikey! What a Beauty."
— Steve Irwin

Intrepid explorers, such as manly exploration geologists, who spend much of their careers in the great outdoors, encounter a lot of wildlife. Over time, they develop a repertoire of stories as wild and fantastical as the animals themselves. I have run into many strange and fascinating critters in the deserts, mountains, and jungles of the wild places I have tramped as a geologist. The two outstanding exceptions are China, where anything that dared move was eaten long ago, and Chile's hyper-arid Atacama Desert, where it is so dry the only living things are microbes that lie dormant in the soil until the once-in-a-decade rain brings them temporarily back to life.

I have many fond memories of awe-inspiring wildlife experiences while working at the "office" – true National Geographic moments: seeing a herd of stately elk emerge like ghosts from the mist after a rainstorm high in the Colorado Rockies; watching the giant tail of a whale sink slowly beneath the waters of Alaska's Stone Rock Bay as we sipped Jack Daniels in our inflatable dingy during our evening fishing break; glimpsing a pair of giant leatherback turtles mating in the Molucca Sea while our outrigger canoe slipped by; gazing up at a pair of majestic bald eagles spinning wildly toward the ground, locked in an amorous "cartwheeling" mating embrace; coming face to face with a rare dugong on my morning swim over Manado's patch reefs; going to sleep listening to the raucous laughter of kookaburras, then waking to the pre-dawn roar of male koalas (eerily similar to that of lions) in the eucalyptus trees above my tent in the Australian Outback – truly magical moments!

No critter exemplifies the wild west better than the coyote. Many years ago, when I first heard coyotes calling to each other as the sun dipped below the horizon, I found their howling chilling. I associated it with a pack on the hunt, closing in on some poor animal before surrounding it, shredding it to pieces, then coming for me. I admit to having been ignorant back then and have since learned a lot about coyotes. The Linnaean name, *Canis latrans,* given to the coyote by taxonomists, is Latin for "barking dog," although barking is only a minor part of their vocal repertoire. Now, I look forward to hearing the melodic high-pitched staccato yips, yelps, barks, and howls of "song dogs," as they are affectionately known, while I am having cocktail hour in my desert camp. I am convinced they are doing the same thing – relaxing, socializing, and giving thanks for another beautiful day.

Desert cottontails and big blacktailed jackrabbits comprise a large part of the diet of coyotes in Nevada. Rabbits are abundant because they breed like…well, rabbits. Driving along gravel roads on moonlit nights, it is impossible to not run over at least a few of the hundreds of rabbits, which dash out of the surrounding sagebrush, do a zig and a zag or two in the road, and fail to get out of the way. Indecision kills. Initially, I felt sad after squishing the poor critters. In the morning, when I drove the same road where the evening's bloodbath occurred, I saw no sign of the carnage. Owls, golden eagles, and coyotes patrol the roads in the wee hours of the morning and make off with the carcasses. I no longer feel so bad about whacking the rabbits. I am just serving breakfast for the local scavengers.

Ranchers are divided regarding coyotes. Many of them hate coyotes, claiming they kill young livestock. Others take a more pragmatic approach. My friends, Tom and Bev Reichert, who own a ranch in Reese River Valley, south of the tiny town of Austin, loved their coyotes.

The Reicherts grow high-quality alfalfa, in demand for horse feed. When they tilled their fields, hundreds of gophers and voles, which eat their crops, got stirred up. At one time, a pack of a dozen coyotes followed behind the tractor, picking off the confused critters as they popped up. The coyotes did their part to keep nature's balance, bringing the rodent population under control.

One fall day, the town of Austin held its infamous "Coyote Derby," a contest to see who can kill the most coyotes in one day. Hunters come from as far away as Utah and California to participate in the hunt and try to win a piece of the $6,000 prize money. There are no rules. Most hunters use calls mimicking a wounded rabbit to draw the unsuspecting coyotes in close. Hunters don't need to leave the comfort of their pickup trucks to do the killing.

The next time the Reicherts tilled their field, no coyotes showed up. The pair of great horned owls, which nest in the giant cottonwood trees next to the stone ranch house, was all that was left to keep the varmints in check.

When we build roads and drill pads for exploration work, we are required to reclaim the disturbance. Giant excavators are used to restore the slopes to their original condition. If the job is done right, after seeding and a few years of growth, it is hard to tell the roads were ever there. I accompanied a BLM geologist, Jim, to inspect reclamation work at White Knight Gold's Indian Ranch project in the Simpson Park Range. We drove down the gravel road in the valley, turned right onto a two-track trail, and drove up the alluvial fan toward the mountains. As we approached the range front, a group of a dozen mustangs on the hillside to my right, including a handsome stallion, stopped grazing and eyed us suspiciously. We continued on, climbing a steep section of the trail. We spied a coyote standing along the side of the trail. I expected to see him run away, but he just stood there. Then he made a short move to the left, followed by one to the right. *What the heck*, I thought, *was this coyote mimicking a rabbit's indecisive zig-zags, or what?*

Once we got closer to the coyote, we learned the problem. The poor fellow's left hind leg was caught in a trap. We stopped and approached the hapless coyote.

I asked the government man, "Is it legal to trap coyotes?"

"Yes."

"Is it illegal to release one from a trap?"

"I don't care." That was the answer I hoped for.

The coyote, exhausted from his last-ditch struggle to escape the trap as we approached, laid on his side and looked up at us, panting. I suspect he was thinking, *Now must be when they put a bullet through my head, or just skin me alive.*

Jim took off the leather jacket he was wearing and handed it to me. He put on a pair of heavy leather gloves. I laid the coat over the coyote and held it down over the frightened fellow. Jim checked the coyote's leg. It was not broken, just skinned badly from the struggle to get free. Jim pressed the springs of the trap, releasing pressure on the coyote's leg. I stood up and pulled the coat off the coyote. He just laid there. Jim reached down and lifted the coyote by the belly and set him on his feet. For a few seconds, he stood there, wondering what the heck was happening. He looked at us, then glanced toward the hillside across the gulch. In a flash, he was off – running with a slight limp down the slope, across the drainage, and up the hillside. When he reached the ridgeline, the coyote stopped, looked back at us as if to say, "Thanks, guys," then disappeared over the ridge.

You are welcome. Have a long and happy life, ole Wile E.

Some wildlife encounters are far from magical; they are just downright annoying. Those often involve spineless lower life forms, such as members of the *Insecta* and *Arachnida* classes, or even lower life forms – politicians. In the line of duty, I have been bitten or stung by bees, wasps, hornets, horseflies, deerflies, blackflies, white socks, gnats, ants, ticks, mites, spiders, leeches, and hordes of bloodthirsty mosquitoes. The worst bugs were the "gononnies" in Indonesia, nearly invisible red mites that love to get into one's trousers to bite you where you really, really don't want to be bitten. Almost as bad are the fire ants of Chihuahua, whose unprovoked bites burn for hours.

Despite many encounters with killer bees in Sonora, I managed to avoid being stung by them. That took some doing since they like to nest in crevices of rock outcrops – the very places geologists love to smack with their rock hammers. Working at Promontorio along the Yaqui River, we often heard swarms of killer bees that sounded like 747s roaring past us. The dense chaparral foliage prevented us from seeing the bees, but we dove for cover, anyway. I was always cautious when approaching outcrops, waiting to see if any bees were coming and going from the site. My partner, Raphael, was not so careful one day. We were investigating the base of a cliff of rhyolite near the town of Tesopaco when Raphael hit the outcrop with his rock hammer. I was about 100 feet to his left and heard him shout, then saw him swatting wildly as he ran down the steep slope. Just then, a couple of scout bees flew up to me. I froze, afraid of making any movement that would piss off the scouts and bring their buddies to attack me. After a few tense minutes, the bees decided I was not a threat and left. I waited another five minutes before I made my way down the hill to catch up with Raphael, who fortunately outran the bees and was stung only a couple of times.

Another amigo, Machaca, had a nearly disastrous encounter with killer bees. He was supervising a bulldozer building a drill road when the blade dug into a nest of killer bees. The bees flew toward Machaca, who ran to the bulldozer, frantically waving at the operator, who was in an enclosed cabin, to stop. Machacha got into the safety of the dozer cabin just as a swarm of angry bees smacked into the windshield. A few years earlier, a Mexican bulldozer operator was not so fortunate. He was operating a dozer without a cab. When he ran into a nest of killer bees, they swarmed him and quickly lived up to their name.

A little higher on the evolutionary scale is the next group of unpleasant wild critters, the scaly legless denizens of the deserts and jungles – vile venomous vipers. Don't get me wrong; I don't mind snakes. I made a pact with them long ago: "I won't mess with you if you don't mess with me." It has worked quite well over the years with a variety of serpents from rattlesnakes to temple vipers and cobras.

We encountered black spitting cobras fairly commonly on our treks on the Sulawesi jungle. They all just wanted to get out of our way – including the one that shot straight down the hill, passing nearly between my legs when we were inspecting trenches at Lobong. Then there were the unidentified snakes. It was prudent to assume all snakes in Indonesia – except the big ones that squeeze the life out of you – are venomous. I was marching along in the middle of a column of Indonesian workers, all about 5'2" to 5'6' tall, when I suddenly found myself eyeball-to-eyeball with a small snake hanging onto a branch over the trail. The boys passed under the snake without noticing. I, however, noticed and quickly ducked to avoid a collision.

My pact with snakes has worked especially well with rattlesnakes, which are the most courteous of snakes, warning you of their presence long before you step on them – well, most of the time.

One of the scariest snake encounters happened a few years ago on a hot September day in the desert of northern Arizona. I was on a geological traverse up and over a hill, then down a steep, narrow canyon. In the canyon was an old prospect, a 20-foot-long, 10-foot-high excavation blasted into solid rock by prospectors more than a century ago. Being old and half-blind, I put on my reading glasses to inspect the outcrop, a silicified breccia, up close. I was fascinated by the jigsaw puzzle texture of the rock, which had been torn apart by a volcanic explosion about 20 million years ago and glued back together by hot, silica-rich solutions, the kind that carry and deposit gold. My hands were stretched out on the wall in front of me, and my nose was up against the rock face, as I inched my way to the left. About halfway along the wall, for reasons I don't know, I looked up. I was eyeball-to-eyeball with a coiled rattlesnake, sitting quietly in a recess in the wall, only inches from my face. I almost landed my left hand on top of the snake. *Holy crap!* Whoever said, "White man can't jump," should have seen me then. I was NBA material for an instant. Fortunately, this was a Southwestern speckled rattlesnake (*Crotalus mitchellii*) instead of a Mohave green (*Crotalus scutulatus*). This fellow did not rattle and, thank God, did not strike. If it had, it would have injected its load of venom into my neck or face. Speckled rattlesnakes are masters of camouflage, exhibiting crypsis variation, the adaptation of populations of snakes over time to nearly perfectly blend in with the colors and textures of the specific rocky terrain where they live. I suspect this one was confident in his disguise and decided to sit still and chill out while it scared the living crap out of me.

Two hours later, I was walking down a dry wash. As I briskly passed a large rock, I heard the distinct buzz of a rattlesnake. I looked behind me to see the business half of a Mohave green rattlesnake lying in the sand of the wash, the rest of it still under the rock. Shit, this sucker was coiled up under the rock and struck at my leg without provocation as I walked by, but missed. Unlike my mellow speckled friend, Mohave greens are very aggressive, even rumored to chase people. The green is unique among rattlesnakes in that its venom contains hemotoxin, the normal venom of rattlesnakes, which destroys muscle tissue, and a neurotoxin – the venom of cobras, which attacks the nervous system. It is a potent combination that makes the bite of Mohave greens the deadliest of all rattlesnakes. Unless quickly treated with antivenom, it can take weeks or even months to eliminate the venom – if the victim lives.

According to CBS News, in 2012, a 6-year-old boy in southern California was bitten by a rattlesnake:

> He was brought to Mission Viejo Hospital in Mission Viejo, where doctors recognized the symptoms as a bite from the especially toxic rattlesnake, the Mohave Green. "It took 42 vials of antivenom just to stabilize him" the boy's father told CBS Los Angeles. "Normally it would just take a few…depending on the snake."

Two snake encounters in one day were a bit much. Had I met the Mohave green instead of the speckled rattlesnake eyeball-to-eyeball, it could have ended badly.

Eye-to-eye with my mellow friend, the Southwestern speckled rattlesnake – Silver Creek, AZ. September 2014.

While I was mapping a silver property in southern Chihuahua in 2015, my amigo, Francisco, and I came across an old mine tunnel that had stibnite (a good indicator of precious metals mineralization) on the mine dump. Excited by the find, I grabbed my headlamp from my pack and scrambled to the portal of the adit, ignoring one of the cardinal rules of entering such holes in the ground – throw in a rock or two to see if anybody is home. The portal was partially caved in, with only about two feet of clearance at the entrance. The caved material formed a pile of rock sloping down into the tunnel. I went in feet first, slowly sliding down into the darkness of the mine. Shortly after my head cleared the entrance, I heard an all too familiar and unwelcome sound, the buzz of a pissed-off rattlesnake (*cascabel* in Spanish). I shot out of that hole like a human cannonball at the Ringling Brothers Circus. My right foot came within a couple of feet of the snake. If I continued my descent, I likely would have been bitten. We decided the adit really didn't need to be investigated after all. I put a notation on my map, "*Casa de cascabel,*" to warn anyone who followed. We moved on.

Only fools mess with rattlesnakes. Nearly all snake bites happen when people torment or try to kill them. There have even been cases of people bitten by the severed head of a snake. I got a little foolish one day at a prospect called "*mucho oro*" ("lots of gold") in Sonora. I was mapping with a young Mexican geologist, Carlos, who, I was about to learn, was deathly afraid of snakes. As we walked along, we heard a distinctive buzz. I walked toward the sound to locate the source – a coiled rattlesnake hidden in the brush. Carlos stood where he was and would not move. "We should go back now," he anxiously exclaimed.

"Nonsense," I said, "It's just a little *cascabel.*" We continued along our way. Within five minutes, we came across another snake, a large diamondback rattler slithering along on the slope about 30 feet from us. Carlos was shaken. "Now we *have* to go back," he stammered. I tried to calm him by explaining the snake didn't care about us and was just looking for something much smaller for a meal. We watched the rattler for a few minutes. It came to a small hole in the ground, probably the home of some rodent. The rattler started to go down the hole, which was a rather tight fit for the fat snake. When the *cascabel* was halfway down the hole, I said, "Watch this." I walked over and grabbed the snake by the middle of its body. I knew there was no way the snake could turn around and bite me or back up and come out of the hole in a hurry. What followed was a lot of loud rattling and tail thrashing by a rather pissed-off serpent – and cringing by a frightened young geologist, who thought this gringo was loco.

Cascabel looking for a meal

... a nice tail to grab.

Friendly tarantula, Promontoria, Sonora

Plants can be nasty, too. One of the worst is catclaw, an innocuous-looking desert bush with spindly branches covered with small, serrated green leaves. Hidden in those leaves are hundreds of backward curved thorns. They are like miniature cat's claws and are just as sharp. The first time I encountered catclaw was in January 1972 when I was a junior geologist for Exxon Minerals, managing a drill rig in the Rincon Mountains east of Tucson. It was a lovely, warm winter day. I thought it would be a good time to get a tan. I took off my shirt and walked up the wash above the drill rig. After only about 100 yards, I brushed against a wispy branch of a catclaw that draped across the wash. Those damn claws ripped into the flesh of my bare chest, stopping me dead in my tracks. Bloodied and somewhat embarrassed by my stupidity, I walked back to my truck to put on a shirt, vowing never to tangle with a catclaw again. Oh, if that were only the case! Working in Sonora and Chihuahua, Mexico many years later, I encountered thickets of *una de gato* (catclaw) that could not be avoided. Every day involved serious bloodletting. The life of a shirt could be as short as a single day in the field.

Another nasty desert plant is the teddy bear cholla, *Cylindropuntia bigelovi*. This cactus stands up to ten feet tall with a central trunk from which extend many close-spaced branches that split into more and more branches. At the end of each branch is the latest growth – a sausage-shaped section about six inches long. The end pieces are loosely attached to the branches and break off with even the lightest touch. They lay on the ground waiting for some unsuspecting geologist to walk by and get one stuck to his boot or pants leg.

Thin spines cover the sausages, radiating in a tight mass from the center of the section. The silvery spines give the cactus a warm and fuzzy appearance, hence the teddy bear appellation. Warm and fuzzy, they are not.

The first time I encountered these horrible sons of Satan, I found a spiny segment attached to the back of my jeans, just above my left boot. I reached down with my left hand to brush it off. My hand became stuck to the damn thing. I tried to pull my hand free, but the spines would not let go. While trying to free my left hand, my right hand also became stuck. In classic Brer Rabbit fashion, I was stuck to the cactus, like a tar baby. Bent over with both hands behind my foot, I hopped back to my truck to retrieve a pair of scissors and pliers. I struggled to cut enough of the spines to very painfully pull my right hand free but left nearly a pound of flesh behind. The spines of the teddy bear cactus are covered with thousands of teeny tiny backward-pointing barbs – like miniature porcupine quills, which make pulling out the spines both difficult and painful. Since the spines radiate around the sausage, they all point in different directions when they enter your skin. Trying to pull out the entire cactus segment at once is impossible. It is necessary to cut each spine and pull them out one at a time with pliers.

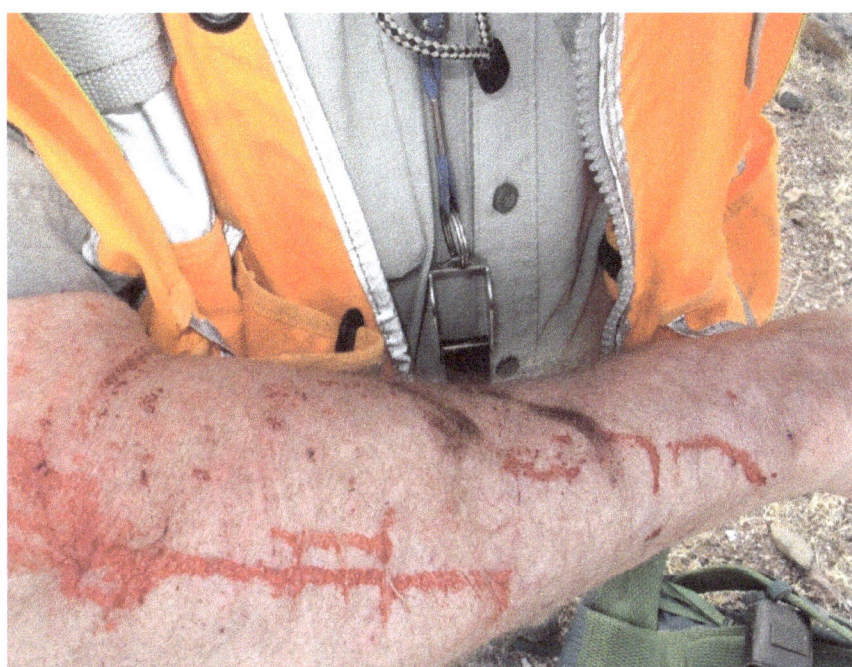

Just another day of bloodletting dealing with una de gato in Chihuahua.

Sticky situation;   *Teddy Bear obstacle course.*

I quickly developed a hatred for teddy bear chollas. Whenever I found them, I threw rocks at them to torment them. It was fun to watch the end segments fly off. Standing on a rock ledge about 15 feet above a big grove of the spiny bastards, I started throwing rocks at my enemy. The assault was going brilliantly until I hefted a big flat rock and sent it crashing down upon a particularly large evil-looking cactus. The impact rocked the teddy bear, creating an explosion of end segments. I stood on the ledge laughing and delighting in my conquest. Then, out of the corner of my eye, I saw a segment come tumbling end-over-end through the air toward me. It landed smack dab on my left shoulder. Damn! Cactus revenge! It was time for the scissors and pliers again.

Chapter 30

LIONS, AND TIGERS, AND BEARS...AND ANOAS

We're definitely not in Kansas anymore.

Forget annoying insect, serpent, and cactus encounters; by far, the most terrifying wildlife stories involve the big players: lions, and tigers, and bears… and, of course, those mysterious anoas.

Lions

North America has its own version of the King of the Jungle – the mountain lion (*Puma concolor*), alias the catamount, puma, or cougar – by any name, one big, mean, nasty, furry pussy. The mountain lion definitely sits atop the food chain. When the Idaho Fish and Game Department reintroduced wolves to the Rockies in the 1990s, mountain lions killed and ate several of the released wolves. In a brash display of nerve and strength, a mountain lion jumped a ten-foot-high fence to enter an enclosure where wolves were being studied. It dispatched a large female that was alone in the back of the 5-acre enclosure. The lion then went back over the fence, carrying the wolf in its mouth, and disappeared into the wilderness.

A similar thing happened near the sleepy Sierra town of Washington, California, along the South Fork of the Yuba River, where I was working in 1990. Washington at the time had a population of about 200 people: an odd conglomeration of recently unemployed loggers (put out of work by the Spotted Owl Scam), gypsy gold miners trying to eke out a living panning gold, pot growers, and pot smokers (a category including most members of the previous categories). Washington consisted of a handful of houses, a couple trailer parks which also doubled as campgrounds (one of which I stayed in), one motel and bar, a couple of boutique shops, and a small school, which dated back to the turn of the century (the 20th, that is). I was working for Panorama Resources, a small Canadian company. We were trying to drill a gold resource at the former Spanish Mine. Not surprisingly, the pot growers ruled the town and the surrounding forests. Why not? They were the only ones actually making money. To prevent conflict while working in the hills around town, we hired a guide, Yankee, a local hippie, who knew everyone in the area and where their crops were located. Yankee steered us clear of the crops, helping avoid a load of buckshot in the ass.

Yankee taught me a lot about marijuana horticulture. I was tempted to join the local farmers. Their line of work sure seemed easier and more profitable than my job. Crops were mainly on USFS land, so the farmers had no investment in land. The Sierra foothills have an ideal climate for growing weed, with one exception – the lack of rainfall in the summer. The biggest expense for the farmers was irrigation line, which they laid for half a mile or more from springs or creeks and buried to prevent discovery of the crops. Their only other expenses were for a handful of good seeds and some fertilizer. The work involved packing a picnic lunch and occasionally hiking to their plantations to check on the plants, and keeping tabs on any other crops in the area. If a fellow grower ignored his crop or was just ignorant and left male plants in it, the other growers would attack it and cut down all the plants. That was the law of the Sierran jungle, for male plants would pollinate the females, preventing them from reaching their peak production of THC.

I learned the incredible value of the marijuana crops from an article in the local Grass Valley newspaper, *The Union*. The DEA had been following one of the growers, 29-year-old Michael J. Michalski, for months. The day for them to pounce arrived. A shitload of armed agents jumped out of hiding to arrest him when he came to inspect his crop on San Juan Ridge, just north of Washington. He was hauled off to jail while the agents cut down his plants to use as evidence against him. Michael went to court and was charged with failure to pay taxes on his crop. What? Yes, Nevada County wanted its cut of the illicit assets. The county Assessor placed the value at $3,000 per plant (apparently based on comparable sales, similar to other real estate valuations) and levied an assessment on Mike's dead confiscated plants based on that amount. But the DEA claimed his plants were worth $9,000 each. The Nevada County Sheriff estimated the value to be $3,500 per plant. Now that caught my attention – between $3,000 and $9,000 for a single plant. Hell, that was 5-to-15 times the value of an ounce of gold. There was no need to own mineral rights, to spend millions of dollars in drilling, to spend more millions of dollars and several years obtaining permits, or to process the ore and refine the gold. Plus, the asset is renewable. What a fantastic business! Michael's crop had 100 plants, which seemed to be the average size of plantations in the area. He had $300,000-$900,000 value for an investment of less than $1,000 and at most a couple of weeks of work. Now, there is a job I would love to have.

Believe it or not, it gets better. Mr. Michalski's attorney argued to the county Board of Equalization that, yes, indeed, the plants *would* have been worth as much as $9000 each *if* they had been harvested at maturity. The DEA jumped the gun and cut them down way too early. Thus, claimed the slick lawyer, the plants were worth no more than $1,000 each. Two of the three Board members, who likely had been smoking some of the defendant's product, agreed. One sober member disagreed with the $1,000 value. The Board reached a compromise– $1,500 per plant. Michalski paid the reduced tax assessment and, as far as I know, was out of jail and planning next year's crop the following day. I never found out if the County sold the evidence or if the Commissioners just smoked it. Why, oh why, was I slaving away as a poor underpaid geologist when there were better opportunities in agriculture?

Yankee told a lot of entertaining stories about the area. One of my favorites concerned a Fourth of July celebration in the 1980s. The townspeople of Washington turned out for the event, beginning with a pancake breakfast, followed by games – like the sack race and the egg toss, played in the main (and only) street in town. The highlight of the day was the Fourth of July parade. In the middle of the parade, a fire broke out in a building on the east side of town, half a mile away. Volunteer fireman jumped onto the town's 1950s vintage fire truck, which was the showcase of the parade. It sped off from the parade to put out the fire, leaving the other two parade entries to finish the parade on their own.

The fire truck never made it to the fire. It ran out of gas halfway there. The firemen forgot to fuel it. It may have been for the better; most of the firemen had been celebrating heartily since early morning and were having a hard time hanging onto the truck as it roared along at 15 miles per hour. They were in no condition to be handling big hoses and ladders in the vicinity of a blazing bonfire.

Enough digression. This section is supposed to be about big kitty cats. Yankee also told stories about Scarface, a huge, mean mountain lion, which was occasionally seen in the hills around town.

The northern Sierra Nevada is essentially a giant westward-tilted mountain block with a steep fault-bounded eastern front. The mountain front is a formidable obstacle to travel (as the Donner Party learned the hard way), looming over the valleys with as much as 6,000 feet elevation gain in only a few miles. On the gently sloping west side of the range, streams and mighty rivers such as the Yuba and the American have nearly kept pace with the rise of the mountain range, dissecting it into long southwest-trending ridges separated by deeply incised drainages. One of those ridges is Chalk Bluff Ridge, just two ridges south of Washington Ridge and the tiny burg of Washington.

Dense forests of mixed pine, cedar, and oak trees on the western side of the Sierra Nevada provide a perfect setting for weed growers as well as wild critters, especially bears and mountain lions. One summer's eve, the residents of a mobile home on Chalk Bluff Ridge turned in for the night, undoubtedly after a hard day of cultivating Sierra Gold, while their 90-pound Rottweiler slept quietly on an old sofa next to the washin' machine on the front porch. They woke up in the middle of the night to a ruckus on the porch. When they turned on the porch light, they saw an enormous lion carrying (not dragging) their dog in its mouth. Mountain lions can't eat an entire deer or even a large dog in one feeding. They bury their kill under leaves and forest litter and come back to feed on it. The next day, the California Department of Fish and Game tracked the lion and found the partially eaten dog carcass several hundred yards from the scene of the murder. They chained it to a tree and waited a couple of days for the lion to establish a feeding pattern. The plan was to come back and shoot it. When they returned, the carcass was gone. The lion snapped the heavy chain and carried its prize to a safer place. Old Scarface was one mean-ass lion.

Historically, attacks on humans by mountain lions have been rare, but they are becoming more frequent as the human population expands into the mountain lion's traditional territory and its habitat dwindles. It is comforting to know there is safety in numbers. Wildlife statisticians have determined the majority of attacks are on lone individuals. There have been no recorded mountain lion attacks on groups of six or more people. That may be because cougars can count and have determined that five is the magic number of *homo sapiens* to safely attack, or perhaps there are not many six-packs of hikers invading their territory. Either way, it is little consolation for an unarmed geologist traipsing around alone in mountain lion country.

As a young geologist working in Arizona in the 1970s, I ran into many cougars. Normally, one would erupt from under a bush or rock ledge and run away as soon as it saw or smelled me. That was not the case with the second category of cougars, the ones who hung out in the bars. They were far scarier and more aggressive than their wild counterparts.

Most of northern Arizona lies within the Colorado Plateau geological province, formed from layers upon layers of sedimentary rocks stacked upon each other over millions of years. The formations have been little disturbed from their original flat-lying orientation. Uplift and erosion over millennia have created deep canyons, including the spectacular Grand Canyon, exposing the many strata. To the south lies the Basin and Range, a very different geologic province, sliced and diced into tilted mountain ranges separated by broad gravel-filled valleys. Between the Colorado Plateau and the Basin and Range lies the Transition Zone, which shares geological and biological characteristics with both provinces. The iconic feature of the area is the Mogollon Rim, a high escarpment marking the southern border of the Colorado Plateau. Here, the cool pine and oak forests of the high plateau meet the sizzling Sonora Desert; desert species mix with those of the high plateau.

The southern White Ledges and Salt River Canyon, Mogollon Rim and Sierra Ancha Mountains in background.

I worked on my Masters thesis in the White Ledges area, in the Transition Zone, northwest of the copper mining town of Globe, Arizona. My thesis work focused on mapping about 20 square miles of Precambrian rocks and interpreting the geological history of the area. The geology of the area was similar to the Athabasca Basin in Canada where there were huge uranium deposits, so I was also investigating the uranium potential for Exxon Minerals (hint – it was zip). The area lies 20 miles from Globe as the buzzard flies, but it is a long, rough 50-mile drive from town. The White Ledges are named for spectacular outcrops of white quartzite, the Precambrian White Ledges Formation, which form a southeast-tilted series of hogbacks, cliffs, and ledges along the east side of the Salt River. Access to the remote core of the area is by a single dirt road, turning into a very steep and rocky jeep trail. The trail ends at a cliff in Hess Canyon, a tributary of the Salt River. From there, travel is on foot over a series of small cliffs and ledges, descending 600 feet to the Salt River.

The river provides a perennial source of water, attracting a bounty of wildlife. This is one of the few places where whitetail and mule deer mingle, grazing on the abundant grasses of the hills. The brush and scattered small trees of the grasslands in the higher elevations slowly give way to the lower Sonoran life zone, characterized by giant saguaros, agaves, cholla, and prickly pear cactus. Down in the lower washes, javelina forage among the cactus and scrub brush. Geese, bald eagles, coatis, and raccoons, seemingly out of place in the desert, follow the riparian zone. Canyon tree frogs sing from pools of water in the side canyons. Cougars prey on anything that looks tasty.

When I walked along the dry washes near the Salt River early in the mornings, I could read the record of the previous night's activities in the damp sand. Tiny footprints made by small rodents were everywhere, as were sets of elongate footprints with a winding line between them – the passing of small lizards. Deer were abundant, leaving the record of their passing in the sand. Sometimes the tracks would be overprinted by the tracks of mountain lions, starting with footprints close together, as the lion tracked the deer, then farther apart as the chase began. The most impressive part of the drama recorded in the sand came when the lion tracks suddenly vanished, only to reappear as much as 25 feet away, after the lion made a great leap.

I was mapping in the ledge country along the east side of the Salt River on a hot spring day, occasionally stopping to massage my leg. I accidentally backed into an agave plant a couple of weeks earlier. Its long black spine broke off just below skin level and disappeared into my calf. The spine occasionally shifted, causing my leg to cramp, bringing all forward progress to a halt.

Walking along a ledge of quartzite, I jumped across a narrow notch where a side ravine cut through, then continued along the other side of the ravine. Looking back across the gap, I saw a frightening site – one of the biggest, meanest-looking, male mountain lions imaginable, lying under the overhanging ledge I was following – a mere 30 feet away and 15 feet below me. I just walked right over him!

My immediate reaction was to duck down and back up to get out of the lion's view, hoping he had not seen me. If he had seen me, I decided there were two possibilities – either he was going to eat me, or he would run away. After a couple of minutes, my curiosity got the better of me. I gambled on the latter possibility. I pulled my camera out of my pack and quietly crept up the ledge to take pictures of the cougar. When the big Tom saw me, he lifted his massive head, glanced at me, then put his head down and tried to sleep. His extended abdomen told the story of his lethargy. He made a kill during the night and had a belly full of venison. His only interest was in digesting his meal in the shade away from the rising heat of the day, not in having a wiry, foul-tasting geologist for dessert.

I got bolder and stood up to get a better picture. The lion didn't react. So far, all I had were pictures of a lazy sleeping lion. Wanting a more dramatic pose, I yelled at the lion to get him to at least raise his head for a moment. He wanted nothing to do with me and could care less about my photo shoot. Eventually, he raised his massive head for a moment. I finally got the picture I wanted. In retrospect, what I did was pretty dumb (which should not be surprising). I was separated from the lion by a narrow ravine, but it would take him only a couple of leaps and bounds to reach me. If you have any brains at all, you don't pull the mask off the Ole Lone Ranger, and you don't pull the tail of a sleepy mountain lion.

The winter of 1988 was a mild one in Nevada. I grew restless in the office and decided to go to the field to check gold prospects in the Shoshone Mountains between the ghost towns of Ione and Berlin, the home of the *Ichthyosaurus*, the official state fossil of Nevada. The ichthyosaur was a 60-foot-long sea monster with a long, pointed snout lined with sharp conical teeth designed for ripping apart just about anything that shared the Triassic Sea some 220 million years ago. It is an appropriate fossil for the wild and rugged state of Nevada.

The afternoon shadows of February were getting long as I was finishing my work. Just before I was about to pack up and leave, I heard a dog barking in the distance. That seemed strange in the middle of nowhere since I had not seen any sign of other people all day. I was concerned that someone lost a dog, and it was barking in hopes of finding a kind soul to rescue it. I hiked up the hill in the direction of the barks, eventually coming to its source – a blue-tick hound standing at the base of a big pinon tree, looking up and barking excitedly. A nervous cougar was perched about 20 feet up in the tree. The hound was not letting it come down.

A few minutes later, I heard voices farther up the hill. I braced myself for the inevitable encounter, expecting to see brave hunters on horseback coming to shoot the poor beast from the comfort and safety of their mount. Instead, a bedraggled-looking middle-aged man and his teenage son came walking down the hill. They were lion trackers from California who were out training their dog, Blue. As they drove up the canyon by Berlin just before daybreak, a big mountain lion ran right in front of their truck. They parked, let Blue loose, and set off on foot after the lion, leaving their horses, lunch, and water behind. They chased lions over hill and dale all day. This was the fourth lion Blue treed.

Soft kitty, warm kitty - pissed off kitty, Ione, Nevada, February 1988.

The lion trackers were regaling me with fascinating stories of tracking lions all over the west, when all of a sudden, the lion saw its opportunity. It launched from the tree, landed about 30 feet down the steep slope, and took off running. The hound was right behind it, baying excitedly. The exhausted trackers took off after their dog, yelling, "Blue, stop! Blue, God damn it! Come back, Blue!" I left to drive back to Tonopah, wishing them luck as they ran down the hill in pursuit of Blue, who was dang good at finding lions, but could definitely use some remedial training.

On a hot summer day, in the mountains northwest of Parral in southern Chihuahua, I descended a steep hillside and walked along the base of a talus pile of big rocks. As I passed by, I heard what I thought was the rattle of a *cascabel*. I froze and looked around to determine where the sound was coming from. The right? The left? In front of me? Directly between my feet (which has happened a time or two)? The sound was coming from my right, about 20 feet away, in the pile of large boulders. The rattling sounded odd. Rattlesnakes rattle for a few seconds, then stop, wait a few seconds, and rattle again. This was more like a rattling hiss. I cautiously approached the boulder pile to listen more carefully. It was indeed a hiss, a prolonged guttural exhale. This was no rattlesnake. It was a puma in its den within the rock pile. A couple of days later, Francisco and I were working in the area. I told him I wanted him to see something but didn't let on to what it was. We hiked to the talus pile. As Francisco approached the rocks, the hissing began. He jumped and exclaimed, *"Cascabel!"*

"No," I said, "listen carefully." We stood there listening, then slowly approached the source of the sound.

Francisco jumped again. "Oh shit, puma!" he yelled.

I marked the map, *Casa del gato*, and we moved on.

and Tigers

I had a close encounter of sorts with a tiger while I was working with a team of Indonesian geologists evaluating a coal deposit near the village of Aurcina in Riau Province, Sumatra. Fortunately, the encounter was closer in space than it was in time.

Aurcina lies along the border of Bukit Tigapuluh National Park, one of the last refuges of endangered species such as orangutans, elephants, cloud leopards, Sumatran rhinos, and Sumatran tigers (*Panthera tigris sumatrae*). Sadly, the refuge is shrinking. Nearly two-thirds of it has been illegally logged. Rubber and oil palm plantations have replaced the primary jungle. It is not surprising that tigers occasionally wander out of the park into adjacent areas. What the hell? There is little difference between being in the park and outside it.

I missed seeing a big Sumatran tiger by several hours but got to see the huge paw prints it left in the mud of the rubber tree forest when it passed in the night. We ventured into the jungle only during the day, so we didn't worry much about tigers. Seeing the tracks from the previous evening was both an exciting and somewhat chilling experience.

I had an even more exciting tiger experience in, of all places, Singapore, the island/country/city all wrapped in one. Although there are over five million people packed onto this small island, there is still a fair amount of untouched jungle just outside of the city. Set in a remnant of primary jungle adjacent to Selator Reservoir is the Night Safari, a zoo that appropriately is only open at night. A tram takes tourists on a big loop through the park. For the more active visitors, there is a footpath that winds through the park allowing up-close views of the animals in their nocturnal environment. The Night Safari is sure to change your mind about animal behavior. Those lazy animals that zoo visitors see sleeping in the daytime are sound asleep for a reason. They are pooped after being up partying all night. Unbeknownst to most people, lions, tigers, bears, elephants, rhinos, hippos, even turtles are mostly nocturnal.

During one of my layovers in Singapore, while traveling to and from Indonesia in the mid-1990s, I decided to check out the Night Safari. I took the 927 bus from downtown for the 50-minute ride to the park, arriving at dusk, about 20 minutes before the Night Safari opened. A crowd of tourists, cameras hanging around their necks, organized into color-coded groups following a guide holding high a flag matching the respective group's color, was hanging around the entrance to the footpath, chatting loudly. Mingling with that crowd did not appeal to me. I studied the map and noticed the path was laid out in a big loop with the endpoints labeled "Entrance" and "Exit." I walked over to the exit and waited there. When the park opened and the mass of tourists squeezed its way into the entrance, I quietly slipped past the exit sign and walked alone, traveling countercurrent to the noisy crowd. The subtly lit boardwalk of the Forest Giants Trail wound through a jungle of towering tropical trees, their massive vine-covered branches overhanging the trail. The only sounds were the rhythmic ringing of cicadas and the trilling of frogs. It was a magical moment – quite different from evenings in our jungle camps. I was walking through the jungle after dark on a smooth, softly lit boardwalk without having to worry about running into hungry wild animals, deadly snakes, or falling down and getting speared by some spiny plant.

After a few minutes, I came to the first enclosure. Under an arch of faux rocks was a big glass wall, about 10 feet high and 20 feet wide. The landscape was softly lit in the front, fading to darkness further back. I peered inside to see what was in there. I saw only a few boulders and low bushes. Then it came – a flash of orange and black, two bounds and a leap – and a huge tiger was standing up against the glass, his massive head above mine and his giant paws nearly at the top of the glass. My heart definitely skipped a beat or two; my sphincter damn near forgot its most important function. I stood there in silent awe of the magnificent beast that came out of nowhere and lorded over me. John Valliant describes the tiger's stealth perfectly in his wonderful book, *The Tiger*, "the tiger who is seen only when he chooses to be seen, erupting, apparently, from the earth itself – from nowhere at all – leaving no time and no possibility of escape." If this had been an encounter in the wild, the end would have come quickly but not necessarily painlessly.

In 2013, near the village of Suban in Jambi, not far from where I saw tiger tracks two years earlier, a 21-year-old Indonesian lad named Sutardi had a similar heart-stopping experience. Sutardi didn't have to cough up the admission fee to the Night Safari but also lacked the protection of the 10-foot-high glass wall.

His heart-stopping experience was a permanent one. He was mauled to death by a Sumatran tiger as he walked just outside his village that night. Odds are he never saw it coming.

and Bears

Any geologist who has worked in the Canadian or Alaskan bush has a bear story or two to tell. Most involve scary standoffs being treed for an hour or so or having to shoot a marauding bear. Sadly, a few geologists never had a chance to tell the story of their bear encounter. They ended up being bear dinner.

I came across many black bears, grizzlies, and a few big brown bears while working in Alaska for Exxon Minerals in the 1970s and 1980s. Thanks to thorough preparations and some good luck, we never had any serious bear problems. We were always careful to camp away from bear trails. We stashed our food where bears could not get it. Burning and burying garbage was a daily ritual. We usually checked out the areas where we worked from the air to see if any bears were hanging around. That worked well for open tundra, but not for brush or forest. Despite those precautions, a few dicey encounters are worth mentioning.

For protection, we initially carried a Colt 357-magnum pistol, which was replaced by a sawed-off police riot shotgun after a geologist on a different project shot himself in the foot. The shotgun was technically illegal. We had a special permit for it, issued to Tom Strong, Exxon's Chief of Security. The shotgun was stored in a safe in the Denver office until we left for Alaska each field season and was promptly returned to the safe when we returned. We received official firearms training at the Englewood, Colorado police firing range. Training consisted of typical gun safety stuff: checking the safety, loading and unloading, clearing the chamber, et cetera, followed by shooting at pop-up figures of men. We ended the day shooting skeet. I did OK shooting targets on the ground but never hit a single damn skeet. Although my colleagues ribbed me for my lack of skeet shooting skills, I figured I had little to worry about – unless grizzly bears learned to fly.

The only time we used the shotgun other than for practice was at a place called Titna in the Kuskokwim Mountains of central Alaska. Left Titna made an excellent spot for a base camp. It was above timberline and provided an unobstructed view of everything around. More importantly, a decent breeze blew across the ridge most of the time. The breeze kept the pesky mosquitoes to a minimum, an essential element for a camp in the Alaska bush. We landed on the ridge, unloaded our gear, and sent the chopper over to Right Titna to drop off food and camp equipment for a fly camp we would hike to later. After that, the helicopter flew away, leaving John and me to our own devices for the next 14 days.

We went to work pitching our tent, organizing our field gear, and setting up our folding dinner table in front of the tent. In ultimate camp dining luxury, we cooked up a delicious dinner of Dinty Moore beef stew, served in our dog bowls, and washed down with a nice bottle of Cabernet Sauvignon. The snow-draped massif of 20,320-foot Denali loomed behind the low hills to the south, providing a spectacular backdrop for the best doggone dinner for at least 40 miles around.

After we finished dinner, we carefully washed the dog bowls and pots, then burned all the garbage, including the cans.

Twenty minutes later, we saw a big black shape emerge out of the tangle of willows about 500 feet below us, Shit, it was a black bear. It was heading directly towards us. So much for bear safety procedures. This big fellow probably smelled the burning garbage and was coming to check it out. To make matters worse, we left the shotgun sealed in its case inside the tent. John went in to get the shotgun while I served as a lookout. I picked up some fist-sized rocks (rhyolite porphyry with smoky quartz crystals, to be specific) to use as a last resort in case the bear got to me before John retrieved the gun, assembled it, and loaded it.

Fine dining at Left Titna camp on day 2 – with our shotgun at the ready; Denali in background.

Our protocol was to load the shotgun with a round of birdshot to be fired as a deterrent, followed by 00 buckshot and finally a magnum slug to stop the bear if it made it to close range. The bear kept coming. When it was about 50 yards away, John fired a round of birdshot over its head. There was no reaction. The stupid bear kept coming. Here in the middle of nowhere, it had probably never seen a human and had no idea what a gun was. When the bear was 30 yards out, John prepared to shoot the buckshot. Shooting a bear is a bit of a dangerous undertaking. The only thing worse than a hungry bear in camp is a wounded, hungry bear in camp. There is also the hassle of all the paperwork and government investigations that follow. As a next-to-last resort, I decided to try the rocks. Maybe the bear would understand what they were. My first throw whizzed over his head (baseball was never my forte). The bear kept coming. The second pitch hit him in the shoulder. He stopped and looked at us. We yelled, and he started to turn around. At that moment, the third rock hit him smack in the ass. He ran down the hill and out of sight.

We assumed that would be the last we would see of that bear, but we kept the shotgun loaded and at the ready. One week later, we finished work at Left Titna and packed up for the trek to Right Titna, chopping and pushing our way through the chlorophyll wall that filled the valley. On the way there, we came across a cute little lynx kitten in a tree just above eye level. Realizing that its mother was not nearly as cute and was undoubtedly watching us, I took its portrait, and we quickly went on our way.

Navigating through the tangle of willows.

Cute kitty

John with the shredded tent.

Upon arrival at the campsite, we found our gear scattered about. Mr. Bear paid our future camp a visit. The bear barrel did its job; it was battered but was intact. Our food was safe inside. Then we found our tent. It had been mauled and was full of rips and tears. This was bear revenge. Somehow that furry bastard knew exactly how to get back at us for the bruise on his behind. Strapping tape came to the rescue. We taped the holes and rips, restoring the tent to a functioning condition. Despite our best efforts, there were still small holes we missed. Every night mosquitos, which were thicker than molasses there, found the holes and terrorized us while we tried to sleep. I curse that damn bear to this day.

Right Titna camp.

The Hogatza gold placer was reported to contain concentrations of uranium, thorium, and rare earth elements. We wanted to locate the source of the uranium, which was likely in the nearby Zane Hills. The gold placer had been worked by a large dredge, which operated from the 1950s until 1975. A privately owned dirt airstrip, the Hog River Airport, was located a stone's throw from the dredge. Whenever we mentioned the name Hog River to people in McGrath, the conversation quickly turned to bears.

"Oh, you're going to Hog River?"

"Yup."

"Better have your rifle loaded."

We contacted Alaska Gold Company, the owner of the dredge and the "airport," to obtain permission to use the airstrip and camp there. Permission came with yet another warning about bears.

Once we arrived, we checked out the landing strip. Being highly trained scientific observers, we noticed bear scat and tracks all over the place. It became painfully obvious the stories were true; bears own this place. That might explain why the local tributary is named Bear Creek. It seems the bears are mean ones at that. Several problem bears were shot when the dredge was operating.

While Rich, Brad, and I set up camp, Paul, the helicopter mechanic, walked over to investigate the old dredge. He came back to inform us there was an old bunkhouse near the dredge. It was unlocked and was in good condition. It would make a cozy and much safer lodging than flimsy tents in the middle of a bear trail. A group discussion followed to decide whether we should abandon our tent camp and settle into the bunkhouse. I must admit the scary bear stories and glaring signs of bears had us all spooked. I made the executive decision to move to the bunkhouse, although we didn't have specific permission to use it.

After a few days, a visitor showed up in camp, Gary, a trapper who lived in a cabin a few miles up the Hogatza River and served as the company watchman. He canoed down the river to meet us. From his radio communications, Gary knew we had permission to be there, but he expected us to be camping on the airstrip, not staying in the bunkhouse. He was not happy with us and told us we had to leave. Reluctantly, and with images of being shredded by marauding bears, we packed up and moved our camp downriver.

There is something about living deep in the Alaska bush that affects a person's brain – adversely. Most bush denizens survive with a crutch, usually a heavy reliance on booze or religion, or both. A lot of them are just plain paranoid. We never discussed religion with Gary and certainly didn't offer him any of our limited supply of Jack Daniels, so I cannot comment on his particular crutch. But he sure was paranoid.

Gary checked the dredge and outbuildings and used his finely-honed tracking skills to find fresh human footprints. He was convinced we had broken into the buildings to steal their contents, which were not much. I learned of this during the next radio contact with my boss in Denver. Thus began the great Hog River Inquisition. I called a meeting with the team to find out if anyone had entered the buildings and if anything had been taken. Well, shit, we had all been to the dredge. It was too cool to not investigate such a classic piece of mining history. Nobody had broken any locks (there were other entry points), taken anything, or damaged anything. I thought we were in the clear.

The gold dredge at Hog River.

It turns out the managers of the Alaska Gold Company were about as paranoid as their hermit watchman. The Hog River Incident, as it was known, became a big deal with the company. My boss, Art Pansze, eventually had to fly to Fairbanks to smooth things over. The problem was solved with an offer to pay rent for the few days we spent in the bunkhouse.

I would have been in more trouble for my poor judgment if it were not for the very bears that influenced the decision. When Art met with the President of the Alaska Gold Company, he was told that the watchman slept in the bunkhouse after we left. On the second night, he woke to a big black bear trying to break into the cabin. He shot the bear. Our own paranoia was vindicated.

While working the Koyukuk region of west-central Alaska, John and I arranged to base our camp at the airstrip next to the small Indian fishing village of Nulato, which lies on the Yukon River about 310 air miles west of Fairbanks. Our helicopter fuel was barged to the site. Camp equipment and supplies were flown from Galena to the strip in a chartered Cessna shortly before we arrived. We were unloading our gear next to the piles of supplies when a middle-aged white man dressed in the classic Alaskan bush uniform of Carhartt overalls and a flannel shirt walked up and asked what we were doing. As we swatted mosquitoes the size of horseflies, we explained we were conducting mineral exploration in the surrounding hills and were setting up camp at the end of the runway to base operations from there for a week or so. He got a worried look on his face and said, "You might want to reconsider that." He explained a barge came upriver a month prior with a load of booze. The entire village had been on a drunk ever since and would remain that way "until it is all gone." Things got out of hand. A State Patrolman flew in the day before to haul off a drunk who shot and killed a fellow Indian.

The Indians of interior Alaska have never been fond of the White Man. One drunk Athabascan Indian is trouble. An entire village on a bad drunk is a death sentence. Not to mention, the Koyukuk Indians have a history of going on the warpath even without the enabling fortification of firewater. On a dark winter night in 1851, Indians from the next village upriver attacked the villagers of Nulato and the Russian trading post there, slaughtering everyone. Only two Indian children, who hid in the snow under overturned canoes, survived to tell the tale.

We didn't want a repeat of the "Nulato Massacre." We packed up our equipment and supplies and had the helicopter sling it to a gravel bar in the middle of the Yukon a few miles upstream. A camp there should be safe from intoxicated natives on the warpath. The odds of bear problems in the middle of the Yukon River would also be a lot less than on the shores.

John and I went to sleep in our pup tent, confident in our decision to move camp to a nice, safe place.

When we woke the next morning, John accused me of snoring loudly. That would be a given today, but back then, I didn't snore. John and I shared tents sufficient nights for him to know that. I unzipped the tent and crawled out to find a set of BIG bear tracks following the shore, taking a sharp right turn, and coming right to the tent door. The "snoring" John heard was the bear grunting with his nose pressed against the door. Good thing we slept through that. It was only Tuesday. We were not due to change our shorts until the ceremonial Burning of the Shorts after our Saturday night bath.

Bear stories are by no means restricted to Alaska. California has bears as well, lots of them. That is probably why the state flag prominently features one.

Shortly after being designated as "excess personnel" by Exxon Minerals in the fall of 1985, I took a short-term job working for Pancana Minerals. I was to fill in for a geologist, Tony Adkins, who was taking a holiday. The job was managing a drill rig at the former Hornet Gold Mine in the northern Sierra, southeast of Oroville.

I drove from Reno, arriving late in the afternoon. I met Tony and the drillers. While Tony was giving me a quick tour of the place, I heard the baying of a hound in the distance. Tony said there were quite a few bears in the area. This was bear hunting season; there were probably some hunters down in the canyon. That was a good enough explanation for me. I thought no more of it.

My lodging was an old one-room log cabin with Spartan furnishings: a wood stove, a small table, and a single bed along the back wall. Not exactly the Hilton, but it would do for a week. I unpacked and made a quick dinner. I turned off the Coleman lantern and went to bed around ten o'clock. It was a warm autumn night. I left the door open to allow air to come through the screen door.

In the middle of the night, I half woke to the sound of the screen door being flung open violently and slamming against the wall, followed by heavy clawed footsteps on the wood floor. There was no time to react before a heavy, dirty, stinky, furry body landed on top of me. I thought it was all over for this geologist. Here I was alone in the dark with a hungry, possibly wounded bear lying on me, its drool and hot, stinky breath hitting me square in the face. Not the way I envisioned meeting my maker.

Then the licking started. I reached up in the darkness and grabbed a pair of big floppy ears. This was not a bear. It was Bella, the hound we heard baying shortly after I arrived. She got lost while chasing a bear. The hunters gave up and went home without her. Fortunately, she had a collar with the telephone number of her owners in Oroville. I was going there to restock groceries and call in a drilling report, but not until three days later. Bella hung out with me as my sidekick and bear deterrent until then. Turns out she gets lost routinely but knows to use her nose to find the nearest humans and stick with them until returned to her owners.

...and Anoas

The Indonesian island of Sulawesi straddles the equator, lying between the island of Borneo on the west and the Molucca Islands and Papua New Guinea on the east. I spent the better part of two years working there as a geological consultant for P.T. Newmont Minahasa in 1994-1996. My job was to manage native crews searching the jungles of the north part of the island, Sulawesi Utara, for gold and copper. Much of Sulawesi remains wild and undeveloped due to its rugged topography and inaccessibility. It is a land of much cultural and biological diversity and intrigue.

At about the same time Charles Darwin was in the Galapagos Islands formulating his famous theory of natural selection, Alfred Russel Wallace, a British naturalist, was developing his own theory of evolution on the other side of the world. Wallace noted differences between fauna and flora from west to east (or east to west if you prefer) across the Indonesian Archipelago. Asiatic species – tigers, rhinos, orangutans – populated Borneo and islands to the west, whereas islands to the east were characterized by Australian species – most notably marsupials. The division between these faunal assemblages was named the Wallace Line in honor of ole Al. Sulawesi lies immediately to the east of Wallace's line in a land called "Wallacea," also named for Al, characterized by a mix of Australian and Asiatic species. Sulawesi also sports numerous odd species that occur nowhere else. Among these is the maleo, a strange bird that thinks it is a lizard and lays a single large egg in the sand next to hot springs, then walks away and lets it hatch on its own. Heck, even the Komodo dragon is a better mother. Female dragons lie on their nests without eating for months to protect the eggs until the babies emerge. The endemic babirusa is a strange cross between a deer and a pig with tusks that curl up and backward until they almost pierce its head. There are several species of bug-eyed tarsiers, cuscus (opossums); black macaque monkeys, which lack tails and look more like small apes; and rodents that are endemic to Sulawesi. And, of course, there is the mysterious anoa.

The explanation for the Wallace Line and the odd species of Sulawesi lies in a combination of tectonics and "climate change" – the very natural cycle of glacial epochs (global cooling) and interglacial periods ("global warming") that shape the evolution of both landforms and animals. During the million years of the Pleistocene ice age, much of the earth's water was tied up in massive continental glaciers up to two miles thick. The seas were deprived of water. Several such ice ages and interglacial periods have occurred since the islands of the archipelago were created by violent volcanic eruptions some six million years ago.

At the end of the last glacial maximum, sea level was nearly 400 feet lower than today. During the Pleistocene (and earlier) sea-level low stands, the islands of the archipelago were connected by land bridges that animals could walk across or were separated by narrow channels the animals could swim across. The exception was Sulawesi, which is separated from Borneo by the Makassar Straights, a wide, deep trench with strong currents that prevented migration to or from the island from Borneo.

Thus, Sulawesi has long been a dead-end to animal migration and a place of isolation where new species like the anoa could evolve.

I was fascinated by stories about Sulawesi's odd endemic species since the time I arrived in Indonesia. I hoped to see some of those rare and endangered animals, especially the anoa. A few days after I arrived, I hiked up to work on the Garini project, high in the mountains above Ratatotok. One afternoon, the survey crew came back all excited. They caught a glimpse of an anoa, which quickly ran away and disappeared into the jungle undergrowth. *Lucky guys*, I thought. *If only I could be so fortunate someday.*

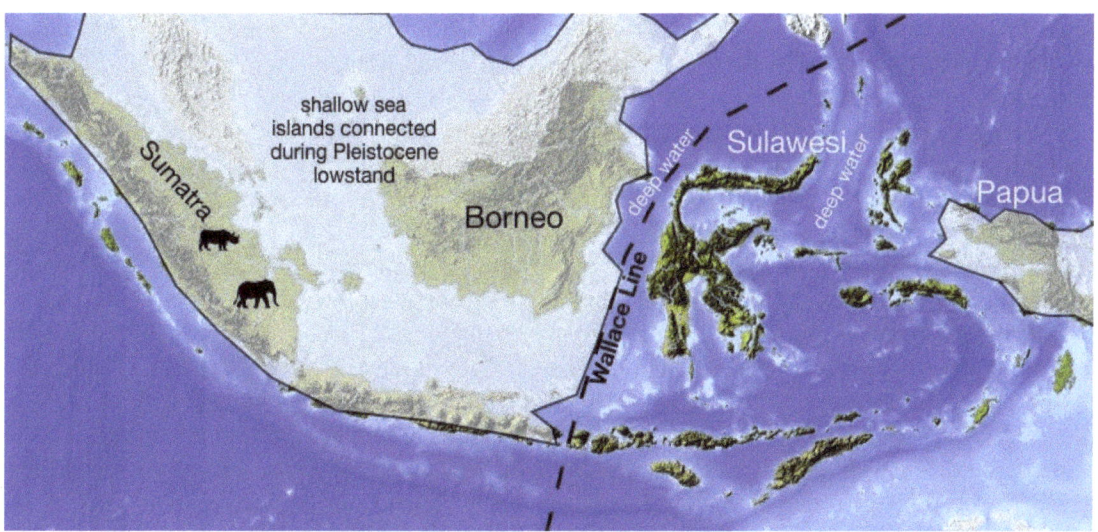

Sea-level at the Pleistocene low-stand; and the Wallace Line.

A year later, the closest I came to an anoa was seeing fresh tracks and anoa scat on a foggy ridge high above our fly camp in the upper reaches of the Dumoga River. Close, but no cigar. In August 1995, my luck was about to change – for better or worse. We were exploring the south coast of Sulawesi Utara, working out of a base camp we built at Torosik Bay. The prospecting crew came back to camp and showed me some specimens of high-grade copper mineralization they found. I needed to determine whether this was part of an extensive alteration system with large tonnage potential or just an isolated vein. The site was far up in the headwaters of the Mataindo River. It would take two days to access the site by trekking up the river from its mouth at the coast. We decided to take a shortcut hiking over the divide, which would be a strenuous all-day hike. The weather was good for a change. Not wanting to wait and risk being rained out and not seeing the prospect, I made arrangements to travel there the very next day. Five of us left camp at the crack of dawn, with Edo behind the wheel of the 1981 Toyota Land Cruiser, Syamsuddin and Charless as guides, and Kasmar as our porter. We drove west along the coast road, then turned north and bounced up a steep, muddy, barely passable abandoned logging road, sliding sideways almost as much as going forward in the thick greasy mud. The road ended at an old collapsed bridge. The main logs rotted, causing the bridge to collapse under its own weight.

We set off on foot, climbing steadily up a crude trail. Even though it was early morning, it was already hot and humid. We were sweating our balls off. After a couple hours of slogging up the hill, the trail leveled out and we followed a high ridge top. At first, I thought a light rain started. Not. We climbed to meet the clouds in a misty forest that rarely saw sunshine. It was much cooler, but very humid. The setting reminded me of the misty forest above Milangodaa, where we had seen signs of anoas. The songs of lorikeets and chatter of macaques playing in the forest canopy surrounded us, but we could not see them through the mist. Thoughts of seeing an anoa, like the one that made the tracks I saw nine months earlier, danced in my head.

We had a long way to hike. The boys skipped their regular breaks to smoke *kreteks*, nasty clove cigarettes. Those were replaced with frequent *linta* breaks. *Linta* is the Indonesian word for leech. There were lots of them climbing up our knee-high rubber boots. Sulawesi is home to two kinds of forest leeches. These were the dark green to brown type, which live in wet leaf litter on the jungle floor. They are basically blood-sucking inchworms. The little bastards hold on to leaf litter on the forest floor, where they stretch their elastic bodies upward, waiting to glom onto any passing critter. They attach to your boot when you walk by and inch their way up. If you don't catch them before they get to the top, they go down inside the boot and suck the blood out of your legs. They inject a dose of anesthetic when they bite, so you don't feel a thing. They also inject an anti-coagulant, ensuring their feast will flow freely until they gorge themselves and drop off. When you pull off your boots and see blood streaming down your legs, you know they have been there. We pulled dozens of them off our boots as we made our way along the trail.

The other type of leach is more ornate, although I would not call them beautiful, with a yellow stripe running down its light green body. This leech hangs out in trees and bushes. They grab onto you as you brush by the leaves. Unlike the leaf-litter leeches, these suckers have a bite that hurts like hell. You instantly know when one has found you. Then the fun of detaching the little vampire bastards begins.

The trail eventually dropped down off the ridge. Charless was in front of me, hacking with his machete at the thorny vegetation that was trying to reclaim the faint trail. Syamsuddin and Kasmar brought up the rear. The slope was steep and muddy. A lot of slipping, sliding, and falling ensued. We marched up and down several hills and crossed a small, clear mountain stream. Just after the creek crossing, a foot-long bright red centipede wiggled its way across the trail, passing almost between my feet. Was this an omen?

A few minutes later, I came across a piece of pink flagging tied to a bush. This was Charless's way of notifying me of an upcoming danger. The Contract of Work the company signed with the Indonesian government was very clear that our exploration work was to have minimal environmental impact. We were not to kill any wildlife, not even venomous snakes. I raised hell with the crew on one of my first trips to the field when I came across a snake writhing in its death throes after being hacked by a machete. The boys thought they were protecting me. They didn't like snakes, anyway. After that incident, we adopted a system of marking snakes encountered along the trail with pink flagging. About two steps past the flagging was a beautiful Temple Viper, its big triangular head and fat coil of light green with yellow bands blending nicely with the jungle vegetation. I took a detour to the left.

The game trail started to climb up a hill. We were entering old-growth jungle. There was much less tangle of vines and undergrowth, typical of the second-growth forest. The trees were much bigger and taller. After about 10 minutes of climbing, I heard a commotion and shouting from the rear. I stopped and looked down the trail but could see only greenery. Suddenly a blur of brown came running up the trail. It took me a moment to realize what I was seeing. Was it a deer, a wild pig? Then it registered. This was a mountain anoa, the rare Sulawesi critter I long hoped to see.

You probably have been asking, just what the heck is an anoa? Well, the mountain anoa, *Bulalus quarlesi*, is strange enough to be difficult to categorize. Most taxonomists consider it a member of the water buffalo family, although it also has some characteristics of a deer. Standing about three feet high, the anoa is considered a "dwarf" buffalo. But weighing in at upwards of 300 pounds, it is a hefty dwarf. Unlike most water buffalo, which have massive horns that sweep out and back over the head, the anoa has straight sharp horns, six to eight inches long, more like those of a mountain goat.

Dang it; my camera was in my pack. The anoa was coming up the trail fast. There was no time to get the camera out. I resigned myself to savor the moment and just watch the anoa run past. I stepped off the path as far as I could to let it pass, spellbound as it ran up the trail. Just before it was about to pass me, it took a hard-right turn. It was coming straight toward me. I took a second to realize that this was not going to be a friendly greeting. I was being charged by a pissed-off miniature water buffalo. My back was up against the wall of jungle. I was trapped, going nowhere. As the anoa closed in on me, it was bucking and bobbing its head up and down in exaggerated "yes" nods.

Then we met. The anoa charged with its head down. At the last second, I instinctively thrust my hands forward. No heroics here; what the hell else was I to do? I will never forget the feeling as my hands met its cold, wet nose, slipped along the coarse fur of its snout, over its eyes, and slid down to the horns. At that point, I literally grabbed the bull by the horns and held on for dear life. And so, we danced. It seemed like forever, but we probably struggled for less than a minute in an Indonesian version of All-Star Wrestling, the anoa trying to raise its head up (sometimes succeeding) and me trying to hold it down. I was unsure how long this wrestling match could go on or what the outcome would be. Out of the corner of my eye, I saw Charless, who knew a good deal more about anoas than I did. He was climbing a small tree about 100 feet further up the trail. With my last ounce of strength, I twisted the anoa's head to the right and pushed it in the direction of Charless. It took the bait and ran off toward him, jumping up and trying to spear him in the butt as it passed under him. Then it ran up the trail and disappeared.

Charless climbed down from the safety of his perch in the tree, babbling away in his local language. I was laughing at the encounter. Not Charless, he was visibly shaken. We walked back down the trail to check on Syamsuddin and Kasmar. They were sitting on the side of the trail, shaking in their boots. The anoa ran up behind them. With no warning, it rammed them and threw them off the trail. They were bruised, had a few scrapes, and their trousers were torn. I missed most of the conversation, but the word *"bahaya,"* which in my limited bahasa Indonesia I knew meant "dangerous," was mentioned repeatedly.

A captive anoa.

Syamsuddin after the anoa attack.

It took a few minutes and a kretek or two before the guys settled down enough to continue our trek. We reached our destination without any further drama. The copper occurrence turned out to be a narrow vein with no size potential. The anoa would be glad to know we would not be returning. So were we.

On the hike back, we passed our pretty green viper friend, still coiled in the same spot and position that it was two hours earlier. Although these snakes are mostly arboreal and nocturnal, we commonly found them on the ground during the day. They have the patience of Job and will lie motionless for weeks until some luckless rodent ambles by. Then they strike with lightning speed. Although seriously venomous, they are typically docile and are focused on catching their next small meal rather than wasting hard-earned venom on hapless humans.

Just when I thought I would make it back to camp with only a bruise on my left knee and some scratches and small cuts on my hands from the anoa encounter (those horns were sharp), plus a single thorn scratch across my hand… oh, and two bites from those damn leaches, I snagged my arm on a thorny vine in the last 100 yards of the trail. With blood trickling down my forearm from two deep, ragged puncture wounds, I arrived at the waiting Land Cruiser. On the long bouncing ride down the muddy 4x4 trail, I daydreamed of having a delicious slushy margarita awaiting me in camp. The only problems being the nearest bottle of tequila was on some other island and ice in these parts is rarer than anoas.

We arrived back in camp just after dark. Syamsuddin and Charless immediately recounted their story with lots of animation and a fair amount of pointing at me. Much chatter in Indonesian about the anoa followed. I really didn't see what the big deal was. It was cool to have fulfilled my dream of seeing an anoa. Not only had I seen one, I even touched it.

It wasn't until the next evening that Sainun, the crew chief who spoke decent English, told me how serious the matter was. He grew up in the next village and knew anoas well. It turns out they are exceedingly ill-tempered beasts. Two of his friends were killed by anoas. I inquired, "How?" He told me the anoa kills by lowering its head, then quickly raising it to jab its horns into the victim's belly and disembowel him. Oh shit, so that is what the beastie was trying to do when I grabbed its horns and wrestled it. Crap, if I had known what was going on, I probably would have freaked out (I like my bowels right where they are) and been done in by the critter. We broke out some warm Bintang beers to celebrate surviving the encounter.

A couple of days later, I traveled to Onggunoi, a short distance to the east, to visit our camp. The rivers were swollen by rain and the roads were impassable. I went by outrigger canoe along the coast. It was a pleasant trip on a warm and sunny day. The sea was relatively calm, with only a few moderate swells. A short distance out of Torosik Bay, I noticed a large object floating directly ahead of us. A big flipper came out of the water. It was a giant leatherback turtle – nearly 10 feet long, the largest and most primitive of all sea turtles. Normally, sea turtles lie low in the water and only poke their heads out to take a breath of air. This one was floating high in the water – most unusual behavior. As we approached the turtle, it was lifted up on a swell. We noticed it wasn't alone. It was on top of a second turtle. We veered to the right to avoid the amorous couple, pleased to know a new generation of turtles would be on the way.

The canoe landed at the designated spot on the sandy beach where the driver, Ferdinan, was waiting. As we drove to camp, he told me everyone heard about my wrestling match with the anoa. News travels fast via bush radio. He was obviously impressed. He smiled at me and said, "*Kami pikir anda sama sama Chuck Norris*" (we think you are just like Chuck Norris). The Indonesians are big fans of martial arts. It was an honor to be compared to good ole Chuck. Sadly, I missed the movie in which he karate-chopped a black-pajama-clad anoa into submission and rescued the damsel in distress.

I suspect that signs of progress, like a paved road and bridges across the many flooded rivers, have probably come to the South Coast of Sulawesi Utara. Despite the changes that come with time, if you were to visit the villages of Torosik, Mataindo, or Ongonnoi today, I bet the villagers would still be talking about the *bule* from America who wrestled an anoa – and lived to tell about it.

Oh, my!

CHAPTER 31

SCATOLOGICAL MUSINGS

Or, Commodes I Have Met, But Wish I Hadn't
Or, Why the Author Carries Copious Quantities of Charmin.

Ask any kid, "Why are farts funny?" – you'll get a lot of giggling. Ask about "number two," and the story is quite different. That is not funny. It is dirty, gross, and embarrassing, especially if it happens when you are only expecting a fart.

Farts, of course, *are* funny. Well, most of the time. There are exceptions, notably wet and lumpy ones. Then there are the SBDs (silent but deadly ones). They are not funny to your dog, who routinely gets blamed for them. "Geez, Rover, that was disgusting! What dead rotten thing did you get into?"

Discussing bodily functions may be an indelicate matter but should not be taboo. Let's face it, shit happens. If it didn't, we would explode. Many of us, myself included, have come close to that happening when things didn't go quite as planned. Come to think of it, it did happen a time or two. Let's not go there.

We Americans are terribly spoiled when it comes to the convenience of modern plumbing. It was not always that way. The modern flush toilet was not readily available until the late 1880s. For basically half of American history, we Yanks had to position our derrieres over uncomfortably hard and poorly contoured holes cut in the splintery wooden boards of the family outhouse, or the "wee hoose" as my grandparents called it in their rich Scottish brogue, to accomplish our daily business. It may have been inconvenient, a tad odiferous, and at times rather chilly, but it was great for the fledgling business of Sears Roebuck and Company, whose mail-order catalog graced the library of many an outhouse and provided free toilet paper to boot.

The sad thing is that the grand flush toilet, which Mr. John Harrington invented in 1596, went essentially unrecognized for nearly 300 years. It took an enterprising English plumber, Sir Thomas Crapper, to make the modern commode popular. Mr. Crapper made a few improvements to Harrington's toilet. Being a clever entrepreneur, he promoted it relentlessly. His billboards at the onramp to the London Bridge and at Piccadilly Square eventually caught the attention of Prince Edward, soon to become King Edward VII. Eddie was ambitious and desired more than just the throne of England. He wanted a special private throne to sit upon. Tom Crapper was more than willing to provide him with one. The King was thrilled with his porcelain throne and appointed Mr. Crapper to be the Exalted Flusher of Royal Turds or some such title. Having sold Buckingham Palace on his improved commode, Crapper expanded his business to promote its many benefits to the commoners. Heck, if it was good enough for the Queen, it should be great for regular shits. Tom's burgeoning business soon became the veritable Flushes R US of the 19th century. The rest is history.

Contrary to popular belief, Thomas Crapper's major contribution to society lies not in the invention of the flush toilet, the credit for which, as mentioned above (if you had been paying attention) belongs to Mr. Harrington, but in his contribution to the English vernacular. Think about it. The toilet could have been nicknamed the Harrington instead of the Crapper. That just doesn't sound right. This is how a conversation on a typical night out could go: "Hey Sam, where did George go?"

"Oh, he's on the Harrington. He'll be back in a few minutes."

Expletives would also not be the same: "Oh, Harring! I left my credit card at the bar." Yes, we owe a lot to dear Tom Crapper.

Despite Tom's advertising efforts and more than 100 years of further improvements in the design and operation of the flush toilet, including the cleverly engineered leaky valve designed to keep plumbers employed, the invention has not caught on in nearly half the world. We call that the Third World, which I think is a misspelling of "Turd World."

The continent of Asia and the string of islands that run through Southeast Asia are the worst culprits. Mr. Crapper's commode has yet to arrive in most of that part of the world. Instead, they have (at least in upscale fancy places where there is some sort of toilet) what is known as the "squat toilet." This is like a regular toilet, except it is sawed off at floor level, which, if you think about it, eliminates almost all of Tom's royal model. The squat toilet is simply an elongated porcelain bowl with a grooved surface on either side. The trick is to put your left and right foot on the corresponding grooves on either side of the bowl to line up your working parts with the hole, then bend at your knees and slowly lower your butt until it is level with your ankles. At this point, if you have been going to your yoga classes regularly, you are successfully squatting and are ready to do your business. If you skipped a few too many classes, odds are you will fall over backward into the porcelain bowl. That's OK. Although a bit damp, this position works. Not to worry, you are about to get wet anyway. Toilet paper is as foreign to these foreigners as is a real crapper. The next step is to take the little bucket of water next to your left hand (unless you knocked it over when you fell into the bowl) and splash water onto your derriere to wash yourself using your left hand. Now you understand why Asians only shake hands using their right hand. The final step is to take the big bucket of water and dump it to flush your contribution down the toilet and off to God knows where. Once it goes down the drain, it is someone else's problem.

While staying at a cute little motel with a hilltop view in Kotamobagu, Indonesia, I put that theory to the test. The *kamar kecil* (literally "little room") down the hall had the usual squat toilet. I did my thing and used the TP that I religiously carried with me to finish the job. I dropped the water from the big bucket into the toittee to flush it. My curiosity got the best of me. I went to the small window at the back of the room and peered out. Sure enough, in a moment, my excrement appeared in an open tiled drain and slid down the hill like a little brown boat with white sails heading out to sea. The journey ended at the bottom of the hill, where I am sure my sailboat joined many others. That is when I realized why Indonesian real estate is cheaper in the lowlands than on the hilltops.

Jakarta has 5-star hotels with state-of-the-art plumbing. But travel to the smaller towns and villages of Java, and you are back in the land of squat toilets, at best. I traveled from Jakarta to the village of Ciberon, on the northeast side of Java, to visit the family of my friend, Enih, and to experience traditional Javanese culture. The family lived in a modest house, which was very clean and had a working squat toilet. Like most villages in Java, Ciberon is a farming community, growing rice, spices, vegetables, and various tropical fruits.

The fields are watered by a series of irrigation ditches, which are multi-use devices providing water, but also serve for garbage disposal and – you may have guessed it – toilets. On the road to the fields, we came across a lovely sight, a gravity-powered toilet suspended over the middle of an irrigation canal. The facility consisted of a small platform perched on four poles about four feet above the ditch. For privacy, a sheet of plastic was wrapped around the platform at waist level. Access was by a skinny log extending from shore to the platform with the added convenience of a single handrail to help prevent a fall into the murky, garbage-clogged water below. Flushing, of course, only happened during a good rain. I am sure the crops grown here are especially tasty and very well fertilized.

The Chinese are big on squat toilets, but theirs are distinctive. They are filthy, like most anything and everything else in the country. Enough said.

Squat toilet in Wenxian, China; the Chinese could use some commode target practice, as well as cleaning lessons.

The Philippines also features adventures in designer commodes. I arrived at the airport in Cagayan de Oro on the island of Mindanao for a flight to Manila after a long drive from the field. I felt some rumblings in the lower part of my alimentary canal and thought it would be a good idea to visit the restroom before boarding the plane. To my delight, there was a modern version of Mr. Crapper's commode. To my dismay, it was only a partial one. There was no seat. What the heck? How do you use that? The choices are to sit on it, or rather in it, float your butt in the water below, and hope that you don't get stuck and miss your flight despite desperate pleas to your fellow passengers to extricate you; or climb on top and try to squat while balancing yourself on the narrow rim, preparing for the brown tsunami that would erupt from the bowl once your contribution hit the water below. I decided nature's call would have to wait. I mused about the possibilities of why the seat was missing. There was a lot of unrest in the area. In retaliation for the ambush and beheading of an army patrol the previous week, the Philippine army was busy battling the Abu Sayyaf Muslim rebels just a hundred miles to the west. I worried that some Islamic radical had taken the toilet seat to hit the pilot over the head and hijack the plane to Cuba, Fallujah, or Mecca. Fortunately, my fears were unfounded. The flight went smoothly, except for my increasingly strong gastric rumblings.

Surprisingly, the grossest experience I have had with excrement was not in a Third World country, but in the supposedly civilized U.S.A. – intellectually superior California, no less. It was New Year's Eve, 1988, when my boss and I were driving across the Mohave Desert on the way to the LA airport after visiting the Mesquite mine. We stopped at a gas station and convenience store in the middle of the desert to get gas and a coffee. Behind the station was a port-a-potty. We still had a long drive to the airport. I thought I might drop off a load before proceeding further down the highway. I opened the door, expecting the usual waft of disgusting odor. What I got was a shock. Oh yes, I got the blast of malodorous air, but a whole lot more. Flies were feasting on the perfectly pyramidal pile of poop that was stacked three feet high above the seat of the crapper. I am not shitting you! I have no idea how people accomplished that. The only thing I can imagine is that a busload of NBA players stopped by on the way back from some basketball tournament after having an elaborate post-Christmas feast laced with Ex-lax, and they were all really desperate. Nobody else could have managed to get his butt that high off the seat. My hat is off to the last player to add to the disgusting pile.

Exploration geologists, being outdoor types, spend the better part of their careers dealing with a dearth of toilets. It is something you get used to. City types can't quite grasp the concept of communing with nature when it comes to bodily functions. Greg Hill, the President of Harvest Gold, and I led the company's CEO and Public Relations Manager on a tour of the Rosebud gold project in northwest Nevada. It was a long drive over gravel roads, culminating in a mile or two of 4x4 trail. We parked on top of a ridge, where there was a good view of the property (and nothing else – we were a long way from nowhere). As we exited the truck, I overheard the PR dude, who probably never stepped off the sidewalks of Vancouver prior to this outing, say to the CEO, "Good grief, what do you do if you have to take a dump out here?" Perhaps he thought we would race the 50 miles back to Winnemucca so he could position his spoiled soft ass over a shiny porcelain commode. I can tell you what to do – buck up, wussy – find a sagebrush to your liking and go for it.

Despite the best intentions and preparations, the time comes in every field geologist's life when he runs out of that life-saving essential – toilet paper. The process begins with rummaging through one's vest and field pack, desperately searching for that last partial roll of TP that just HAD to be hidden deep inside, only to come up empty-handed. Time to improvise. The next step is to inventory any items that may resemble toilet paper, either in shape, size, color, material, or absorptive qualities. Cloth sample bags, which are usually in abundant supply in an exploration geologist's pack, are a favorite alternative – a little stiff and rough, but they do a reasonable job with minimal irritation of delicate parts. I gave up carrying plastic sample bags years ago. They work fine for rock samples but are rather slippery for other purposes and only make a messy situation messier. Pages from a field notebook work OK, but it is unacceptable to use ones that already have notes on them. It is a mortal sin to use your field map. I am not even going to mention using pages from holy books. That subject is covered elsewhere.

Natural materials are also viable alternatives. Mother Nature has provided lots of vegetation that works with varying degrees of success. I am more than happy to share the botanical knowledge I have gained through many years of practice. Big maple leaves work pretty well; in autumn, you have the added bonus of choosing your favorite color. I am not fond of oak leaves. They are too stiff and have spiky edges. I love aspen trees, especially when their leaves turn golden in the fall and quake in the breeze. But the leaves are just too small to get the job done. The big soft leaves of mule ears, *Wyethia millis*, are covered in soft wooly white hairs that aid in wiping. The leaves are thin, so it is best to double up. Another good choice is skunk cabbage, which grows splendidly in moist mountain areas and has strong giant leaves. The pinon-juniper forest of Nevada does not offer much to use. I must admit I have attempted to use pinecones under dire circumstances, but I do not recommend them. They are rough and abrasive, and the sap can glue your undies to your butt. My final advice is to read a good botany book before heading to the great outdoors. The ability to recognize and avoid poison oak and poison ivy can save both your rear end and your love life.

In areas lacking leafy vegetation – for instance, the barren Sonoran, Mohave, or Gobi deserts – rocks can substitute for the Charmin you left at home. Shale is a very common rock, which splits into nice thin sheets. There is even a variety called "paper shale," which sounds like a perfect substitute for TP, but it is not sturdy and will crumble in your hand as you wipe. However, add a few thousand feet of overburden, heat the rocks for a few million years, and voila, shale has been metamorphosed to produce slate, which splits into stiff, strong sheets. Then you have something quite functional. Add a little more pressure and temperature, and you have schist, which is always good for a shit. I try to avoid igneous rocks. They break with hackly or uneven fracture, leaving very rough surfaces. And by all means, don't even think of using vesicular rhyolite and basalt. Those glassy rocks can shred your hand just by picking them up. They don't belong anywhere near more sensitive parts.

There is one more way to get around a lack of toilet paper, but it is rarely encountered. In fact, I have only come across this unique technique once. Thus, we arrive at the story of the shit-eating hound.

I was introduced to this rather talented canine while on a whirlwind trip across the Gobi Desert. Dusk was approaching; we had a long drive to reach the nearest soum where we could spend the night when we spied a couple gers and a herd of goats and sheep. We met the occupants and were invited to spend the night with them in their ger.

I woke at the break of dawn to the call of nature. I stepped out of the ger. There to greet me was the family's big furry Malamute-shepherd dog, who had spent the night sleeping at the threshold of the ger. He was a healthy young fellow, about three years old by my estimate and obviously well fed.

The shit-eating hound – lying in wait for breakfast to be served.

I stepped outside and looked around in search of an outhouse or some reasonable facsimile thereof. I expected to find at least some sign of human elimination activities near the ger but didn't see shit. Mongols are tough. Although it is never advisable to crap in your own nest, even Chinggis Khaan's toughest warriors would not have gone far from the warmth of their ger to take care of business in the brutal winter they had just experienced. Befuddled by no sign of human scat, I hiked up the steep rocky hill behind the gers, searching for a suitable site to mark my territory. I don't want to give away the punch line, but you have probably already figured out that my new best friend, the big furry family dog, followed me on my journey. I found a suitable outcrop of equigranular biotite granite that blocked the chill south wind. I settled down to let nature take its course. The reason why there was no outhouse and no sign of human waste was about to be revealed. The reason why man's best friend was not allowed in the ger was also about to become abundantly clear.

I unbuckled my belt, dropped my drawers, and assumed the Asian squat position – sans squat toilet – in preparation for presenting my little offering to enrich the impoverished soil of Mongolia. Just as I started gophering, I was almost bowled over. My big furry friend was anxious to get his warm breakfast. He was persistent as hell. I had to shoe him away several times before finishing my business and burying it beneath big rocks. It was not easy. Those kinds of rude interruptions tend not to facilitate the process of defecation.

I still marvel at what an efficient method of waste disposal that dog demonstrated. But no doggie kisses, please!

Chapter 32

APPS, PLEASE

Pretty please!

In classic baby boomer form, I have never been in sync with computers. That may have something to do with the horrors of Fortran IV and punch cards way back in the 1960s. I avoided using computers for several decades. Modern times finally caught up with me. Now I have all kinds of computer skills, like typing these stories in Word and cussing as I try to insert a picture where I want it. I can even send and receive emails. Sometimes I wish I either couldn't or, better yet, could do so properly.

I really, really need someone to develop a special app for email. I would call it the "Don't Do That" email app. Its function would be to delay sending emails with certain nasty words – especially if the boss is in the address field. If you try to send such an email, it appears to have been sent but is hung up in your outbox. After a few hours, you get a notice asking if you really want to send the message, edit it, or hit the "Bleach Bit" button. That would avoid so many of the "Shit, I wish I hadn't hit 'Send'" regrets. Believe it or not, Omar Khayyam, the 12th-century mathematician and poet, warned about this problem long before email existed, writing in *Rubaiyat*:

> The moving finger writes; and having writ,
> Moves on: nor all your Piety nor Wit
> Shall lure it back to cancel half a Line,
> Nor all your tears wash out a word of it.

I could have saved a lot of tears if I heeded his sage advice.

On a similar note, I would love to have one more app, the "Really – To Everyone?" app. This one would block the evil and dangerous "Reply to all" button. If you hit that, bells and whistles go off. The computer asks, "Do you really want to include ALL the recipients?" That would have saved my ass more than once. I will give a recent example.

I was working with a mine geologist to develop procedures for check assaying at the mine. We had been sending emails back and forth on the subject for several days. Late one night, I got home (after enjoying adult beverages with friends) to read an email in which the mine geologist mentioned the mine was considering hiring a certain assayer. I recognized the name but will refer to him as "Scummy" to avoid any potential litigation. Feeling my oats from perhaps more than one beer, I considered it my duty and obligation to respond immediately, providing my secondhand knowledge of a litany of improprieties committed by the assayer (rumors, but well-sourced ones), including theft of high-grade gold samples. My tirade concluded with this gem, "A scumbag, for sure." I hit "Send" and retired for the night, feeling proud of myself for ratting on such a scoundrel and saving the company from dealing with him.

Early the next morning, I got a phone call from the mine geologist.

"Your email has created quite a stir."

"Huh, what do you mean?" I asked.

"Scummie was very upset. He wants 'appropriate discipline' to be taken."

"How the hell did he find out?" I inquired, genuinely puzzled.

"He was copied on the email. You must have hit 'Reply to all'."

"Oh shit. Well, I guess I will be hearing more about this."

And did I ever hear more! Scummie sent an email to the company accusing me of "harassment," "bullying," and "an extreme frontal assault" on his reputation. He threatened to sue the company for "liable" (sic) and demanded an apology as well as "some significant and appropriate gesture." I thought maybe flipping him the bird would be a significant and appropriate gesture. The Project Manager responded with a one-sentence apology, noting the email exchange was intended to be between two individuals, not for Scummie or the rest of the company. Scummie shot back with a long letter denying all the accusations but using language suggesting to anyone who could read between the lines, there must be some basis for the charges.

My response was if Scummie wanted to pursue libel charges, he would have to prove my accusations were false – a can of worms I was confident he did not want to open. Fortunately, the scumbag decided pursuing such matters might just prove his scumbagness. He dropped the threats. In the end, he was not hired, and I was not fired (at least not right away). I dodged a bullet on that one. I now double-check the reply field whenever I send an email.

But I would really love to have the "Don't Do That" and "Really – to Everyone?" apps, just to be safe.

Chapter 33

WANDERING STAR

> There's a race of men that don't fit in
>   A race that can't stay still;
> So they break the hearts of kith and kin,
>   And they roam the world at will.
> They range the field and they rove the flood,
>   And they climb the mountain's crest.
> Theirs is the curse of gypsy blood,
>   And they don't know how to rest.
>
>   "The Men that Don't Fit In"   Robert Service (1907)

If your reading comprehension is up to speed, you may remember that way back in the Introduction, I proposed leaving the reader deep in thought at the end of this tome: pondering philosophical conundrums – specifically who wrote the *Book of Love*, why are farts funny, and what is the meaning of life? There is no need to spend a lot of time pondering the first two questions. The answers are obvious. Who wrote the "Book of Love"? Duh, the answer is in the Monotones' lyrics. Yes, of course, Who wrote it (when he wasn't busy hanging out on first). As for farts, well, they are funny simply because they ARE funny. I know; my boys told me so. There is also sociological support behind that premise. Passing gas is something we all do. Regardless of how prim and proper we may be, farts happen, whether they be manly blasts or girly tweets by the Queen. And they often do their thing at the most embarrassing times. Flatulence is a fundamental bodily function that brings unity to humanity, reminding us that down deep we are all the same – much like cattle, actually.

The last question, "What is the meaning of life?" remains a point of deep philosophical debate, best left for scholarly discussions among students of ethnotheology. On a personal level, I have been pondering the question for a long time. I thought long and hard about my odd career since it has occupied more of my life than anything else. Analyzing it may provide some insight into the conundrum…at least for me.

Exploration for mineral and energy deposits requires understanding and applying several fields of science: geology (yes, a real science), chemistry (geochemistry), physics (geophysics), and biology (paleontology). It also requires the ability to merge high-tech theories and applications with practical old-fashioned mapping and prospecting methods. Unlike academic studies done in classrooms and laboratories, exploration geology requires physical work in challenging terrains. Self-confidence, resourcefulness, and basic survival skills are essential. A touch of craziness is also helpful. It is not for the weak-of-heart – or legs. Few careers require such an unusual blend of disparate specialties.

Exploration geology has provided me with a satisfying and rewarding career. It has also been a love-hate relationship. There were many days when I looked around the spectacular surroundings of my outdoor "office" and thought, *Wow! I can't believe I am being paid to be here.* Those glorious days were offset by times when conditions sucked, and things were not going well – and I thought, *Damn! They couldn't pay me enough to do this again.* Despite the hardships of fieldwork, I have always been drawn back to the field. There is something special about geological fieldwork. Spending days in the great outdoors is part of it, as is experiencing different terrain, geology, wildlife, people, cultures, and customs. The real draw is the challenge of solving the mystery of where Mother Nature hid her ore deposits. Every day in the field is a treasure hunt – hiking over hill and dale searching for the geological clues that will help solve the puzzle. I can't think of any career more interesting and challenging.

Over the years, I have spent my fair share of time stuck in an office: staring at a computer screen, analyzing data, writing reports, preparing budgets, applying for permits, and promoting companies and projects to raise funds to keep the exploration programs alive. There was also the challenge of trying to manage a staff of geologists and geophysicists. Herding cats would be easier. I served in executive positions and sat on the Boards of several junior mining companies. But I am not good at sitting. Like a caged tiger, I hated being a captive in an office. I missed fieldwork and the thrill of the hunt. It didn't take long before I grew restless, became "Bored of Directors," and returned to the basics of exploration geology and the satisfaction of a good day spent in the field.

The biggest problem with traipsing the globe as an exploration geologist is that it requires spending a lot of time away from home and family. Finding the right balance between work and family is difficult. I fear I never did. When my first son was three years old, he clung to me and cried whenever I left for the field, making me feel guilty as I left. A few short years later, after I was home long enough for the novelty of my presence to wear off – and to assume the role of disciplinarian – he turned to me and said, "Dad, isn't it time for you to go back to the field?" No guilt on that trip.

I must admit, at times I thought about changing careers – especially when my highly compassionate spousal unit, frustrated by having to take the garbage out too many times during my absence, screamed at me, "Why don't you get a REAL job?" Gee, I thought I already had one, at least one that paid the bills. Perhaps I was wrong. I did a lot of soul-searching, had lengthy consultations with my friends Jack D. and Johnnie W. late into the night, and realized I was indeed deficient at garbage detail, among other domestic duties. I took the issue to heart. Maybe I should have pursued a "real" career, hanging out in an office working as a doctor, lawyer, or Indian Chief, while being slowly strangled by a cursed necktie. I struggled to come up with an alternative career. *What else could I do well enough to make a living? More importantly, what could I do that I might enjoy?*

The answer was always, "Nothing."

I have been infected with what John Steinbeck, in his classic travelogue, *Travels with Charley*, called "the virus of restlessness." Steinbeck explains:

> When I was very young and the urge to be someplace else was on me, I was assured by mature people that maturity would cure this itch. When years described me as mature, the remedy prescribed was middle age. In middle age I was assured greater age would calm my fever and now that I am fifty-eight perhaps senility will do the job. Nothing has worked.

I suffer from the same restlessness. Whereas most people hate change and cling to what is familiar and comfortable, I need change. I thrive on it, especially if it brings challenges to overcome. Crazy, I know, but that is me. From the time I was a child, I have had difficulty sitting still. I enjoy outdoor venues, such as concerts, but I hate it when attendees bring chairs and plop their asses down for the duration of the event. I much prefer to walk around, observing the venue, meeting and watching people.

I tend to become bored with my surroundings, even when they are spectacularly beautiful. Eventually, I succumb to the urge to move on, to explore and experience new and different places.

My childhood hero, Roy Chapman Andrews, said it perfectly:

> I was born to be an explorer. There never was any decision to make. I couldn't do anything else and be happy."
>
> *This Business of Exploring* – Roy Chapman Andrews

Like Roy, I was born under a wandering star, a misfit destined to explore....and to have many adventures and misadventures to share. I guess that is what has given meaning to my life.

I am often asked if I would recommend a career in exploration geology to young college students. Reflecting back on the long workdays, the physical hardships, the extended absences from home, the perilous travel, brutal work conditions, and the job instability, the answer is, "Heck no," – definitely not for today's spoiled, soft snowflakes.

Then I have to address the question, "If you had the choice, would you do it again?"

"Hell yes!"

www.ingramcontent.com/pod-product-compliance
Lightning Source LLC
Chambersburg PA
CBHW041233240426
43673CB00010B/323